The Garish Sun

The Garish Sun

by

Jose Manuel Quintero

Book One, Earth Series

Also by Jose Manuel Quintero

The Earth Will Groan

The Cosmic Lotus

Heaven's First Home

copyright 2011

All rights reserved. Not to be used or reproduced without permission of Jose Manuel Quintero

category@bellsouth.net

Photographer: Philip Talleyrand

Email: philip@philnicecreative.com

CONTENTS

Notes on the text 6

Beginning Chapters 7

Chapters 1 through 6

Middle Chapters 106

Chapters 7 through 29

Ending Chapters 483

Chapters 30 through 37

The Garish Sun

All characters in this book are fictional. Any resemblance to persons living or dead on this planet or any other is purely coincidental.

NOTES ON THE TEXT

Any resemblance to documents, public or private, on this planet or any other, is purely coincidental. The annotated version of this text is not typical of this or any other planet except as screen captured or as electronically amended text. This argues against an otherwise tedious word processing exercise and is consistent with the study and cross-references of multiple readers. By default, the latest reader's annotations are indicated with binary numbers e.g. 1000... The corresponding note is in decimal numbers e.g. 1. Readers may wish to refer to Chapter 15.

Not all electronic versions of this document include annotations or cross references. Some hard copy versions may include screen captures of the annotations and cross references. Not all electronic or hard copy versions include the bilingual English/Bemyem dictionary as the final chapter. The dictionary in all cases reference words that are native to the attached document.

The English employed is conspicuous both for its abundance of cliches, malapropisms, parochialism, and absence of contractions not only in the narratives but also in the dialogues. This is most typical of practitioners of English as a second language. This suggests the existence of a first language. This supposition is consistent with references to Bodumnloi and Bemyem. While no examples of Bodumnloi are included in this text, its romanized version, Bemyem, is extensively employed in annotated versions. This environment is consistent with the existence of Bemyem as an independent, organic and living language.

BEGINNING CHAPTERS

1

1 1

1 2 1

1 3 3 1

1 4 6 4 1

1 5 10 10 5 1

1 6 15 20 15 6 1

Chapter 1

Chapter 1000....

Chapter A

Chapter $(1 + x)$

Chapter $(a^1b^0 + a^0b^1)$

Chapter 2^0

Episode 1

Intuition informs us that sometimes the things that unite us can be as dangerous as the things that divide us. It has always been this way both here and on Earth. The planet which Dr. W-abe and Dr. Mim-sey call home, is a robust version of Earth but Earth it is not.

Its largest gulf, rivers and continents are larger than any on Earth. This climate raises and drops more water than Earth's with the help of a little under four pounds of sunlight each day.

To Earth, Clidim, is a legend that is invented and reinvented as a perpetual fantasy. Almost everyone on Earth considers the existence of another inhabited planet to be nothing but a fiction.

For Clidim, Earth is a legend with a purpose. The legend of Earth is cultivated to prepare its population against the day the alleged discovery of Earth is revealed to be real.

On Earth, English is common and embedded with words from many languages. The least of all is Bodumnloi. Bodumnloi is significant to only a few with a penchant for fantasy and the few, on Earth, who are not native to Earth.

On Clidim, Bodumnloi is the common language for many billions of its inhabitants.

Bemyem is the romanized form of the language adopted as a prophylactic by the planet and is embedded with many words from English. On the most ordinary level English is the language that gives the Clidimite's legend of Earth its credibility.

But English is essential in the highest circles of Clidim because of the inevitable discovery of the discovery of Earth or against the unlikely event that Earthlians approach the planet. Earth's discovery of Clidim is unlikely given the cosmic ocean that separates them.

Doctor Mim-sey has reason to believe that those distances can do nothing but disappear. The gulf that fills Dr. Mim-sey's horizon seems anything but ephemeral. It is hospitable for its 877,650 square miles. The southern half simmers in tropical temperatures.

Its share of sunlight that falls on the planet raises water temperatures and thousands of cubic miles of water vapor every day. Of that water vapor, 22 percent returns through one great river in the form of rain and melted snow.

This river is pacific for its size but resisting its flow is out of the question. There are few dams in the 1,560,000 square miles that it drains and the insurance companies make sure the flood plains are respected. So there is no need for levies. The river endures bridges but their huge cost keep them to a few.

Where the river meets the gulf there is a granite ledge that in some eras is insulated from the river and the gulf by a delta. But even without the insulation of the delta the ledge is immune to the prerogatives of the of the water. Here the city of So-Wemtolo has its foothold.

This city is popular despite its size. Many millions live and work here and in buildings that are cities themselves. Dr. W-abe and Dr. Mim-sey work in one of those buildings. It is large but not one of the largest. Still it hosts 399,000 workers and like its larger neighbors it has floors reserved for suburbs. Visitors represent another 250,000 daily occupants. The building-burbs for this building are occupied by an average of 132,000 tenants.

It is unlikely that anyone not working in the building would live in the building-burbs. Almost everyone not working in the building usually does not make the money or has the need to live there. Anyone not working in the building likes to say he does not want to live in the building-burbs even if he could.

This may be true since almost everyone living in the building also has at least one home outside the city. Dr. W-abe lives in the building's building-burb and has one home outside the city which allows him to avoid the hot weather. The burb-home allows him to avoid the cold.

His rational and his investment are a contradiction to his motives since the weather is the last thing to impose on his choice of which home to use. His wife and children avoid the building-burbs as much as possible but somehow they all see enough of each other to call themselves a family. He is a longtime friend of Dr. Mim-sey.

Dr. Mim-sey, too, is a `-burber and has two homes outside the building-burb. He can easily afford them. But he likes the weather around his building most of the time and ignores it or tolerates it the rest of the time. His wife lives in the house in the mountains most of the time and they see little of each other. He works a lot. She travels a lot.

They see very little of their daughter who is single and works for a publisher. That is the way Ali-ce wants it. The Mim-seys would like to see her married but that is all they would change. After many years of friendship, Mim-sey and W-abe like most of what they know of each other and ignore or tolerate the rest.

Dr. Mim-sey could enjoy his job without compromise if he did not have to deal with a problem that only he seems to appreciate. Part of the problem involves sciences that are beyond his expertise. Part of his solution is to convince Dr. W-abe that there is a problem.

Dr. W-abe is not impressed with his colleagues, so he is not planning on being convinced of anything by them or Dr. Mim-sey. Dr. W-abe cannot be told much, if anything, by anyone even though he almost always listens genuinely.

So conversations with him are not really conversations, even with his friend Dr. Mim-sey. But this is where Dr. Mim-sey has the advantage. He can tell when he has Dr. W-abe's attention no matter how he disguises it. Regardless of how Dr. W-abe likes to bring the issue to a moot point, Dr. Mim-sey brings it back from a dead end. He is indifferent to how many times or how many ways he has to do it, Dr. Mim-sey does it.

For his part, Dr. W-abe likes to chisel at the foundation of a conversation by asking questions that are not far from rhetorical. He watches his friend settle into an oversize chair and asks Mim-sey, "Why eight dimensions?"

"It does not have to be eight," responds Dr. Mim-sey, "Mr. Abbit says the best they can do is eight, if not sixteen, for now. But there is no limit. You can have a 'Theory of Everything' for any number of dimensions but it has to be a power of two."

Dr. W-abe smiles patiently. "What is wrong with eleven parsed as 1, 3, 6, 1 dimensions? That is popular on Earth, at least for the time being, even though it falls inelegantly between 2^3 and 2^4."

"Abbit is okay with the one, the three and the last one of the eleven dimensions. It is the group of six of the 1, 3, 6, 1 dimensions that present a problem."

"What is wrong with the group of six dimensions?" W-abe asks.

"He says that three of the six are "redundant, recursive, self-similar" or something like that," says Dr. Mim-sey looking at the other chairs with the urge to choose a better one.

"Which one do you suppose it is?" W-abe mutters rhetorically.

Dr. Mim-sey moves to a chair that is closer to the bookshelf and continues, "He does not know. He is approaching the issue from the outside. But I need to know if are you willing to concede his point?"

"What is to concede? Am I to grant the assertions of this middle-aged, retail cashier, this Roy Abbit, who has not passed a math course in forty years? Since then, he imposed on the library solely to write his one and only manuscript. Which is unpublished, thanks to you," says Dr. W-abe tensing as he sees the studied order of his books become a ransom for the conversation.

Dr. Mim-sey stands and starts to reach for a book but stops and says, "He is not an Einstein, however, he is willing to acknowledge quantum mechanics as basic to the theory which their Einstein was not."

"Is this where and how Abbot questions the 'six dimensions'?" W-abe asks.

"He says that the six dimensions probably include three already given," repeats Mim-sey, "The real spectrum of dimensions is 1, 3, 3, 1 and not 1, 3, 6, 1. The first two sets of dimensions and the last one are unchallenged by Mr. Abbit. The set of six dimensions count the previous three again and artificially loads the six dimensions.

"And if anyone wants more than eight dimensions for a Theory of Everything that is okay but it has to be a power of two. That eliminates eleven dimensions but 8,589,934,592 dimensions are okay?" W-abe asks.

"You are uncharacteristically modest. How about a google-power of two?" asks Mim-sey, not expecting an answer either, while he pokes his pocket calculator before putting it away.

In an effort to end the conversation, Dr. W-abe asks, "Has he done the math for eight dimensions?"

Dr. Mim-sey responds with, "No, but he can do the math Chu Shih-Chieh did in fourteenth century Earth."

Dr. W-abe is not clueless but asks, "And that is?"

"You, like many here and on Earth, know the scheme as Pascal's Triangle. Each row begins and ends with the same digit - one. The rest of the digits represent subsets like cards, or for Mr. Abbit, they are the dimensions that satisfy a Theory of Everything," says Mim-sey.

"That is arithmetic. It is not even algebra. What does he know?" W-abe asks.

"Quantum physics is intimately associated with those numbers - probability. Quantum physics and anyone's Theory of Everything must be consistent with those numbers," says Mim-sey.

Dr. W-abe is relieved to see that Mim-sey has not found fault with the bookcase but braces himself for Mim-sey's almost predictable myopic examination of his seascape. Hopelessly, he restates his objection, "Why do the dimensions of a Theory of Everything have to be sorted into eight subsets?"

"That is row three of the triangle but that is not what Roy Abbit is saying," Dr. Mim-sey goes on, "He is saying that if a Theory of Everything cannot be defined in eight dimensions then the next quantum of dimensions must be sixteen. The next even power of two and the next row of the Chu/Pascal triangle."

"If sixteen dimensions are not enough for a Theory of Everything, then thirty-two dimensions must be?" asks Dr. W-abe trying to draw Mim-sey away from the painting.

Still inches from the surface of the painting, Dr. Mim-sey says, "Adding dimensions just provides mathematical room for the Theory of Everything to include more. Increasing the number of dimensions 'smooths-out' the theory. The theory can be explained with eight dimensions. But there are gaps. Your colleagues, those with your mathematical skills, can jump through the holes and eventually eclipse your theory with with a better one."

"So no one can afford to ignore the issue. Anyone, who relies on a theory with too few dimensions, is set for a fall," begrudges Dr. W-abe at least for the moment.

"Eleven dimensions are more than eight. But it is not a power of two. It is vulnerable in two respects," Dr. Mim-sey imposes, "Anyone, using eight dimensions trumps the eleven dimensions but cannot defend the theory with less than sixteen dimensions."

Dr. W-abe forgets his painting enough to say, "A Theory of Everything with sixteen dimensions is 'safe' until one with thirty-two dimensions comes along."

And Dr. Mim-sey restates his fundamental focus with, "Yes, and with each power of two the theory becomes more airtight and surpasses any theories with fewer dimensions."

Dr. W-abe consoles himself with the supposition that he has lost nothing but logical ground and asks, "And this middle-aged nonmathematician says so?" as if to be done with the issue.

"Seven hundred years of Earth's arithmetic says so," says Mim-sey, as the way to tell Dr. W-abe he might be assuming too much.

Episode 2.

Arlington, Virginia is not in the District of Columbia. But Roy is too young to comprehend municipal boundaries. So, for him, Arlington is part of Washington D.C. and it stays that way for him long after leaving Virginia.

This is where young Roy lives with his parents, Edgar and Jenny. They live in an apartment complex that is a mix of military and civilian residents who work in and around the District of Columbia. It is a relic of military housing that predates the second world war. They live there because it is what they can afford. Mr. Abbit makes little more than a janitor and Mrs. Abbit makes less.

Things were better in Italy but they are in the United States now. Here is the promise of a middle class life for the family someday and something better for the children. But for the time being, there is a roomy apartment with a minimum of furniture for the bedrooms and kitchen and none in the livingroom. The rest of the household is in storage in Italy waiting for the day Edgar and Jenny have the money to ship it here.

These things take time but this is not comprehended by Roy. Christmas is coming. It always does. But why the furniture? Furniture does not come once a year. Everyone he knows has furniture. None of them have furniture waiting for Christmas.

Furniture is not as important as toys. Toys are what you wait for, not furniture. And then there is the baby. We are waiting for the baby too, but no one knows if it is a boy or a girl. If you know a baby is on its way you should know if it is a boy or a girl.

At the moment Roy is outside under a grey sky in a grey canyon of angular apartment buildings near a cavernous stairwell that goes up to the mocking emptiness he calls home. There are no friends and nothing to do. It occurs to Roy that all you have to do is to go through the neighborhood and find someone to be your friend. He is not sure he will see anyone but he is sure that he must remember the way back. It is important to remember the way back. Without knowing or even wondering why, Roy finds no one. He walks as far as the cemetery and comes back home. The only curious thing is the rise in the sidewalk. Sidewalks are supposed to be level. When the ground is flat, the sidewalk should be level. Someone is not being responsible.

Before dinner Jenny asks Roy about school. At one point she asks him to spell "bread". But Roy is aware of the bread commercial on television. From his earliest years to his last, Roy's apparent mindlessness is derived from his penchant to focus on tangents. He does not respond to the spelling question before it is eclipsed with the thought, "*Is this a test? Am I going to be called a cheater, if I look at the TV?*" This supplants his concern with spelling anything right or wrong.

His mother says, "It is right there on the television and you cannot even spell the word bread."

Roy does not respond to that, either, because he is still preoccupied with cheating.

Jenny is concerned to the point of distraction about her nonresponsive son. Roy for his part only makes things worse by not trying to guess what she is thinking. He sits and waits for Jenny to do something. She goes into the kitchen without saying anything. Now, Roy has to decide on his own whether he should stay in his chair or not. There is no good reason to stay in the chair. But what is a good reason to get out of the chair? Whatever he decides to do, he prepares to answer the question, "Why are you just sitting there?" or "What are you doing out of your chair?" accordingly. What he finally decides to do depends entirely on which question he can answer best. Roy settles for crossing the empty livingroom to look out of the window.

After dinner Roy feels free to walk to the cemetery. It is a short walk through Fort Meyer. He is almost certain, Miss. Meyers, his first teacher, lives there. Consequently, Fort Meyer becomes Fort Meyers (with an 's'). But he is more certain that he has to be back before dark. At the edge of the cemetery, Roy hears someone. But it is not Miss. Meyers or his mother. It is quite ordinary to hear people who are out of sight for part of the time. It is as ordinary as someone in another room or in dreams.

The voice projects pleasantly, *"Roy Abbit, do you remember me?"*

"I cannot see you and your voice sounds like mine," Roy reflects. He tests the voice with the question, "Is this a dream?"

The voice offers no focus of image or location. There is only the voice. The voice projects, *"There are differences but visits like this are remembered like a dream. Do you remember me?"*

Roy reflects, *"Ali-ce? We have talked a few times before."*

Ali-ce is pleased that Roy remembers and does not need the support of ordinary clues. "*We have chatted many times,*" projects Ali-ce, "*but most of the conversations were when you were older.*"

Roy reflects without trying to find Ali-ce, "*I know I am seven. I remember all the way back to kindergarten, Miss. Meyers and a few things before. I have not been older. We have talked a few times, not many times. When I am older, I suppose we will talk some more.*"

"*Much more. I will talk to you when you are younger too,*" Ali-ce projects.

Roy is trying to chip some paint off a fence and reflects, "*I am not going to be younger.*"

"*You have been younger,*" projects Ali-ce, "*and I will talk to you when you are younger.*"

Roy is pleased to have the company of Ali-ce but the idea of being younger is impossible. He gives up on finding loose paint and is giving up on Ali-ce but reflects, "*When I am younger do I go back to the hospital?*"

"*Do you remember the hospital?*" Ali-ce projects.

Roy reflects, "*No, Mom says that is where I came from and that is where I will get a brother or a sister soon.*" Roy wishes Ali-ce is more than just a pleasant voice. He reflects, "*This has to be a dream, I do not see anybody.*"

Ali-ce reflects, "*You never had a dream where you did not see any of your body?*"

"*I was in my underwear. I could not see myself but I knew I was in my underwear. I could see my classroom and I hoped no one would see me or at least notice that I was in my underwear,*" Roy reflects, "*I remember that they did not notice. I was glad about that.*"

"*We will make people and clothes for the visits after this. When I visit before age seven I will make people and clothes if you want them,*" Ali-ce reflects, "*We do not make dreams but we made this pavilion.*"

Roy glances at the pavilion and reflects, "*Washington has lots of this stuff. Some of it is bigger.*" Roy remembers Ali-ce but she is not like Miss. Meyers. He reflects, "*You will you visit me when I am younger?*"

Ali-ce projects, "*Yes and when you are older.*" She waits while he tries to focus his memory.

Roy is beginning to think she has not been to the school. "*Have you visited me before when I was younger?*" reflects Roy, "*I do not remember you from first grade or kindergarten.*"

Now Ali-ce understands his question and his doubt, Al-ice projects, "*I have not visited you then, yet. You remember me from when you are much older.*"

"*Older than my parents?*" Roy reflects with the confidence.

"*You do remember,*" Ali-ce projects innocently.

Roy reflects, "*I was trying to trick you. I do not remember being older than my parents.*"

Ali-ce projects, "*You just tricked yourself. You have not been older but you will be older. I have visited you many times when you were older and I will visit you a few times when you are younger. Remember this pavilion and remember me and you will see. This is not a dream. Now the sun is setting and you ought to go home before it is dark. Good-by.*"

"*You are right,*" he reflects, "*Good-by,*" and he goes home like it is any other day.

Episode 3

The building, which Dr. Mim-sey and Dr. W-abe, share is separated from the river by a small park. The park is separated from the river by a low railing. This patchy patch of green is an historic site which saves it from encroachment. Urban land is virtually priceless and this little plot is easily more priceless than most.

But for hundreds of years it was a batpark. A few hundred years after it fell out of use, it went from city property into private hands. From then on, the owner dedicated his remaining years to securing its historic designation and building a monument to 'batball.' Batball has become big business. The great size of the batball business and the unprofitability of this park's size combined to preserve it.

It is now used to commemorate greatness, current greatness, to give some patina to the game. A great batter or an accomplished pitcher is retired here along with his number, glove or bat. Later, if he is inducted into the Chamber of Fame, it is from this same park that he retires along with whatever equipment is retired with him.

For those who do not know the game, they would guess that batball is played on one of the three green squares that touch at the corners and define a triangle in the center. Home base, first and second bases are located at the vertices. Home base is at the vertex with the angle of 90 degrees. Green squares always vary in size among parks with one exception.

The area of the square that borders the lines between home base and first base together with the area of the square that borders the line between second base and home base must equal the area of the square that borders the line between first and second base. For a park with 90 feet to first and second base, the square opposite home base must be 16,200 square feet.

For this park, it is 70 feet to first base and 120 feet to second base. First base and second bases are separated by 138.924 feet with the square adjacent to this line and opposite home base of 19,300 square feet. There is a pitcher's mound for each base and a fielder for each square.

A batter and pitcher face-off, first at home base. The batter who makes it to first and second base faces the pitcher again at each base. Any ball that is caught by the fielder before it touches the ground is an out and the batter is retired. Any ball that is caught outside a green space is a base hit. If the batter can get to the next base before the fielder throws the ball to the defending baseman, he is safe and faces the pitcher again.

The same is true for any ball that touches the ground without being caught first. A ball that touches the ground behind the base-at-bat without being caught is a foul ball. Four foul balls or miss-strikes are an out and the batter is retired.

Miss-strikes are failed attempts to hit the ball. This includes no attempt at a ball in the strike-zone. Miss-strikes and are called by the strike-judge. After four outs, the teams trade positions, up to twelve times, and collect points for every batter that completes the circuit of bases.

Each park takes its pride in not being the same size as any of the others. The differences in distance between bases is deliberately unique for each park. These differences affect the number of points that can be had in a game but in the end there is only one winner.

The fundamental proportion of $a^2 + b^2 = c^2$ is common to all the batball parks. Those squares that are adjacent to home-base are a^2 and b^2. The square that is opposite home-base is c^2. It is a common, leveling, proportion that subsumes all debate about the game.

It becomes an anchor. An exception without exception. A cosmic touchstone. A vein of purity. The proof, somehow, that there is a rule, with the gravity of a law, that eclipses not just individual excellence but the excellence of a team. This fundamental proportion rises above everything.

It is even at the root of curves and curved dimensions. Without limits on the shapes, numbers or dimensions of spaces, chemistry is born in the form of particles too small to see. A physics emerges without limits on time, dimensions and even the means to measure its constituents. Without such limits comes cosmology. With physics and cosmology comes an extraterrestrial commerce and culture along with uninhibited space travel.

The significance of batball and the park is not completely fathomed by Officer Gim-bal but it is not lost on him either. It is preserved not as a sense of awe but as a sentiment. A batball park is a righteous place and righteousness is Gim-bal's most sympathetic perception.

When he wants to confirm his rightness or support it, he immerses himself in a batball park. Even this crude approximation, this empty park at this behemoth building on this titanic river plucks at his sympathies. So this is where he prefers to meet Dr. W-abe, at least this time.

Dr. W-abe surveys the desiccated, uninspiring landscape and grumbles, "I cannot be tempted to believe that two million years of civilization is going to come to an end at the hands of the Earthlians."

Gim-bal wonders if the tattered gravity of this park is the source of Dr. W-abe's subdued mood. "It is not going to happen overnight. I prefer to believe that it will never happen but it is not as remote a possibility as it used to be."

"They have to discover us, get to us and then erode us," says Dr. W-abe looking across the river. "They have to do that without our cooperation and before we simply fade into our own history."

Gim-bal asks, "I suppose 'fading into our own history' is the natural decline of our civilization?"

Dr. W-abe's gaze is fixed. He says, "Yes, if we face an end, that is the one to be preferred."

"So the task becomes working against Earth discovering us?" Asks Gim-bal.

"Maybe not," notes W-abe, "it is one thing to suppose that the visible universe is, say, 12 billion light-years across and it is another thing to cross it in, uh, less than 12 billion years."

Gim-bal can see Dr. W-abe begin to relax. Perhaps it is the river. "Overcoming the vast time barrier between us and them still leaves the task of overwhelming our civilization."

Dr. W-abe asks, "So how much of a threat can Roy Abbit's category mathematics be?"

Gim-bal deliberately understates his point, "There is a chance he could be no threat at all."

If Dr. W-abe notices, he ignores Gim-bal's suggestion and continues his list of possibilities. "But the larger danger is that we could make Roy Abbit the pivot that turns history against us."

Gim-bal considers Dr. W-abe's point. "The simplification is an unrealistic temptation. Suppose the Earthliens are convinced by circumstances that he is the object of assassination by us as extraterrestrials. Is that enough of a grudge to launch ships across space in the name of revenge?"

Dr. W-abe loosens his gaze from the river. It seems to have served its purpose. He asks, "So the huge gap of treasure and time could exhaust the motive of revenge?"

"At this stage of the game, yes," says Gim-bal, "it would take many thousands of years of coexistence and proximity before any assassination could turn the tide of Earthlians against us."

"So the real task becomes that of not accelerating events by murdering Roy Abbit which would promote his mathematics. Our task becomes one of killing his scheme or ideas?" questions W-abe.

Gim-bal says, "There is a chance even that much will give currency to his mathematics."

Dr. W-abe turns and surveys the dry, stony park. "A failed attempt or a less than perfect one to suppress his mathematics would focus unwanted attention on it and hasten history against us."

Gim-bal is ready for this conclusion. "Even without the excuse of harm to its author, Roy Abbit."

"The conclusion at the onset, is that we must protect both Roy Abbit and his mathematics from our own interference," says W-abe.

Rus-ty Gim-bal's professional experiences and mental reflexes require him to ask, "Are you convinced of that?"

"Not completely," says W-abe, "but it is consistent with my prejudice that Roy Abbit is not the real danger and that tinkering with ancillary issues could just make things worse."

"That suggests that your friend, Dr. Mim-sey, could do, or provoke more harm than good," says Gim-bal.

That is just one of Dr. W-abe's private fears and he suppresses them with, "It is too early to tell. I have to leave the city for a few days. Will you be here when I get back?"

"No," says Gim-bal, "but I will look you up when I am." He knows it is the end of the conversation for now but it is not the end of the issue. Dr. W-abe is not avoiding him and will not.

$$1$$
$$1\ 1$$
$$1\ 2\ 1$$
$$1\ 3\ 3\ 1$$
$$1\ 4\ 6\ 4\ 1$$
$$1\ 5\ 10\ 10\ 5\ 1$$
$$1\ 6\ 15\ 20\ 15\ 6\ 1$$

Chapter 2

Chapter 01000....

Chapter B

Chapter $(1 + 2x)$

Chapter $(a^1b^0 + 2a^0b^1)$

Chapter 2^1

Episode 4

    Dr. Mim-sey is in the flight-port waiting for Dr. W-abe's return. The flight-port is near the top floor of their building. Dr. W-abe does not travel a lot and he does not like commuting. He cannot get any significant work done during a commute and while he has little regard for anyone he knows, he has less for those he does not know. Commuters should stay home.

But it is the flight-port that tips the scale in favor of buildings that host as many as his does. Its effect is not obvious at first. From the beginning, flight-ports on the top of buildings were promoted with great fanfare but avoided by hoards.

Acreage, which building tops have, landscaping and every material incentive that can be provided fails until the old, ground-level flight-ports begin putting in their own second-story runways. Nobody uses them either. But passenger volume on the ground-level runways remain high and volume increases in some cases.

It is as if second-level runways do not exist. Except the riders do take any flight that is scheduled for a ground-level landing and refuse to take any flight scheduled for the upper-level runway - takeoff or landing. An emergency landing that uses an upper level is an extreme emergency in the minds of everyone. Media and attorneys make the most of it. Scale is irrelevant. It is like landing on a pinhead.

It is an historic debate whether it is genius or an accident of necessity but one builder put his flight-port three floors below the top floor. The top floor tenant wanted the top floor, the flight-port and insulation from the flight-port environment.

So the client gets it. The builder has to do it without selling any real estate on the flight-port floor or on the ones just above and below. The builder can only hope that this does not catch on because he is not going to get rich that way.

But he does get rich. Passenger volume takes off at the maximum and the builder even has to add a second runway. Shippers and air transport related industries cannot pay too much for the few acres that exist above, below and next to the flight-port. That is when the lights go on. Passengers want to see architecture when they look out of the flight-craft's windows.

Just because the passengers are pinheads, it does not mean they want to land on one. A building facade near the runway makes the takeoff and landing psychologically secure. The builder has learned his lesson. Now, buildings are never too big. Acreage is never too expensive. As long as the flight-port is not on top of the building, success is certain.

Dr. Mim-sey is still preoccupied with this when he notices Dr. W-abe has brought up the other subject. "If a Theory of Everything can have any number of dimensions...," says W-abe.

Dr. Mim-sey says, "...as long as it is a power of two...."

"...then why just eight dimensions or sixteen and not some higher power of two?" W-abe asks while going into the Port Lounge without hesitating or indicating whether a drink is his idea or Mim-sey's.

If Dr. Mim-sey wonders about W-abe's motive, it is not for long. He quickly expands his own purposes to include the drinks and conversation. "Just more blanks to fill," responds Mim-sey "We have the three dimensions of 'regular' space and the fourth dimension of time . A one and three of the sequence - 1, 3, 3, 1."

"Four dimensions are common to our experience," says W-abe.

While Dr. W-abe finds a seat, Dr. Mim-sey says, "A Theory of Everything with sixteen dimensions would have to satisfy twice as many dimensions. That is a lot more work."

"We aren't limited to Newton's and Eienstein's rules," says W-abe.

"We have not been limited to ordinary experience for a long while. Chemistry and physics brought many of us to molecules and atoms," says Mim-sey. "Fewer of us are familiar with Quantum Mechanics and even fewer, still, are familiar with string theory."

W-abe knows his choice of seats is fine if Dr. Mim-sey gets a view of the door. "So how does this cashier jump into this ever shrinking circle?" W-abe asks disparagingly.

Dr. Mim-sey is trying to decide if his vantage point is optimal. "We do not have to go back very many centuries to forget everything we ever knew about Pascal's triangle."

"I left it behind with the binomial theorem," counsels W-abe, "with that anyone can calculate any number of coefficients."

And with that you do not need to know anything about the family of coefficients you are not using," demurs Mim-sey while deciding he does not like the Port Lounge.

"The family of coefficients I am not using?" W-abe asks while feeling like he is being ignored by the wait service. "The three dimensions of space have a long and familiar history."

"But time has joined the family only lately in the history of math and physics," proffers Mim-sey with an anxious search for a waiter.

Dr. W-abe transfers his concern for the wait-service to the math. "So time is allocated to the first or last digit of the series one, three, three, one? And three-dimensional space is allocated to the second or third digit of the series? And a Theory of Everything only needs to define the remaining set of three dimensions and the odd one dimension? It has never been that easy"

Dr. Mim-sey welcomes the relief that comes with the approach of a waiter and coaxes, "But a Theory of Everything is nothing if it cannot account for every possibility that its components suggest. It has always been that difficult. And Pascal's triangle simply tallies the number of possibilities."

Dr. W-abe orders drinks with a marginal interest in Mim-sey's spin on the math and then asks, "If 2^3 combinations are not enough must we use 2^4 components?"

Dr. Mim-sey is not disarmed by Dr. W-abe's condescension. "The price to be paid is that sixteen combinations or dimensions must be accounted for," says Mim-sey.

"The universe follows this rule whether we know it or not?" W-abe hectors.

Mim-sey takes the opportunity to challenge Dr. W-abe, "The universe does not avoid any possibility just because it is rare or sublime," maintains Mim-sey.

"So when we describe the universe with 2^2 combinations of possibilities it is the same universe as one with 2^3, 2^4, or 2^n combinations of possibilities?" W-abe asks.

Now, Dr. Mim-sey makes his point with more substance. "Yes, but with more combinations you get finer details. And with finer details, we get more of what the universe is about."

Dr. W-abe finds himself just where Dr. Mim-sey wants him. "So it does not matter if time as a dimension is the first, last or any number of a row of Pascal's triangle?" W-abe asks.

"Time can be part of any of the numbers in between and in any row," says Mim-sey.

Dr. W-abe on getting Dr. Mim-sey's point mutters, "Just do not count the same thing twice."

Episode 5

For Roy this is not Ali-ce's second visit. He is forty-seven years old. For Ali-ce this is her second visit and it has not been forty years since her last visit. She finds Roy in the state of Florida and mentally exhausted.

He has finished seven years of working for a vocational school that trains aspiring emergency medical technicians and pharmacy technicians. It has been his job to translate the curriculum to a computer based medium and teach the basics of computer operation to get at the course material. This results in five thousand multiple choice questions in two hundred chapters. The multiple choice questions are written to be parsed by the computer into five thousand fill-in-the-blank questions, five thousand essay questions and twenty thousand true/false questions. These thirty-five thousand questions directly support the two courses of instruction during a period where relevant software is the rarest commodity.

But for those seven years the material is officially supplemental in nature and therefore optional for student use. This option is a low priority on the list for people who have full-time jobs and family concerns to compete with the six-month vocational course. No matter how easy it is to use the computer-based training programs on the topic, it is easier not to use them. So much easier that when improvements are made to the computer lab, they do not include Roy or his course material. Ali-ce finds him working as a cashier and putting his sparse hopes and skills into a mathematics manuscript.

They are crossing the remoteness of Virginia Key, near Miami, to get to her pavilion. The key is used for waste processing and is no longer a place for recreation. It is a sunny walk from the causeway. The taxi driver knows why they are there. Since they are not dressed for the beach and carrying nothing, he only has to remember the beach's history as a nude beach to know their story.

Ali-ce, plainly unimpressed with the sandy road suddenly asks, "What is that charmer?"

Roy says, "It is a marble," guessing at what she is looking at on this dusty featureless road.

She picks it up but seems more captivated by the object than it merits. "Marbles are forms of limestone. Its charm is in its polish and the plurality of its colors," says Ali-ce, "The colors and patterns are sublime and captivating but they are fixed as they say 'in stone'."

"This is a different marble. We have round glass toys. They are called marbles and my favorites since before your visit when I was seven. They are manufactured and cheap. So cheap you never buy just one. I never asked my parents to buy me any because they did not have the money. It would not be right to make them alibi their place as parents with phoney indifference in order to avoid saying they could not afford marbles," says Roy.

"You did not think that way at the time," challenges Ali-ce. "You were too young."

Roy responds, "Not in those words. Our house was upstairs. The room by the front door had no furniture and no toys. We had a table and some chairs in the kitchen. But we had to wait for Christmas for toys. It was not an unwillingness to wait on my part. I did not know when Christmas was but if it was coming it had to be soon. So not wishing to anger my parents and a willingness to wait for Christmas, it was enough to settle for admiring the marbles that I saw from time to time. It was not a matter of feeling jealous or deprived because I knew that I would have my own."

"Did you get marble for Christmas?" Ali-ce asks diligently wiping the dust from the new treasure.

"I do not remember but I do remember adoring marbles without regret. I enjoy them, now, without longing for any," says Roy looking for some water to polish Ali-ce's road-gem.

"But this marble is a charmer" says Alice focusing on her find.

"That one is not a marble. It has dimples. Marbles are perfectly round," clarifies Roy.

"That is a requirement?" Ali-ce asks examining the object as much with her fingers as with her eyes.

"Yes, they are used in games," explains Roy, "Fairness requires that they be similar to other marbles. Small marbles are similar to other small marbles. And large marbles are similar to large marbles. When you meet someone with marbles, you can game with them and hope to win the game with talent since the marbles are similar."

"So they have to weigh the same and have the same shape so that gamesmanship is the only difference. The winner is the better gamer, the better," grants Ali-ce.

Roy wonders, "What is the charm in this dimpled, ball of glass?"

"I can ignore the dimple and the uneven shape but the inside focuses ordinary light into a sharp violet light," bubbles Ali-ce as she hands Roy her objet d'art.

Roy takes a close look at the violet nugget. "I imagine the color of the light comes from the color of the glass but it is sharp and purple and you are right, now I see it is quite intense," says Roy.

"Of course this glass is not typical," Ali-ce judges retrieving her curio.

"I do not know. Except for chemical differences and surfaces most glass is the same, I suppose," says Roy as they turn east to a path three-quarters of a mile from the causeway.

"Then most glass is melted and formed for use," supposes Ali-ce.

Roy says, "We have some glass that can be used in hot and cold situations for cooking and freezing but most glass is just manufactured by melting and forming for use."

"Here is my pavilion. Do you remember it after all these years?" Ali-ce asks.

"It looks very much like I remember it," says Roy stepping out on the beach and putting the pavilion between himself and the trees.

"Does this pavilion look like most glass?" Ali-ce asks trying to gauge Roy's impressions.

"Yes, but I do not see any sections or seams so, somehow, it must be one piece," stepping from the right to the left to change the effects of the background on his view of the pavilion.

"I know you have lasers in your industry. Do you know what a laser is?" Ali-ce prompts.

Roy says, "Yes. They have ordinary and exotic uses from hanging pictures to warfare."

Ali-ce walks with Roy as he moves around the perimeter of the pavilion looking and touching, "Then you will not be too surprised to know that we use lasers to make this pavilion."

"Some edges are sharp. Are they carved by lasers?" Roy asks.

"This glass is carved and cooled by lasers," responds Ali-ce.

"I guess, if the air above the melted glass is too hot, it will not cool," says Roy

"So we just pull some of the glass into a cooler slip which fixes the shape of the glass," says Ali-ce as they step closer to a hatchway.

Roy argues, "I suppose the slip floats over the glass. But lasers heat the glass not cool it."

"Lasers maintain the heat at the surface of the pool of glass," Ali-ce relates, "The slip will be cooled and circulated over the glass. But the laser is focused to a narrow line. By directing the laser along the edge of the pulled glass the cooling can be delayed.

The cooling can be accelerated by slowing the vibration of the metal atoms in the glass with the laser and the viscosity of the glass. The glass adopts the shape that cools just above the line of the laser which is just above the pool of melted glass."

"Like Michael Angelo's 'David'," says Roy touching the pavilion.

Ali-ce steps on a ledge of the pavilion to get a better look at a niche. "Who is David?" At the same time, Roy gets a better look at Ali-ce. "The 'David' is a statue. It was so big that Michael Angelo made a model and put it in some water." Roy conveys, "As he progressed, he lowered the water on the model and knew what to carve next."

Ali-ce follows, "The water and the laser line outline the object."

"But this pavilion must be ten times the size of the 'David'. Can this pavilion or a copy of the David be pulled from a pool of melted glass?" Roy wonders.

"The accuracy, along with the symmetry and polished finish of the surface of the final object, depends on coordinating the heating and cooling effected by the laser. This is more efficient than cutting and polishing cold glass," says Ali-ce in an effort to respond to Roy's question.

Ali-ce hops back to the sand a little to Roy's disappointment. "I suppose some of the surfaces could be manufactured, chiseled, ground and polished as a finishing process," says Roy.

"But for artifacts this large, the process would take much more work and treasure to manufacture," Ali-ce observes, "So the finishing process is accomplished by more lasers to control the drawing and cooling of the artifact along with the final surface as well."

"Can you justify all the lasers it takes to finish an artifact like this pavilion? The cost of energy and other capital must be very large," says Roy as he walks a little to the north for another view of the hulking pavilion.

"It turns out that there are some benefits that do not meet the eye," Ali-ce offers.

Roy says, "I suppose the sheer clarity of the structure is not all there is to it."

"That much we both appreciate. The few grams of a marble and the kilotons of this pavilion are just as transparent," enthuses Ali-ce without concern for the contrast in scale.

"This pavilion is thousands of tons of glass?" Roy asks trying to comprehend the size of the pavilion from the north side.

"Yes," says Ali-ce walking along with Roy and wondering what is so remarkable.

Roy tries to see the top of the pavilion but its curve and the sun frustrate his effort. He says, "But there is no reinforcement to support the glass. The transparency of the top and the bottom of the pavilion is the same. The lower reaches should be thicker and much less transparent," as he continues his trek around the pavilion.

"You are right the 'lower reaches' as you call them are thicker. But the lasers can be used to align the molecules of the molten glass and enhance the strength of the glass as it cools. Once cooled, the atoms of the constituent glass cannot flow and do not compromise the strength of the artifact," says Ali-ce with a view of the beach and its water through the pavilion.

"Ordinary glass cannot survive a lot of weight or stress," Roy grants.

Ali-ce puts a hand on the pavilion as if to confirm her thinking and says, "The glass you are talking about is like peanut butter and jelly. Stir them together and they make layers and folds that only mingle. The mix is only as strong as the friction between the peanut butter and jelly."

Roy says, "If you press the bread together, the jelly and peanut butter travel. They squirt out of the sandwich. The force of the squeeze on the sandwich is transferred to the peanut butter and jelly and they transfer the squeeze by flowing out of the sandwich."

"This glass is not just homogeneous. Its molecules are aligned with the help of the lasers and a more complex structure is achieved by adding metals. Now with the physics of electromagnetism, stress does not have to overcome the friction of the molecules, it has to overcome the electrostatic cohesion of the molecules," says Ali-ce.

Roy says, "This is like the difference between cement and mud," still studying the pavilion.

"Yes. Except the constituents of glass and its impregnated metals are molecular. This is important. Lasers comb out the molecules to an optimized electrostatic/friction pattern. Now, the flow of the glass has to overcome electrostatic forces if it is going to fail as a structure," Ali-ce says from the south side of the pavilion.

Roy says, "I suppose 'aligned molecular glass' is more transparent and stronger."

"This pavilion is taller and stronger than any structure that can be made with ordinary glass," announces Ali-ce, "It is stronger than metal except for transparent (aligned) metal."

"The difference is as subtle as the difference between a brick and diamond," says Roy.

Ali-ce looks at Roy. "These colors, we see, are from the metals that are added to the molten glass," adds Ali-ce, "Gold makes the glass red, cobalt makes it blue and uranium makes it yellow."

Roy wonders what the uranium does but does not ask. "Why is the color important?"

While the colors are what Ali-ce likes the most, she responds to Roy's question, "These colors are just a side effect. It the arrangement of the metals in one end of the pavilion that permit the storage of huge amounts of electricity. It is like a capacitor. The metals in the other end are like a huge coil. There are metals and glass connecting the coils and capacitors to control the flow between them," says Ali-ce while facing the east side of the pavilion and without supposing another trip around.

"Even if this pavilion is a big tank circuit. What is the advantage of circulating electrons back and forth?" Asks Roy vaguely recalling the principle from his years in the navy.

"When the flow it is fast enough and in the right direction we overcome local gravity and we can move away from any source of gravity," recites Ali-ce drawing on her training as well.

Roy jumps back. The sand gives and he slips. "This is a ship?"

Ali-ce is surprised at his surprise. "It does other things too," Ali-ce says defensively. "We can use massive objects to speed up or slow down. We do not have to shift electrons so much. We can use the coils to tap magnetic sources and resupply electrons that get lost in the process."

Roy says, "If you move as many electrons forward as you do backward, there is no change in inertia. It cancels out," as he steps a little further from what is now a ship for a better view.

Ali-ce is pleased that Roy does not take too much for granted. "That is the trick. We move electrons quickly to create the inertia we want. The larger molecules of the ship try to flow with the large current," discloses Ali-ce, "So the mass of the ship tries to flow in the direction of the current. Then at a slower rate, a lower current flows from the capacitors to recharge the coils. There are many layers of coils and capacitors that discharge quickly and recharge slowly."

Roy prompts Ali-ce with a smile, "Are you planning on taking me somewhere?"

Ali-ce considers the options but says, "No," looking for some way to say yes.

"Then why are you telling me this?" Asks Roy still inspecting the ship from top to bottom.

Ali-ce responds, "You were only seven on my first visit. The subject barely came up."

"But I am forty-seven and we have not had many visits," says Roy.

That much she grants and says, "We will have many visits. Some of them will be after you are forty-seven; some of them will be before you are forty-seven. This is our second visit."

"You are going to visit me when I was younger? That is impossible," contends Roy.

"It is not impossible to visit you in the future or the past. You say you remember visits in the past," says Ali-ce with the surrounding foliage growing darker in the approaching sunset.

Roy says, "I feel like I have seen you many times before," as the ship also loses definition in the fading light.

"Then I will visit you in the past. If you remember it, it has happened or it will happen," concedes Ali-ce as they approach the ship in near darkness without street or other lights.

Roy says, "I do not remember any future visits. I do not know anything about my future."

"But you know you will live in the future," says Ali-ce raising some of the ship's lights as the darkness and quietness blanket the empty beach.

Roy grants, "I have to say yes."

"None of the future visits have happened yet," conveys Ali-ce while opening the entrance hatch to a well-lit room inside the ship, "Besides your brain is selective. It will not allow you to remember any future visits until you have lived through the intervening years from now to then."

Roy says, "That is impossible," to both the apparent substance of her remarks and the apparent size of the cockpit.

"Do you remember any of the months before you were born?" Ali-ce asks.

"No, No one does," maintains Roy sitting in a chair too comfortable to be simple glass.

"Everyone does. No one knows they do," asserts Ali-ce.

Episode 6

Officer Gim-bal does not go directly to his office on returning to Earth. Instead, he goes to New Orleans to get a look at the Mississippi River. The location he has in mind is similar to the riverfront where he left Dr. W-abe. It is in front of the Aquarium of the Americas. Here, there is a large, flat, paved area that fronts the river. There, he wants to do two things.

He wants to focus on the similarities of this river to the one outside Dr. W-abe's office. And he wants to confirm a certain story before he leaves New Orleans. Here, the river is wide, calm, as warm as the gulf waters. It seems inconsistent with the story about the drunks who tried to swim across to the other shore. These swimmers never made it and were never found. Strong currents are said to have exhausted the swimmers and they drowned.

He has a problem with that story just looking at the river. As slowly as the river seems to flow, it seems that there is nothing to threaten a swimmer. It seems like the swimmer would make it across but get to the south bank way down stream. Perhaps, it is a dangerous and ignorant thing to do but it seems possible.

This location is shared by boaters who make their living on the river and from tourists. Boaters need the mooring space and the tourists find the boats atmospheric. Tourists never say much, just mostly look. The boaters are here to repair, replenish and return to the river and what makes money. Boaters act like they have seen tourists forever while the tourists act like they have never seen boats before.

Gim-bal ventures, "Excuse me, I am Rus-ty Gim-bal. I wonder, Do you work on the river much?" standing farther from the boat than the usual tourist.

"That is funny. I know you do not live on the river, not with that name," guesses the sailor while faking the loose end of a line.

Gim-bal asserts, "I never thought about that. But you are right. Anyway, I suppose you live or work on this river," still without approaching the boat.

"All my life," attests the sailor finishing his line and noticing Gim-bal's distance.

"Have you ever heard of anyone trying to swim across this river?" Gim-bal asks.

"Two or three times," says the sailor moving to the bow and away from the gangplank.

"Did anyone ever make it?" Gim-bal asks matching the sailors move without closing the distance to the boat.

The sailor concedes, "I never heard of them making it across the river down at this end."

"They get hit by a boat?" Asks Gim-bal a little beyond the bow.

"Not as I hear," he says satisfied that Gim-bal is not attempting to board.

"What do you hear?" Gim-bal asks a little grateful for the shade of the boat.

"Big catfish," says the sailor resting his foot on a winch.

Gim-bal asks, "Do catfish get that big here?" As he steps a little closer to the boat.

"They get mighty big but I do not know if they get that big," grants the sailor.

"Recently, I have seen pictures of big catfish that were caught in South America and Africa," declares Gim-bal while a small boat pushes a large barge past the boat.

Without acknowledging the traffic or the heaving it causes his boat, the sailor asks, "How big was it?"

"The one I remember, four men were holding it for the picture," says Gim-bal.

Taking his foot off the winch, the sailor says, "That is big enough to pull someone down."

"No doubt, where there is one, there are two or three," says Gim-bal.

"You may be right," says the sailor as his boat settles into calm.

"Do you suppose there are enough swimmers to feed the big cats?" Gim-bal asks.

Putting a foot up on the winch again, the sailor says, "That does not seem likely."

"What do they eat, most of the time?" Gim-bal asks.

"I have not though about it but the river is more than a thousand miles long," says the sailor, "There must be lots of stuff falling in all the time."

"It ought to be ready to eat by the time it drifts this far," maintains Gim-bal.

Glancing at the water with a new eye, the sailor says, "You've got that pretty much right."

"In spring, a flood in Iowa is going to kill a few cows, horses and pigs," lists Gim-bal.

"By the time they get down here they ought to be ready to eat," says the sailor.

"I guess two hundred years ago they had buffalo and bears to eat," proposes Gim-bal.

"Maybe," grants the sailor, "During the Civil War, they had their share of Yankees to eat."

"Slaves too," offers Gim-bal taking his turn at looking at the muddy river.

"There is where you got it wrong," says the sailor looking down on Rus-ty.

Returning his focus to the sailor, Gim-bal asks, "Why not slaves?"

"Seems like they would know better," says the sailor, lifting his cap slightly and replacing it. "Anyone who lives with this river has to know the river."

Wondering if that gesture signals the end of the conversation, Gim-bal asks, "So you would not swim in this river?"

"Not on your life," hawks the sailor without any other clues as to his mood.

Turning away from the bow a little, Gim-bal asks, "Because of the big cats?"

"I never think about the big cats. I just think, if you go in the water, you are going to die," explains the sailor.

Facing the stern, Gim-bal says, "It does not matter what gets you."

"You are just as dead whether it is a big river or a big cat," counsels the sailor.

"I guess you are right. Thanks," says Gim-bal as he walks away from the boat.

Getting back to work, the sailor says, "Anytime."

1

1 1

1 2 1

1 3 3 1

1 4 6 4 1

1 5 10 10 5 1

1 6 15 20 15 6 1

Chapter 3

Chapter 11000....

Chapter AB

Chapter $(1 + 2x + x^2)$

Chapter $(a^2b^0 + 2a^1b^1 + a^0b^2)$

Chapter $2^0 + 2^1$)

Episode 7

    The `-burbs, like the flight-ports, are gold mines for the building's owner/builder. And like the flight-port, the building-burbs contribute to the viability of the building.

More than four million gallons of water a day are used and recovered in the building where Dr. Mim-sey lives. The green spaces of the building-burbs are at the terminus of water recovery and reuse system. All of the waste water for the building is used to make a slurry of the trash and garbage from the tenants and guests.

That slurry is pressurized and it is heated to a temperature of over seven hundred degrees. Pressure keeps the water from turning to steam. The heat decomposes the slurry to carbon compounds, salt and metals. Then at normal pressures and temperatures the components are separated. The sterile water is used to keep the building-burbs green and reused by the buildings occupants.

In order to maintain water and green space quality the landscaped areas are accessible only to `-burbers and maintenance workers. This way anyone with access can also be held responsible for any abuse or pollution of the space. There is no public green space. This exclusivity is touted as a selling point and of course raises the value of the '-burb homes.

So building-burb inhabitants with access to the green space, pay handsomely for the luxury of the location. The green spaces range from as little as 20,000 up to 110,000 square yards. Then they are arranged in an archipelago of green spaces that transect the building. The concomitant canyon provides light and a microclimate that serves air conditioning.

Dr. Mim-sey likes the view from Dr. W-abe's balcony as much as his own. Dr. W-abe does not think about the view. He likes the insulation of isolation and the short commute from the office. W-abe does not mind the company of Dr. Mim-sey as long as that is limited and infrequent.

He likes Dr. Mim-sey but he cannot concentrate on his current conundrum in the company of others. Dr. Mim-sey needs and wants to convince W-abe of some kind of threat posed by Roy Abbit and Earth. Dr. W-abe is remotely aware of Earth beyond its legendary existence.

"Imagine that you have one black marble and some white marbles," coaxes Mim-sey.

W-abe objects, "Are you going to tell me something I already know about probability?"

They take seats on a too small balcony. "No. Well, not yet," Mim-sey says, "I'm going to remind you of what you dropped when you picked up the binomial theorem."

"Then I do not have to worry about breaking into a cognitive sweat," quips W-abe.

Mim-sey follows a R. A. T. gondola across the atrium. "Your reputation as a mathematician is safe," Mim-sey intones, "But what you need to know is how that cashier, who is not a mathematician, insinuates himself into that contracting circle of mathematicians who are working on a Theory of Everything."

"His place in that circle is a quantum leap no one will ever explain," upbraids W-abe.

Pushing back in so many words, Mim-sey says, "He is not trying to insert himself and no one is trying to make room for him. But there is always room for anyone who has something to contribute."

W-abe is more interested in his ice cubes but says, "Okay, I have my imaginary marbles."

"Remember, it is what Roy Abbit knows, not what you know that is important," Mim-sey notes, "Let the white marbles be zeros. Let the one and only black marble be the number one."

Dr. W-abe pokes an ice cube with one finger, "I know this one. W-abe relates without much interest, "I can use the black marble to represent the decimal value of one by putting it first in the row. Decimal value two must have one white marble first then the black marble. Zero, one is binary two."

"What are you doing with all the other white marbles?" Mim-sey asks.

Dr. W-abe finishes his drink and mumbles, "I do not need them."

Mim-sey works his drink a little faster, so he will be ready when Dr. W-abe decides to get another one. "It is a good thing you are not running the universe," Mim-sey needles. "Making those little things disappear would release a lot of energy. It would produce an infinite amount of energy, in fact, and be the end of us."

W-abe looks at the bottom of his glass. "I could line them up after the black marble," he stipulates. "The decimal value of that number, along with all of us, would be safe."

Dr. Mim-sey, still sipping the way to the bottom of his glass says, "Then you agree that a binary number can have any number of zeros at the end of the number, without effect."

Dr. W-abe gets up, returns with a bottle and says, "No one will dispute that."

"A few billion white marbles or an infinitude of them will have no effect?" Mim-sey asks.

W-abe fills both glasses with ice and a shot. He grants, "There are even some infinities that are larger than others but that will not affect the decimal value of the binary number."

They move to chairs just inside. Dr. W-abe leaves the balcony doors open for the view. "But it does mean that one universe can be larger or smaller than another," Mim-sey shares.

"Yes, but binary two in any universe is still binary two," W-abe proposes. "It may be useful to distinguish a binary two from one universe as opposed to a binary two from another universe."

Dr. Mim-sey helps himself to a little more ice and asks, "Keeping other universes out of it, is binary two unchanged by the number of zeros that are attached to the end of the number?"

"Binary two is a decimal of two regardless," rehashes W-abe.

Mim-sey takes a sip and says, "I will get back to that. Binary three will need another black marble. Now, the first two marbles will be black and all the rest will be white."

"And for binary four we only need one black marble, again, but it must follow two white marbles." W-abe asks, "Are you creating black marbles and making them disappear?"

Mim-sey swirls his drink and says, "Or we can say we are rearranging an infinite number of black balls and an infinite number of white balls but that is getting ahead of the point."

"Are you getting to the point?" Asks W-abe with reproach.

Mim-sey replies, "So far, we have used three sets of numbers."

"Not with binary numbers. They include the set of ones and zeros only. If you insist on the set of decimal numbers there are two sets of numbers - not three," W-abe says as the complexion of the atrium grows harsher as the environmental lights begin replacing the sunlight.

Dr. Mim-sey turns on a table light. He says, "One, two and four represent one set of binary numbers - those with one black marble. Binary three needs two black marbles. Zero needs no black marbles at all. These are the second and third sets."

"More counting will require more black marbles," remarks W-abe.

"Keep counting and again you find numbers with *one* black marble. Binary eight only needs one black marble." Mim-sey repeats, "Each power of two only needs one black marble."

Dr. W-abe studies a sip and asks, "My 8,589,934,592 only needs one black marble?"

"And 32 white marbles since that number is two to the thirty-third power," says Mim-sey.

While Dr. W-abe moves to the kitchen but not with the thought of food, he says "And along the way you can have as many extra zeros as you want at the end of each number. What does that prove?"

Mim-sey openly taunts his friend, "Now there are thirty-four sets of numbers. One set has no black marbles. There are thirty-three rows with one black marble. Five hundred twenty-eight rows have two black marbles. Five thousand four hundred fifty-six rows have three black marbles. Forty thousand nine hundred twenty rows have four black....."

Dr. W-abe steps into the pantry with the emerging hope of spotting something appealing. "And how do you know?" W-abe scoffs from just inside the door.

"The same way our moldering cashier knows." Mim-sey says, "It is row 33 of Pascal's triangle."

"Any number in the triangle is the sum of two numbers from the previous row but that is a lot of counting," remonstrates W-abe on returning to the kitchen with his selection of cans and jars.

"Or just a little arithmetic since 'factorials' were invented or discovered," Mim-sey reveals.

Episode 8

Florida does not have a philharmonic orchestra. Attendance is too low. No money. The sympathy for the symphony is more than a little lacking. Miami does not have a philharmonic either. But it does have its own multimillion dollar civic center just in case a philharmonic turns up.

However, if by chance an orchestra arrives and you wish to be seen in the company of one, you do not want to go to the Civic Center. Those patrons arrive by ambulance and paddy wagon. You want to go to the Carnival Center.

The Carnival Center is just next to the Miami Arena. The Miami Arena is for basketball and ice hockey. But it is not used for either. For reasons that are opaque to developers ice hockey never lasted in Miami. As soon as people started attending the basketball games in the Miami Arena, it became obsolete owing to its surprisingly inadequate size. A new, bigger sports arena now stands conveniently close to the old Miami Arena and the Carnival Center.

The old Miami Arena is the host to the circus most winters. Some say they are just taking advantage of the weather. Others say it is the only chance to see elephants on ice skates. Both sides may be right. Remember, The Phil' when it is here, is at the Carnival Center not the Civic center. The Carnival Center is not the Miami Arena and good luck on explaining that to a taxi driver.

Today, the Boston Symphony is performing Mozart's *Requiem* at the Carnival Center. Two of the soloists are women and there is a tension that seems out of place for these veterans. One of the women is dressed sumptuously in black with white accents and matches the tuxedo clad male soloists. The remaining soloist is in blazing white with the thinnest black trim on her neck and sleeves. Did someone misunderstand the dress code?

The *Requiem* is performed without a flaw and tensions, real or imagined, seems to have no effect. Neither Roy nor Ali-ce is concerned with the agony of spirit that was Mozart's or the parochial agony of one soloist or another. It is a thrill for both of them. For Roy, it is one choral performance that is as grand as the music. For Ali-ce there is no comparison. This is a first for her. They do not talk until they get to her hotel. Neither one notices.

Maybe it is the time of day but they are alone on the elevator. Roy says softly, "Ali-ce."

The elevator stops at their floor. "Yes, Roy," says Ali-ce absently.

"Which visit is this?" Roy asks following her down the hallway.

"It is our third. But that is not the question you want to ask," coaxes Ali-ce.

"How do you know?" Roy protests as they enter the room together.

Ali-ce disappears into the bathroom but leaves the door partially open. "You are almost fifty-seven and most of our visits will be before your next birthday," clarifies Ali-ce.

"Yes, and I will be much younger then, in most cases," Roy concedes, "But it will always be the present when you do. I cannot remember past or future visits until they happen."

"And you want to know why," says Ali-ce as Roy picks up the tv-cable listing and checks the digital clock.

"I think I know why. In all cases, I cannot remember anything until it has happened. And, like you say, most of our past visits have not happened yet," depicts Roy. "I will not be able to remember any future visits until I am older simply because that is in the future. But I am wondering, Why do you visit at different times like this?" He hears her washing her hands. He knows it is his turn and takes it.

"Oh, you want to know why our visits are not chronologically ordered," surmises Ali-ce.

"Exactly," says Roy.

"I do not know all the details," says Ali-ce. "I am not sure I can explain what I do know."

"Try, I promise not challenge your explanation," entreats Roy. He wets his hands before unwrapping the soap. The wrapper is fighting him.

Ali-ce watches the war. She says, "You know my pavilion travels at very high speeds."

"I like to call it a ship but yes." Roy asks, "Do you travel faster than the speed of light?"

Ali-ce unwraps some soap and gives it to Roy. "Yes, but those details need a lot of explanation," She explains. "I can give you the example of electrons to begin with. They do not travel faster than the speed of light."

"Okay," says Roy. "I imagine that most of what you can tell me about electrons I can review in the library in between our visits."

"When an electron travels from point A to point B," Ali-ce begins, "You can see it at point B if it is a phosphorescent screen."

Ali-ce stays to hand Roy a towel. He smiles and says, "Thanks. There is scintillation where it hits. If something is in the way the screen stays dark."

They settle in the kitchenette which is their favorite equalizer. "When enough electrons hit the screen you get a dot for the duration," informs Ali-ce.

She puts some water on to boil. Roy gets two cups and interjects, "This dot appears sharp if the electrons are not too scattered," while reaching for the tea.

"Yes, but a barrier with a hole or slit in it," Ali-ce says, "causes a physical change in the electrons."

"Electrons interfere with each other and the dot or slit cannot be focused," Ali-ce says.

"But you still get the interference even shooting one electron at a time," interrupts Ali-ce.

"This pattern is built up one scintillation at a time in either case," Roy continues.

Ali-ce likes the way he makes the tea so she lets him do it. She prompts, "But if you put two slits in the barrier you get more than two stripes on the screen. That much I know."

It is cool enough on the balcony, so they move the tea there. "If it were just two stripes," Roy says, "you could suppose that the barrier caused the interference and the ultimate pattern."

"So, you see, the interference has to happen after the barrier," says Ali-ce.

"This kind of interference is linked to the way certain waves behave and it can be described by using mathematics," says Roy as they watch eastbound planes or at least their lights.

"But I do not arrive here as a wave," Ali-ce insists. "I and my pavilion, uh ship, would be a bright fuzzy dot. No doubt producing serious heat and only a one way, one time, trip."

"So no matter how fast you come and go, you do not want to transform yourself into some kind of wave like an electron can," interprets Roy. The tea is cool enough to drink and he gets some cookies.

Ali-ce pours some more tea. She says, "An electron can move from point A to point B and always remains a particle as long as it takes every possible trajectory simultaneously."

Roy rather likes the cookies. "This might be easier to believe, or learn to believe, if you could tell me why it is necessary to take every possible trajectory simultaneously," musters Roy.

"The average or combinations of all the possible trajectories come to the same result as the wave forms," Ali-ce offers. "All, but one of the possible trajectories cancel each other."

"From an infinity of paths only one matters for the motion of an object," says Roy.

"What is true for the electron is true for my pavilion," pitches Ali-ce.

"Ship," utters Roy setting the cookies next to Ali-ce.

"Ship," mimics Ali-ce standing up just as Roy sits.

Ali-ce moves to the bed forcing Roy to choose her or the cookies. Bringing the cookies, Roy asks, "How does that explain why our visits occur at different times in my past and future?"

"I have only given you part of the explanation," Ali-ce responds, making room for Roy, the cookies and a huge smile. "Perhaps you need to assimilate this before I tell you the rest."

"But no doubt most of the following explanation will happen in my younger years," Roy postulates. "I will not be able to remember this conversation then since it is in my future."

Ali-ce helps herself to a piece of cookie. "You will remember it but not in a time frame."

"It will seem familiar but I will not be able to remember where I learned it?" Roy asks.

"If by 'where' you mean where-in-time then you are correct," replies Ali-ce.

Roy puts the cookie plate on the night stand. "Some of this will just seem familiar?"

"You might even remember me and this conversation," Ali-ce predicts. "But you will not be able to put into a future perspective. You will have to fix it in the past somehow."

"If I cannot, I will not believe I had the conversation?" Roy asks.

Ali-ce turns off the bed light. "Or you will think it was a dream," she beckons.

It amuses Roy to say, "This does not feel like a dream."

Ali-ce is more than amused to say, "This is not a dream."

Episode 9

Everything in Officer Gim-bal's life is a mystery. What leads him to this New Orleans motel, he does not ponder. Except that he is in New Orleans. He has to stay somewhere for the night. This one is handy. So is the gas station food store. For the price of a steak dinner he buys some 'imported' saltine crackers and some domestic margarine. This is better than a mealy, cold, sliced bread sandwich or a charred chili dog. He speculates on how they serve chili frog.

He hopes the motel room TV has something on public televison. If it is a fund raising week there will be the perennial parochial experience programs punctuated by impossibly redundant pleas for you to pick up your phone and make your pledge now. He wonders how many 'pledges' are made with the hope or expectation of ending the plea promotions. If it is cable, the adult entertainment (for a fee) programming is salted with enticements that never include hermaphrodites. What kind of enticement is that?

What saves the day is that the ice machine works. (When it does not, it is because there is a convention in town. It is the only reason, ever, that an ice machine is out of ice.) It does not matter to him that ice is a dollar a bucket, 'correct change only'." The ice is for his guilty pleasures. An airport scotch is his licence.

There is every reason not to drink, but 'not to drink' has to be forced by circumstances. He has no excuse today. So he has his scotch, at least that is what the label says it is and all the ice he needs. One of his favorite treats is a saltine cracker with margarine.

Butter is okay but margarine melts better in his mouth. The flavor is foreign and kind of perverse. The plain crackers are a conspicuous disappointment every time he eats one.

Somehow, they are supposed to taste like something, but they do not. Rus-ty wonders, with each cracker he eats, if he would like them better if he was from Earth. His wondering settles on the question, "Why is a no-name scotch, crackers and margarine more entertaining than cable TV?"

It is nearly ten in the evening. Officer Gim-bal has waited long enough. He picks up the phone and calls the desk clerk. After enough rings to build a front desk, the clerk answers, "Front desk, How can I help you?"

Rus-ty is considering what kind of planet the clerk should wake up on but says, "This is Gim-bal - Room 111. Can you tell me if there is a large grocery store around here?"

"There is a Higgly's Piggly, west of here on Metaire Road," advises the desk clerk.

Gim-bal blurts, "A Higgly's Piggly? That is a funny name."

"I was thinking the same thing. Mr. Rus-ty Gim-bal," the desk clerk chuckles.

Rus-ty is thinking, "*This one will like the planet with an abundance of nitrous oxide and the acid rain.*" But his priorities require him to ask, "Is it far from here?"

"It is less than a mile, just past Bonnabel Boulevard, on the right," natters the desk clerk.

Gim-bal asks, "Do you know where I can get a five gallon bucket?"

"I think all the hardware stores are closed until 6:30 A.M.," ventures the desk clerk.

"That will be too late for me." Gim-bal asks, "Do you have any other ideas?"

Once he is cornered, the desk clerk sincerely tries to help. "We do not have any big buckets but you might try a Mike & Dee's Hamburgers," he replies. "See if they have any empty pickle buckets."

Rus-ty thinks, "*He is trying to get rid of me or he used to work at a Mike & Dee's.*" Gim-bal quips, "I bet you know where one of those is."

The desk clerk has just about plumbed the depth of his people skills but manages to respond, "Just east of here at Fairmont."

"Thank you," says Gim-bal cheerfully - feeling fortunate that he did not have any more need for help from this clerk.

"No problem," mutters the desk clerk absently.

1

1 1

1 2 1

1 3 3 1

1 4 6 4 1

1 5 10 10 5 1

1 6 15 20 15 6 1

Chapter 4

Chapter 001000....

Chapter C

Chapter $(1 + 3x + x^2)$

Chapter $(a^2b^0 + 3a^1b^1 + a^0b^2)$

Chapter 2^2

Episode 10

    Dr. W-abe's house is nothing like his office. The concierge is the equivalent of his receptionist but that is it for similarities. His home has a view and greenery. It is spacious, well lit and commodious. He has places to sit and places to set things. The furnishings serve his comforts.

He has music instead of phones. There are foods and refreshments that are not allowed in any of the offices. At the moment, Dr. W-abe is not hungry enough for a full meal and it does not occur to him to wonder if Dr. Mim-sey wants to eat. Mim-sey does not want to eat because he does not want Dr. W-abe distracted by dinner. Dinner will mean the end of the conversation, a preemptive end of the evening. He will want to go home and Dr. W-abe will want him to go.

Dr. W-abe prepares hors d'oeuvres and fresh drinks but with a caveat, "This conversation is already over if you drag out factorials."

"This conversation is not about telling you anything except how that register wrangler could have anything to say about a Theory of Everything," advises Mim-se y.

Dr. W-abe signals his assent by taking a seat. "Okay, but there is no connection to 1 x 2 x 3 x 4 x 5 x 6 x 7 = 5040 with 2 x 2 x 2 x 2 x... or any power of two," W-abe argues. "Except that you get, rather conveniently, to numbers that are usually inconveniently large."

Mim-sey realizes the end of the conversation means the end of the drinks so he hastens to say, "When it comes to Pascal's or Chu's triangle, it is the small factorial that proves my point."

Dr. W-abe stops his glass on the way to a sip. "Now you may have my attention."

Mim-sey disguises his satisfaction by asking, "How do I do that?"

W-abe laughs, "By explaining to me how you turn telescoping numbers into a microscope."

"If you were hoping for a calculus you will have to settle for arithmetic," says Mim-sey with a slight mocking tone.

Dr. W-abe gets back to his drink and mutters, "The medium of curmudgeons."

Dr. Mim-sey pours himself a new drink without asking for permission as a token of his advantage. "You are the one that is fending off an embarrassment of mathematical riches."

Dr. W-abe, knowing he can turn the advantage his way any time says, "All right, I am listening."

Mim-sey poses, "5 x 4 x 3 x 2 x 1 equals 1 x 2 x 3 x 4 x 5."

"I am waiting to be wowed," W-abe quips.

"When one product is the numerator and the other product is the denominator, the fraction is always equal to one," Mim-sey proposes. "No matter how you telescope the numbers."

Dr. W-abe, getting up to close the balcony doors, quibbles, "And the microscope?" Mim-sey turns without getting up and says, "Take away one number from both the numerator and the denominator. What is the quotient of 5 x 4 x 3 x 2 / 1 x 2 x 3 x 4?"

Dr. W-abe partially closes the drapes and recites, "120/24 is 5."

"What is the quotient of 5 x 4 x 3 / 1 x 2 x 3?" Mim-sey questions.

W-abe, on returning to his drink, utters, "60/6 equals 10."

"What is the quotient of 5 x 4 / 1 x 2?" Mim-sey quizzes.

Dr. W-abe, wondering what is waiting at the end of this exercise says "20/2 is 10. That leaves 5 / 1 which is 5. The telescope goes out and back." W-abe baits, "Where is the microscope?"

Mim-sey watches W-abe take his seat and pick up his glass. "The set of quotients is 1, 5, 10, 10, 5... and these are almost all of the numbers for row 5 of Pascal's triangle," says Mim-sey as a bid for W-abe's interest.

"All of the rows of the triangle begin and end with the number one. The remaining number must be one," blusters W-abe. "What is that probative of?"

Dr. Mim-sey's anticipation mingles with disappointment as Dr. W-abe puts the stopper in the decanter. "That number is also equal to 0/0 which is the last possible *factorial* in the series."

"Is this how our math nonentity proves $0! = 1$?" W-abe decries.

"Yes, because the other 'one' of any row is that proper fraction we started with," Mim-sey demurs. "The series of quotients and the final quotient imply that '0' is equal to one."

W-abe deplores, "So now anyone can *prove* $0 = 1$ and keep his curmudgeon credentials?"

"If you want to know what that has to do with the black marbles and the white marbles, it is too soon to pickup your marbles and put them away," Mim-sey says. "We have seen the binary numbers that can be represented with all the white marbles (0), with only one black marble (2^n) and any number of black marbles in between. There is still one more set to consider...."

Episode 11

Roy Abbit has an excellent view of the shopping plaza. He is renting a small condo in what used to be the U. S. Immigration and Naturalization building. I. N. S. moved to Coral Gables.

Gables residents pretended to put up a fuss but it was just for show. The rest of Miami put up a fuss and it was real. The population that needs I. N. S. cannot afford traveling and parking and eating in Coral Gables. So they are more cut off than ever.

But the population that pretended to oppose the I. N. S. office in Coral Gables made every parry, riposte and promise necessary to get it there. Now it is prestigious to be seen in and around the I. N. S. building. You are working to get your dearest first cousin's daughter by a second marriage to this country. You are saving her from the fate of being born where she was. No more working in obscurity and tedium. Everyone in Coral Gables can see you there and wonder at your sacrifice.

Roy has wondered about this for so many years it is easy to think about other things. So he is ready when Ali-ce distracts him with conversation.

"On our last visit I made a promise that I do not think I can keep," discloses Ali-ce.

"I usually remember our visits very well, at least for a long while, but I do not remember a promise," says Roy. "It seems like that is something I would anticipate and remember."

"We talked about particles. Electrons were the example, and how they travel at high speeds," recounts Ali-ce. "I said that by traveling at all possible trajectories, a particle could remain a particle and travel like a wave including traveling as fast as a wave."

Roy and Ali-ce are sitting in matching, $7.00, molded, plastic, chairs.

"It sounds familiar but I think I read it somewhere." Roy quizzes, "What was the point and what was the promise?"

Ali-ce watches shoppers load their cars below but Roy finds watching his pineapples grow more interesting by contrast. She says, "The point was to explain why our visits were not chronological. You were forty-seven then. How old are you now?"

"I am forty-two now. Still, I cannot remember a conversation and a promise that have not happened yet," answers Roy.

"But," Ali-ce says, "if I do not keep my promise you will remember when and after you are forty-seven."

"How can I hold you responsible?" Roy asks.

Ali-ce cannot decide what is more amazing, buying all that stuff or getting it all into a car. She wonders if all these people live in empty houses.

"I am not responsible if there are no adverse effects," says Ali-ce. "But if my broken promise causes some terrible consequence you will not forgive me."

"There are a few reasons why you are wrong," says Roy. "First of all I adore you. I will not let, even myself, especially myself, punish you."

"How can you be sure?" Ali-ce asks.

"Because of what I have just said. It is a curse and a blessing of being human," attests Roy. "We can modify or manufacture reality often and without working at it."

Roy is convinced that pineapples need to keep their roots wet. He requires each pineapple to have its own pot with the bottom half filled with gravel. Each pot sits in a shallow dish of water. But standing water breeds mosquitos.

Ali-ce sees toys, appliances, clothes, food, drinks, furniture heaped in a constant parade of shopping carts.

"Intellectual and emotional reality can be forced," argues Ali-ce. "There are other realities."

"Intellectual and emotional realities are the ones that count with us. Because the others are mechanical or accidental," clarifies Roy.

"Like the sun overheating and destroying Earth?" Ali-ce asks.

"Yes, I cannot hold you responsible for something like that," Roy contends. "No one can be responsible for accidents, chance."

"Probability?" Ali-ce asks.

Roy fills each pineapple dish with fine gravel. The dish holds less water, true, but no mosquitos. The soil drains into the gravel and over flows the dish.

To Ali-ce's amazement she sees whole families, pushing two or three shopping carts. And taxies; people are loading taxies and paying taxies to carry all that stuff.

"Say, something terrible happens to me," says Roy, "specifically because you did not keep the promise that you have made to me in my near future, how much of that is deflected or focused by all the events that occur and can occur between now and that fateful day?"

"Your fate and the significance of my promise, broken or kept, is a combination of chance and machination," frets Ali-ce.

"Simple honesty prevents me from blaming you. Unless I want to." Roy insists, "I do not want to and I will never want to."

Ali-ce sips her iced tea. "I do not think there is anything harmful in what I promised to tell you but my father said that it could have the worst possible consequences," she says.

Roy refills her 99¢, plastic, pilsner, drinking glass. "Did he explain?" asks Roy.

"He could not or would not," notes Ali-ce. She sets her glass on her $1.99, no drip coaster.

Roy persists, "But I do wonder more why our visits occur when they do."

She likes to set the glass so the light from the $3.99 citronella candle, shines through it. "All right, I guess my father has to do a better job of intimidating me before I break a promise," Ali-ce affirms. "You have made it easier by reminding me that I cannot be responsible for the part that chance plays in life."

"Are there any other reasons why you countervail your father?" Roy presses.

Ali-ce swirls her glass to spin the ice. "You know and make me say it?" She banters.

Roy puts charcoal in his $14.99, double handle, cast iron, hibachi. "Yes," he nods.

"You know I adore you," Ali-ce says. "I do not like to refuse you, even when I must."

"I need to know that and want to build on that," admits Roy while lighting the charcoal. "The little time we spend together must mean the most for us."

"There is so little that we will ever share," says Ali-ce as she assembles some kebabs.

Roy adds ice to the drinks. "What little there is must be as much as possible," He replies.

"How will it help to know why our visits are not chronological?" Ali-ce asks.

"It will help to know whether your visits are deliberate or accidental," asserts Roy.

Ali-ce brushes the kebabs with olive oil and asks, "How will that help?"

"If they are deliberate and life is good to me I will give you the credit," Roy contends.

"And if they are accidental?" Ali-ce asks while putting some salt and pepper on the meat.

"Then you cannot be responsible for the worst that happens to me," coaxes Roy.

"I am afraid that our visits are both deliberate and accidental," confesses Ali-ce.

"There is no such thing as 'many accidental visits'." Roy asks, "So why are you here?"

Ali-ce put some dried basil on the tomatoes. "My father's life work is in language and mathematics. Mine is literature and language." Ali-ce concedes, "Our shared interest in language makes me useful. I collect literature that he studies for its use of language."

Roy adds cinnamon to the baked beans. "That sounds a little innocent," He challenges.

Ali-ce slices some Italian bread. "You may be right," she says. "But I do not know his motives and interests in mathematics. So I have to trust his justification for my visits."

Roy sets two places with his prized Reed and Barton stainless steel flatware. "And if he knew that your motives and interests were more than literature and language?" He asks.

"I will always be his daughter but if his motives and interests were conflicting he would make sure you would never hear from me again and there would be no explanation," Ali-ce insists. "I am more important to him than anything he can achieve in language and mathematics."

Roy tears some romaine lettuce into bite size pieces. "So his motives and interests would have to be deep indeed if he is willing to risk his daughter," he judges.

Ali-ce gets the chopsticks. "Right now he is confident that he is not risking me. So his motives do not have to be deep," she maintains. "Right now he is just serving his profession."

They start with the chop sticks on the salad with anchovies, oil, wine vinegar, salt, pepper. "So how do you explain these accidents - meeting time after time in my future and past?" Roy asks while slicing and buttering Italian bread for them both.

Ali-ce turns the kebabs. "To begin with, have you ever heard of the principle of uncertainty?" She asks.

"You cannot know both the position and velocity of an electron for example," recites Roy doubtfully.

Ali-ce fills the glasses to the top with ice and then adds the tea. "In fact they are inversely proportional. The more you know about the velocity the less you know about position," she remarks. "The more you know about position the less you know about velocity."

With so much ice in his glass Roy drinks half the glass of tea easily. "I imagine that there is some connection with this and the conversation we will have years from now," he says.

Ali-ce refills his glass so it can get cold before he wants some more tea. "In the conversation we will have then, I explained how I traveled as a 'particle' but at very high speeds," she says.

Roy pushes the anchovies closer to Ali-ce's plate so they are more convenient. "You cannot be sure of your speed and location? That is too dangerous to be believed," he adds.

"I can be sure of my speed or location any time but not simultaneously," synopsizes Ali-ce.

"So you could be here, say, every four years," says Roy.

Ali-ce takes the kebabs off the hibachi so they can get cool enough to eat. "But there is a price to be paid for that. You have a very definite location in space and time," Ali-ce vouches. "The more specific I am about the time to visit the more *indefinite* I am about the location."

"You can begin your trip from a specific location and travel at a specific speed but you cannot be certain that you will get there?" Roy ventures in his effort to grasp the point.

Ali-ce cuts some more Italian bread from the loaf and gives Roy a slice. "I will get there. The problem is how close I will get," she answers. "I can expect to get there very close and right on time."

"Once you are close to here most of the trip is done," advises Roy.

"But how close is close?" Ali-ce asks. "If I am only four light years away when the trip is nearly done, it will still take four years traveling at the speed of light to finish the trip."

Roy says, "I can see why you would want to get closer than that."

"And the closer I get to this specific location the more uncertain my velocity must be for the trip," says Ali-ce as she unskewers the kebabs and puts the pieces on a platter.

"And the more uncertain your past, present, or future time of arrival is," says Roy as he positions the dipping sauces for Ali-ce and himself. He likes the soy and sweet sauces. Ali-ce likes the hot sauces.

"It is the price I pay to be near you," says Ali-ce.

"Because you adore me?" Roy asks putting down his chopsticks.

"Because I know you love me as well," says Ali-ce.

"I feel like I have known you all of my life," says Roy following with a bite of buttered bread.

"Maybe longer," says Ali-ce before a bite of tomato and basil.

Episode 12

You navigate among galaxies. You shuttle to Earth. Then things get difficult. The Higgly's Piggly has all the liver you could want but it is frozen and prepackaged. The manager does not ask why Rus-ty wants 30 pounds of liver even though it means putting down his cell phone to get it.

The Mike & Dee's has plenty of pickle buckets but none of them are empty. Buying and dumping the pickles is easy. The manager is incredulous but he unlocks the dumpster and watches Rus-ty dump the pickles and keep the bucket.

Officer Gim-bal takes Esplanade Avenue to the French Quarter then Decatur to Ursulines Avenue where he checks into a small hotel just around the corner from Bourbon Street. This hotel is not important because of its proximity to Bourbon Street. The Ursuline Hotel has a hot tub and a clientele that will not ask why he is floating thirty packages of frozen liver in it. He is able to get the room that is next to the hot tub and sets to thawing his livers.

It is nearly 2 A.M. when Officer Gim-bal and his livers get to the Aquarium of the Americas and the river. The same small boat is there and he wonders if they set a watch.

"Hello," calls Gim-bal, standing as close to the railing of the boat as he can get.

"Just a minute," carps the sailor, verbally taking control of the situation.

"Sorry to bother you," says Gim-bal, satisfied that he cannot be ignored for the moment.

"What do you want?" Quibbles the sailor as he positions himself at the head of the gangplank.

Rus-ty puts one foot up on his bucket of livers and asks, "Did I talk to you earlier today?"

"Are you Rus-ty Gim-bal?" The sailor asks as he unlatches the chain giving access to the deck.

Rus-ty takes his foot off his bucket of livers saying, "Yes." He asks, "Do you have the Midwatch?"

"Yes. But I have the day off in the morning," says the sailor self-consolingly and with some anticipation.

"I need to ask you a favor," says Gim-bal taking a step back as if to give the sailor room.

The sailor remembers his duty to the boat and challenges, "What do you need?"

Rus-ty picks up his bucket of livers and says, "I want to do some chumming off your bow."

"I guess that explains the bucket," responds the sailor only partially convinced of Rus-ty's motives.

Rus-ty volunteers some information to make it easier for the sailor to say yes to his request. "It is beef liver."

The sailor jokes, "You are not going to get on the boat with a bucket of pickles."

"I want to see how big the catfish are around here," says Rus-ty without much more in the way of the truth to offer.

The sailor asks, "You don't need to go anywhere else on the boat?"

"No. You can hold the flashlight if you want to," Rus-ty offers in order to reassure the sailor.

The sailor steps aside to give Rus-ty access to the deck saying, "Come aboard. I have got a good light."

Rus-ty guesses the sailor might have a little time on his hands but keeps the observation to himself.

$$1$$
$$1\ 1$$
$$1\ 2\ 1$$
$$1\ 3\ 3\ 1$$
$$1\ 4\ 6\ 4\ 1$$
$$1\ 5\ 10\ 10\ 5\ 1$$
$$1\ 6\ 15\ 20\ 15\ 6\ 1$$

Chapter 5

Chapter 101000....

Chapter AC)

Chapter $(1 + 3x + 2x^2)$

Chapter $(a^2b^0 + 3a^1b^1 + 2a^0b^2)$

Chapter $2^0 + 2^2$

Episode 13

    Dr. W-abe is beginning to think about dinner. Everything in the kitchen looks new, unused. Because it is. The wife and relatives are good at finding ways to stay out of the building-burbs. This is fine except when it comes time to decide what to eat. It is easier to think of anything else. So, Dr. Mim-sey's conversation will do for the moment.

Dr. W-abe unstoppers the decanter and pours a fresh drink for Dr. Mim-sey and himself with the remark, "...Except, we have not seen the set of numbers that include only the black marbles? How difficult is that?"

"There is the set of all white marbles. There is the set of all white marbles with one black marble and there is the set of black and white marbles. So there has to be a set of all black marbles," alleges Mim-sey while he puts some more ice in his drink after the refill.

Dr. W-abe is used to this. Dr. Mim-sey does not like to short his drink with ice. "That is as easy as falling off a cash register. It does not take a Newtonian methodology to do that," says W-abe.

"But how often do all the black marbles occur among the others?" Asks Mim-sey taking the smallest possible sip, this time, for maximum appearances.

"It is interesting that there is only one row with all white marbles no matter how high you count. The next easiest observation is the rows with only one black marble - shown to be equal to a power of two. But each power of two has more rows, in fact twice as many rows of marbles in between. They can have two, three, four black marbles, etc., together with so many white marbles. Sometimes there are more black marbles than white marbles, sometimes less. Even equal numbers of black marbles and white marbles in a row," W-abe reasons.

"It is possible to calculate how often these subsets occur," Mim-sey tells his friend.

"The occurrence of the black marbles is not random?" W-abe asks.

"No. But like the powers of two, the groups of two or three, etc., are different for each power of two," Mim-sey answers as he forgets about sipping for effect and drinks with satisfaction.

Without missing Mim-sey's growing enthusiasm for his drink, Dr. W-abe says, "But to get from eight to sixteen, say, you still have to count - 9, 10, 11, 12, 13, 14, 15. So the calculation must be simple."

"Yes," conveys Mim-sey. "List or stack your binary numbers like you were going to add them. Each number in the stack is in a row of ones and zeros of its own. After a few rows you can see that there are also so many columns. Label each column with, say, letters of the alphabet."

"You're going to get so many subsets of letters instead of marbles," W-abe concurs. "The number three includes the letters 'A, B' because there is a binary one in column A and B."

"Sort the subsets like this - Group all the subsets with one letter in one group," Mim-sey counsels. "Group all the subsets with two letters in another group and so on."

Dr. W-abe makes a bet with himself whether Dr. Mim-sey will get to the bottom of his glass before he gets to the bottom of the conversation. W-abe entreats, "Do we have something yet?"

"You do when you compare the groups to the Pascal or Chu triangle. They are the same," Mim-sey explains. "Row five of the triangle says that there are five rows in the stack with one black marble or one letter. There are 10 rows in the stack with two black marbles or two letters and 10 rows in the stack with three black marbles or three letters. There are five rows in the stack with four black marbles or four letters and one row in the stack with five black marbles or five letters. You can go to any row in the triangle and each number tells you the coefficients that occur for each subset in the stack of binary numbers," explains Mim-sey.

"Numbers in the middle of any row of the triangle must be the subsets of mixed black and white marbles," says W-abe taking a sip to encourage Dr. Mim-sey to do the same.

"The first entry of any row of the triangle represents the subset for zero. This is entry zero. This is the one row in the stack that has no black marble. The next entry of any row of the triangle includes those rows with only one black marble. This is entry one. Those entries represent powers of two," Mim-sey instructs while taking a quick glance to measure how much of Dr. W-abe's drink remains. "There are six of them by the time you get to row five: 2^0 2^1, 2^2, 2^3, 2^4, 2^5."

As he adds the last of the ice to his drink, Dr. W-abe says, "Each row of the Pascal or Chu triangle increases the number of rows in the list or stack with one or more black marbles."

"With two exceptions. You know the first one," says Mim-sey.

"Entry zero of any row of the triangle represents the row in the list or stack with no black marbles," says W-abe.

"The other exception is the entry of any row of the triangle that represents the row in the list or stack with no white marbles," informs Mim-sey while he silently vows to get an ice server like Dr. W-abe's.

"It must be the last entry for any row in the triangle," says W-abe. "This entry terminates each row of the triangle and is always the number one," Mim-sey comments. "There is only a single occurrence for each row of the triangle where all the marbles are black, all the digits are ones."

Dr. Mim-sey takes a gulp of his drink betraying his impatience with pretense and the suspicion that he may never own a piece of foamed glass, antique or otherwise.

W-abe maintains, "Where a power of two is represented as 2^n and the number with all ones or all the black marbles must be 2^n-1 because it is just one digit short of a power of two."

"Seven is one less than eight and is binary 111," says Mim-sey.

"The number 8,589,934,591 is one less than 2^{33} so all the marbles are black. It is the last entry in row 33 of the triangle and all the binary numbers are ones," says Dr. W-abe. "That entry is a one just like the first entry of that row. This much is obvious but what good is it?" He asks as Dr. Mim-sey finishes his drink.

"That becomes more obvious when you consider this question - Is the row with all the black marbles a subset of the next row? Is three the subset of four or is three a subset of two? Is 2^n-1 the subset of 2^n? Is 2^n-1 the subset of 2^{n-1}?" Mim-sey asks as Dr. W-abe finishes his drink.

"Is 8,589,934,591 a subset of 2^{32} or 2^{33}?" W-abe parrots while pouring fresh drinks.

Episode 14

This used to be a poor neighborhood. But the worlds largest retailer bought the largest parking lot on the boulevard and built a store that could offer the largest possible volume of consumer goods at the lowest possible price for the greatest possible profit. The blessings do not end there.

There are scores of new jobs. None of them pay enough in a month to make the rent. But there is an answer for that. Fortunately, they only hire workers as part-time. This gives the employee ample opportunity to earn more on a second job.

Roy is a part-timer but he is not in retail. Not now at least. So for the time being he is an instrument of gentrification. His rent is paying the landlord's mortgage. He is doing his part to raise his poor neighborhood to a lower-middle-class one.

The neighborhood restaurants, of course, cross the spectrum - from the most pedestrian to the most exclusive, regardless of the local economic climate. Roy has chosen a Chinese restaurant for dinner.

Ali-ce has some work to do for her father. At least, Roy can count on her returning to him. So the parting is understandable, however unwanted. The conversation is less one of a good-by and more one of a continuing dialogue woven through a familiar routine.

"You know time travel stirs a certain debate," says Roy while they wait to be seated.

"If the debate is whether it is possible," says Ali-ce, "you, at least, know there is no debate."

"The debate," says Roy, "is about whether or not the time traveler can change history."

The sign to 'Please Wait...' is posted on a billboard for Tsing Tao. It is a beer they both like. Some of the patrons think Tsing Tao is Chinese for "Please wait to be seated."

"Do you mean," Ali-ce asks, "Can the time traveler change history or is the change he effects really a change?"

"I do not know what you have in mind but the usual argument is that if a time traveler killed my grandfather, I would not be here, as his descendant," acknowledges Roy. "History would have no effect from me or any of my descendants, since we are, all of us, also removed from history."

"I have to ask you this to put the question in perspective," coaxes Ali-ce. "Who among your ancestors has had a significant impact on history? No offense."

"Under the circumstances, no offence taken. Their lives have been ordinary in the historic sense," says Roy. "What says none of my ancestor's progeny cannot be significant historically?"

"Nothing, but the point is most of what changes with your grandfather's death is not historic. It is, so to speak, local in time," clues Ali-ce as they are lead to a table that is set for four. The table is fine for two but she wonders how four could fit at this table this size.

"I would not be here certainly," says Roy while the waiter removes two of the settings.

"Not as certainly as you may think," allows Ali-ce. "In order to eliminate you from even local history your grandfather would have to meet his end before he fathered your father."

"So the earlier the better," adduces Roy, "if the purpose was to change history."

"Even then," professes Ali-ce, "the streams of events are so vast and mixed that history would flow past the event of his death and not change its course."

"History has a kind of inertia that is not overcome by relatively smaller events?" Roy asks.

"Like my infinite paths that bring me here, history has infinite paths," says Ali-ce. "Much of what happens one way could have happened in infinitely other ways to the same end."

"You mean," Roy asks, "that there are an infinite number of ways events could bring me to this day or any other day?"

The drink special is 'Rum and Coke' but they have hot Oolong tea.

"What if your grandfather was adopted?" Ali-ce asks while she looks at the menu.

"He would still be my grandfather," replies Roy wondering if a 'Rum and Coke' will still be called a 'Cuba Libre' after Cuba is free.

"But that is just another way your line of ancestors could change," contends Ali-ce.

"My grandfather could have been my step-grandfather," Roy quibbles. "That would complicate, even frustrate, the assassin's intentions."

He orders Mu Shu Pork which they will share as the appetizer.

"The entire fabric of life and all its events direct your life," claims Ali-ce. "You could be born to different parents as a consequence of the assassin's temporal success."

"He would only succeed in making me a different branch of the family tree," supposes Roy and deciding on the Cashew Chicken.

She orders Hunan Scallops with mushrooms and carrots.

"So going back in time and changing things would not change much," warrants Ali-ce while handing the waiter the menu.

"What about the historic record? There are certificates, books, sometimes statues and monuments. These things last thousands of years," says Roy still holding his menu. He adds Double Yellow Mandarin fried noodles with shrimp, chicken and vegetables to their order.

"Suppose one of your ancestors was a pharaoh," presses Ali-ce. "Suppose he disappeared one day."

"There would be a historic record," says Roy pouring some hot tea.

Ali-ce suggests, "No doubt there would be a funeral and a tomb as a memorial."

"It could be discovered centuries later - preserving his record," says Roy.

"But there may be no body," says Ali-ce. "The explanations would challenge the record."

"The hundreds of thousands of people who knew of the pharaoh would be gone," utters Roy. "But a few scholars could preserve the fraction of history that remained."

"What is preserved would be subject to more differences of fact and fewer and fewer scholars to preserve them," says Ali-ce.

The waiter brings more tea and the Mu Shu Pork.

"His historic significance would be lost, eventually," claims Roy.

"And it would matter less and less whether some time traveler played a role," says Ali-ce.

"That is just for the few that are famous," agrees Roy while rolling some Mu Shu in a pancake for Ali-ce. "There are billions of human lives interacting in historic and local time."

"You are easier to convince than my father," presses Ali-ce while she enjoys her appetizer.

"I will take that as a compliment," volunteers Roy as he rolls some Mu Shu for himself.

"It was meant to be," admits Ali-ce before they focus on the food for a moment.

"I suppose you have discussed this with him before," Roy resumes.

"Many times," confirms Ali-ce, "and with no success in changing his mind."

"What is wrong with that?" Roy asks.

"It puts him in agreement with those who justify murder as a means to manage history," warns Ali-ce.

The waiter returns with the entrees.

Roy asks, as they admire the food, "Do you mean through war or assassination?"

"I think he is against war and war is remote. But I think he would support an assassination to avoid war," urges Ali-ce while she gives Roy a generous portion of her Hunan Scallops.

"What about you?" Roy asks returning Ali-ce's gesture by giving her plenty of his Cashew Chicken.

"I do not think that even war can change the course of history," charges Ali-ce.

"War is used all the time to change history," advises Roy taking some Double Yellow.

"Scores of millions are killed," enjoins Ali-ce while she takes some Double Yellow too. "After the war billions resume their lives in some way and the war fades into history itself."

"The war that would change history," says Roy, "is the war that becomes a way of life."

"For many lifetimes," Ali-ce agrees, "and for billions of people."

"You would know if he was in support of that sort of thing," utters Roy.

Ali-ce protests, "But no assassination will promote or prevent that kind of war."

"So if his intentions were to prevent such a war," Roy says, "his actions would be meaningless."

"Unfortunately, the stigma of a murderer would become his and that would change his life for the worse," laments Ali-ce.

She helps herself to more steamed rice.

"What would his punishment be?" Asks Roy.

"He would live out his life but he would never be trusted," Ali-ce deplores, "and he would always be scorned."

She takes a sip and seems to study the pale color of the tea.

"What daughter would not want to prevent that?" Roy asks as he studies Ali-ce.

"I could not make myself responsible if, I did fail," Ali-ce says. "I can only do so much."

"Then you do not need to oppose him," remarks Roy with some relief and much hope.

Ali-ce weighs, still focusing on her tea, "In one sense I can do nothing but oppose him."

"How is that?" Roy asks.

"What if my father thought he could justify your murder?" Ali-ce asks.

Episode 15

There are no boats tied to the starboard side. Officer Gim-bal and the sailor take a position on that side of the bow. This water is muddy and slow moving. There is no prospect of seeing the fish in that water. The hope is that a prowling catfish will stir eddy currents. These catfish will be chasing the sinking, bloody, liver and perhaps competing among others.

The port side of the boat is on a concrete pier. So it is unlikely that any big catfish reside here. But it is possible that the water will carry the scent of the blood and liver down the river. Whatever liver catches on the bottom will mark a trail upstream to the boat.

Officer Gim-bal has 90 pieces of liver. By throwing in a piece every minute and a half, he can chum for almost two hours. By then it will be four o'clock and the end of the watch. He will not have to plan on the cooperation of the next watch.

"What makes you think you will see any big cats?" Asks the sailor.

"Well, when a cow or a bear falls in the river it washes down the stream," Gim-bal illustrates. "It tumbles along and eventually sinks."

"That could be hundreds of miles from where it fell in," marks the sailor who is becoming engaged but is leaving room for his doubts.

"From there it probably rolls along the bottom and gets broken into large pieces," states Gim-bal as he begins to recognize the speed and relentlessness of the river's flow.

"So the cat does not have to be able to eat a horse," says the sailor.

"No, but it is still probably too tough to eat. But big chunks are pushed down the stream," invokes Gim-bal not wanting to get ahead of the sailor's willingness to believe the idea.

"So these chunks are getting beat up under water and getting softer and smaller as they move down the stream," advocates the sailor while Rus-ty throws meat into the spot created by the flashlight.

"You can imagine," Gim-bal evinces, "that bacteria help too."

"The meat has to get pretty far down the stream before it is edible," pads the sailor.

"That is what I think," surveys Gim-bal leaving room for the sailor to contribute a little.

"The biggest fish are where the most food is," snorts the sailor from experience.

"I am betting on that," prompts Gim-bal pumping the dialogue.

"And catfish do not need to see the food to find it," the sailor volunteers his knowledge with a little more enthusiasm. "So they have a better chance at being at the head of the line."

"Those whiskers are quite an advantage," Gim-bal construes more as an encouragement than as a matter of fact. "They can eat night or day whenever the food goes by."

"Have you seen anything yet?" Asks the sailor as a way to say nothing has happened yet.

Gim-bal, throwing in some liver a little sooner than he wants to, says, "Not yet."

"There is an eddy," crows the sailor. "Throw some liver in that eddy."

"Look. The disturbance is lasting," declares Gim-bal who wants to succeed for his own reasons.

"Eddies do not do that. Throw in some more," urges the sailor hoping for some big fish.

"Even if there are several of them," Gim-bal injects, "they have to be big to make eddies like that."

"Throw some more in," the sailor bristles. "Maybe they'll surface."

"Move the light a little up the stream so it is not so bright where I throw the liver in," huffs Gim-bal having an environment in mind where the light is not as bright as this.

"Look at the size of that" A fish breaks the surface and the sailor squawks, "Nobody is going to believe that."

It is difficult to tell how many fish are there. But the agitation in the water suggests large and strong fish that are sure about the food and getting it.

"The head must be four feet wide," rants Gim-bal who is not ready for such a large cat fish. He is throwing in liver freely now. He knows he does not have to reserve any for the rest of the watch.

"We just saw a part of the head over the mouth," marvels the sailor moving the light to the center of the activity.

Gim-bal affirms, "I guess plenty of food gets down here."

He has achieved his purpose beyond his expectations. So he does not require the sailor to adjust the light anymore. Now he is just chumming to finish his supply of liver. It is time to move on and return to his Virginia persona.

1

1 1

1 2 1

1 3 3 1

1 4 6 4 1

1 5 10 10 5 1

1 6 15 20 15 6 1

Chapter 6

Chapter 011000....

Chapter BC

Chapter $(1 + 3x + 3x^2)$

Chapter $(a^2b^0 + 3a^1b^1 + 3a^0b^2)$

Chapter $2^1 + 2^2$

Episode 16

While Dr. Mim-sey and Dr. W-abe work in the same building, they have not worked together for more than twenty years. Their brief time working together comes during the ultra-secret 'Project Bump'. They are recruited because of their specialties. Dr. Mim-sey is trained in language and mathematics. Dr. W-abe is trained in math and physics.

They do not have any leading positions in the project. Their main job is error checking. When the project ends, their careers go in different directions. Now, half a lifetime later, Dr. W-abe is an engineer and Dr. Mim-sey is a civil servant.

They do not talk shop because no one, not even Dr. W-abe, finds engineering a topic of conversation outside of the workplace As for engineers, Dr. W-abe finds they know enough to either agree with him or not. In either case there is little to discuss. While Dr. Mim-sey is a long time friend, he has nothing to contribute to issues in engineering. He is obliged not to talk shop.

His shared vocational turf with Dr. W-abe is off the horizon. Project Bump begins with an unexpected and unexplained interruption to communications. There is the need to know "What went 'bump' in the night?" The official conclusion of the project is that the interruption to a remote communication is due to a star that exploded and the released energy produces the one and only interruption of the kind. Only part of that story is true.

Dr. Mim-sey never discloses the fact that the project is still ongoing. Dr. Mim-sey and Dr. W-abe take on new careers soon after it is found that an inhabited planet is the source of interrupting energy. It has been and is Dr. Mim-sey's job to learn if and when this planet becomes a threat. He wants to do that without revealing his connection to the project.

Specifically, Dr. Mim-sey has to know what kind of threat someone like Roy Abbit is and keep the effort as confidential as possible. So Dr. Mim-sey always returns the conversation to the amateur math of Roy Abbit which could be the key to when and how there is an escalation of the threat Earth portends. Dr. W-abe is Dr. Mim-sey's litmus test.

"I don't like using large numbers," huffs W-abe as he waits for Mim-sey to insert his access card at the matrix elevator.

"So why do you?" Mim-sey asks.

Dr. W-abe pockets his matrix card when he is with Mim-sey and along with it his history of their activities. Cards leave a trail and Dr. W-abe likes to leave gaps.

"The big numbers have to work like the small ones. So I cannot avoid them," W-abe drones. "And I use big numbers to fend off amateurs."

"The professionals are just as fend-ed. They just refuse to admit it. But in this instance the case is just as obvious with small numbers," Mim-sey pitches. He routinely watches those card trails.

W-abe asks, while he watches Mim-sey study him, "Is this obvious to the nabob of no-sales?"

"Just as obvious," casts Mim-sey.

The matrix elevator arrives very slowly for this time of night.

"I think I am going to be offended by the obvious," marks W-abe wondering if it is fewer cars or a system operator that accounts for the system slowness. "Hopefully, you are going to do it in the simplest terms."

"Do you remember all those years ago," Mim-sey asks, "when you were learning to count?"

"I remember learning to count," W-abe typifies. "You draw a line and put numbers on it... 1, 2, 3, 4...."

When an operator identifies certain passengers using a matrix card, he slows down the matrix to examine the database on related origins and possible destinations. It is the possible destinations that could signal significant activity and history of the passenger of interest.

"You learn that numbers come in order and it helps you create the next number," elucidates Mim-sey while speculating about the advantage of a slow transport for more time as Dr. W-abe's parasite.

"I can do it now with binary numbers," W-abe elects to volunteer, "as well as with decimal numbers."

"With binary numbers in a list, each number represents one row," says Mim-sey returning his matrix card in his pocket. "You have to add more columns as the rows require more ones and zeros."

"We did this before," recalls W-abe. "We labeled them A, B, C...."

"So your number line, with 1, 2, 3, 4..., includes letters also," suggests Mim-sey.

"One is 'A'. Two is 'B'," W-abe indicates. "Three has binary ones in columns 'A' and 'B' so decimal three must be 'AB'. Four is 'C'...."

A matrix car arrives not much later than usual, Mim-sey estimates.

"...Okay. Notice that decimal three has no letter 'C'," marks Mim-sey as he keys the keypad.

"You can build a fire on a frozen lake but the ice has to be a little thicker," apprises W-abe.

Row 2 of Pascal's triangle consists of the numbers 1, 2, 1 which is a sum of four. Entry zero is the first number and represents the null subset. The second number is entry one and represents the subsets with single coefficients - subsets like 'A' and 'B'. In this case they are the only subsets with single coefficients," reprises Mim-sey while keying-in for the matrix station near his office. "Entry two is the third and last number, It represents the sole subset with two coefficients, in this case 'AB'."

"In order to include 'C' in that row of the triangle," W-abe inserts, "you would have to have three subsets with single coefficients."

The detour to Mim-sey's office on the way home or to dinner is the price W-abe pays for his penchant for anonymity.

"Three cannot be counted as a subset of four until row three of the Pascal or Chu triangle," Mim-sey enjoins. "This row has the sum of eight. There 'three' is a subset of eight."

"So after row two, three is a subset of four" interposes W-abe as they exit the elevator near Mim-sey's office.

"Not so fast," Mim-sey warns. "When you add 'C' to the list, you also add the subsets. 'AC', 'BC', and 'ABC' to the list of subsets."

The detour allows Mim-sey to get another matrix card.

"Does that mean that decimal five, six and seven are also subsets of decimal four?" Asks W-abe.

"Yes," Mim-sey fleers. "It also means that for any power of two all the decimal numbers after that power of two are subsets of that power of two until the next power of two."

"All decimal values up to the next power of two," says W-abe.

"And then you have an avalanche of subsets to add to the next group," says Mim-sey.

"So since 8,589,934,591 is one decimal value short of 2^{33} that number is a subset of 2^{32} which has a decimal value of 4,294,967,296." W-abe says, "You are saying that eight billion is a subset of four *cleye*."

"That number is 50% smaller," says Mim-sey.

"Wow" says W-abe.

"Is that your professional opinion?" Asks Mim-sey who is now an intern from his office.

Using one of his employees' matrix cards as an alias-card will not raise any flags. If it does, the questions will come to his desk, first but not ahead of the answers.

"Thirty-three percent of the subsets for 2^1 on the number line, occur after 2^1 and 50% of the subset for 2^{32} occur after 2^{32}," says W-abe who, of course, knows nothing about Mim-sey's subterfuge.

17.

Margaret's hair style is sacrosanct. She seizes on it before she is a teenager in Puerto Rico. Along with her adopted hairstyle, she co-opts English and American Rock and Roll. Family and friends are constantly amused by her predilections and affectations and she is buoyed by the attention.

She cultivates the affection for her affectation and it satisfies her like a kind of masturbation. It wins approval for her heavy jaw, deep voice and her emerging mustache which races ahead of her youth. All of which is instinctively forgiven, somehow, by everyone.

Long bangs and shoulder length hair becomes her adopted trademarks. Everything else adopts her. Her fortunes inch by pennies and miles until she finds herself with six children and a mortgage in Miami. Somehow she survives every curve a churning corporate employer throws at her.

Every day has different hours. She works them and then some. Each and every manager has different corporate demands designed to cull achievers and she survives each and every swing of the ax. Every time they find a way not to pay a bonus she finds a way to feed the family and pay the mortgage.

She can afford neither the beer, beef, dentists nor breast implants to keep a husband. In the end, she cannot maintain the accouterments that keep her children at home and in the style to which they are accustomed. She is left estranged and without a choice.

Margaret launches Mark Garret's Rock and Roll Band. As the lead singer she has shoulder length hair, long bangs, skinny arms, no chest, a strong voice and chin and a deep memory of the genre.

Angst and a lack of accent make Mark a hit. Mark's endurance of the industry is attributed to masculinity, drugs, alcohol and dedication. Nothing is known, in fact, about gender, a lifetime of drudgery, moderation with alcohol, immoderation with caffeine and hypertension.

So by some consequence of history Margaret becomes Mark and the Grand Canyon becomes the Grand Canyon. The Grand Canyon cradles the Colorado River which no longer reaches the sea. Margaret cradles the children by every proxy, all the money, each telephone call and every excuse to connect with them short of real embraces. And somehow the confluence of Mark and Margaret escapes them all.

The children adore both Mark and Margaret without confusing one with the other. The tide that raises Mark's success never delivers Margaret to a spouse or returns her to her children except by proxy.

Both Ali-ce and Roy revel in Mark's music in their own way. Their shared appreciation connects them. It assuages their estrangement and redeems their reunion.

Ali-ce is back in town after eleven weeks of research at the Library of Congress. Roy has to cook. Ali-ce says she is like the microwave. She warms. She does not cook. Roy does not believe her. She knows it.

Roy persists, "We could have children."

Ali-ce puts the teapot on the heat.

"If I did not know you were a man before this," Ali-ce beams, "I do now."

"You had your doubts?" Roy asks as he cuts a box of spaghetti noodles in half.

"Not about your morphology or your capacity," says Ali-ce as she sucks on an uncooked noodle.

Roy asks, "What else is there?" as he puts a couple quarts of water on the stove.

"Your innocence...," grins Ali-ce stirring a tablespoon of powdered turmeric into the pasta water.

"I never pretended to be innocent," contends Roy while browning a pound of ground beef.

"But you underestimate what having children means," says Ali-ce cutting an onion in half.

"And that confirms my gender?" Roy defends as he stirs the ground beef.

"In an simplified way, yes," claims Ali-ce rendering the onion into one inch pieces.

"I know that having children is profound," Roy professes as he opens a large can of chopped Italian tomatoes. "I just thought that children would frustrate your father's plans."

"For murder?" Ali-ce taunts while opening a small can of tomato puree and one of tomato paste.

"For history," says Roy as he pours the ground beef into a colander.

"He might forget his plans for history and settle for mere murder," Ali-ce puffs, "if he thought we were having children."

Roy returns the pot to the stove to brown some ground pork. "Then we must be careful," he murmurs.

Ali-ce selects the ground nutmeg and cloves from among the spices. "You do not have anything to worry about," says Ali-ce while Roy gets the Worcestershire sauce and hands it to her.

"Our species are not compatible?" Roy supposes as he takes the teapot off the heat.

"In the ordinary sense, yes." says Ali-ce. "It would take many generations of social assimilation to produce the few progenies that could mature enough to mate in the current sense."

"In the meantime?" Roy asks.

"Gene splicing or RNA-interference could bring our species closer together," apprises Ali-ce as Roy adds the pork to the colander.

Roy returns the pan to the stove and adds oil and the chopped onions. "Society would have an obligation to provide support and protection to the progeny that survive in the meantime," he clues.

Ali-ce adds the noodles to the boiling water while Roy returns the cooked meat to the pan with the onions. "What makes you think that there is social support on either world for such a thing?" She asks.

"I imagine that after many, many years of coexistence of our two worlds that interbreeding is inevitable," comments Roy.

"More than many years and in the meantime what is the insurance against the annihilation of one or the other of our species?" Ali-ce asks while stirring the spices into the chopped tomatoes.

"Interbreeding is the only thing that will end the competition between our species," observes Roy while he stirs the chopped tomatoes and puree into the meat and onions.

"The interbreeding multiplies the historic paths among individuals," offers Ali-ce as she moves the colander to the sink.

"And they do not need to be limited to one or a few planets that can be destroyed with only some effort," poses Roy as he fishes a noodle out of the water to see if they are ready.

"More planets more effort," says Ali-ce putting the tea bags into the teapot with hot water.

"And at some point more effort than can be mustered," says Roy draining the pasta into the eternally useful colander.

"Endless petty wars," judges Ali-ce while she stirs the meat sauce.

"Meanwhile annihilation becomes more remote," says Roy picking up a large wooden salad bowl.

"But could we cohabit in the meantime?" Ali-ce asks.

"You and I do," Roy holds. "Why not entire communities? At least the communities that are large enough to support the gene splicing and RNA-interference that is intended to culminate in unassisted and mutual procreation."

"You and I cohabit, blissfully," persists Ali-ce, "but only because of some choices that I made before I met you." She seeds and chops a sweet pepper and half a cucumber.

"It is not color. You can change that if you wish," abides Roy. "You do it from time to time."

"It is not easy but, yes, I do. It would help if I could get it right," frets Ali-ce.

"You cannot decide which color you like?" Roy asks pouring oil and vinegar into a blender.

"I see many colors I like," prompts Ali-ce. "But gender is easier."

"You can choose your gender" says Roy adding anchovy paste.

"Do not be alarmed," Ali-ce entreats while putting a spoonful of feta cheese and some pitted, black olives in the blender. "It is a lifetime choice that comes with public confirmation and ceremony,"

"Can you change your choice?" Roy asks while he removes the tea bags from the celadon teapot.

"Not without a lot of public scandal," Ali-ce asserts. "People can tell if you are new to your choice. So they infer you choose to change."

Roy adds sugar and lemon juice to the blender. "So it takes most of your life to become a man or a woman?" He asks turning on the blender.

"We live most of our lives after we make the choice but that is because we spend most of our lives raising our children," Ali-ce extols. "They are ours even after they choose their gender."

"Why does it take so long?" Roy asks pouring the salad dressing into a carafe.

"Believe it or not it is history again. We could remain fully functioning hermaphrodites for our entire lives," Ali-ce presses as Roy cuts some slices from a loaf of fresh artisan bread. "Most of the species we know are hermaphroditic but the simpler species mutilate each other after mating."

"Why?" Roy asks as Ali-ce puts the lettuce, cucumbers and peppers in the wooden salad bowl.

"It eliminates the competition," Ali-ce asserts. "A potential mate is less interested if the intercourse is not symmetrical."

Roy and Ali-ce move the salad and pasta to the table before returning to the kitchen.

Picking up the dressing and turning to the table, Roy asks, "So you have a history of mutilation after mating?"

"Not me but more than a million years ago a male could keep many females that way and not worry about competition and a female could keep males the same way," offers Ali-ce picking up the teapot and following Roy to the table.

"What put an end to that practice?" Roy asks.

"The harem keepers would compete or otherwise fail the harem and no one wanted its mutilated members," explains Ali-ce.

Roy and Ali-ce move the bread and ice water to the table.

"By choosing to be male or female," Roy asks, "was it an affirmation of potency for that choice?"

Ali-ce selects a bottle of red wine and Roy opens it.

"One would not choose to be female if they were better at being male," Ali-ce says. "For almost everyone that sorts itself out while we are still adolescents."

"An affirmed male will have the most progeny with an affirmed female," speculates Roy.

"That is another bias that supports choosing," states Ali-ce as she gets the wine glasses. "Another one is that your affirmation is suspicious if you exercise your alternative option."

"If you were known to compromise your affirmation," Roy imputes, "there would be others who could promise more success in a marriage."

They move the glasses and wine to the table.

"So by the time we make our choice it is for life and everyone knows it," credits Ali-ce.

"We could marry," speculates Roy.

They survey the table for completeness.

"We could never have any children," says Ali-ce looking at Roy.

"We would always have each other," urges Roy looking at Ali-ce.

"My father would disown me," ventures Ali-ce as they sit to eat.

18.

Although he is groomed for a role among the 'Earthliens' with firm resolve to be undetectable as a counterfeit, Officer J. R. Gim-bal is ironically isolated from his public. In large and small ways his isolation is imposed. While he is in New Orleans, he does not visit the N. O. Glass Museum simply because it is not open while he is there.

As C.E.O. of Glasscastle Publications he must work out of Lynchburg, Virginia even though the office is in Bethesda, Maryland. The airports in Lynchburg and Washington, D.C. make it possible. Extraterrestrial travel makes it necessary. This way he can come and go with fewer people knowing when he is gone. Office visits are on a pretext and intended to support the impression that he is in week to week interaction with the business. Otherwise, he 'works out of Lynchburg'.

The company is unaware that he has two assistants. They are not paid through the company but are trained and groomed to stand-in for him. They are indistinguishable from him in knowledge of the business and physical appearance.

When Rus-ty is out, 'he is in Lynchburg' in the form of one of his assistants. J. R. Gim-bal makes it a practice to begin and end his trips with a stop in Bethesda. If any unspoken impressions are made about his coming or going, he is coming or going from Lynchburg.

"Including the side trip to New Orleans you were 'out' 18 days," says J-R One.

"How many of them were working days?" Gim-bal asks as the sun sets without notice.

"Thirteen," J-R One allows. "Today is Memorial Day."

The house lights ease into service for the evening.

"Is there anything in the folio for those days that I might not appreciate?" J. R. Gim-bal quizzes his assistant while he closes the street-side curtains.

J-R One is prepared to answer. He only has to wait for Gim-bal to ask the question.

"The redundancy is layered, as usual," J-R One says as he hands Gim-bal his briefing papers. "You can get as deep into the thirteen days as you like."

"What is not in here?" J. R. Gim-bal asks with every intention of doing his homework.

"Ali-ce Mim-sey's time is hers," J-R One says, "but she spends a lot of time in Florida." He takes a seat without presuming anything about Gim-bal's prerogatives to sit or stand or even to stay in the room.

"All of her company assignments go through me and I do not see a compromise," asserts J. R. Gim-bal, "But as Dr. Mim-sey's daughter she will never be 'just an employee'."

"Yes," J-R One quibbles. "But Roy Abbit lives in Florida, actually Miami."

"We have an unsolicited manuscript from him and we have a year to accept or reject it," cajoles J. R. Gim-bal mostly to buy time to think.

"It is expected to be rejected," reveals J-R One giving the significance time to focus.

"So what does this have to do with company business?" J. R. Gim-bal asks.

"If Abbot resurfaces as company business it most probably will not involve the publishing pursuit," says J-R Two, "and it will most probably involve Ali-ce Mim-sey, as an employee."

"It is not likely to be an adverse 'company' concern. Is it?" J. R. Gim-bal asks.

"If by adverse you mean some hostile action or transaction by her father or our host, this Earth, she will do everything to avoid that," grants J-R One checking this time for a reaction. "But she may not prepare you as soon as required regarding all of her father's intentions or initiatives."

Gim-bal wonders if convenience ever is an option.

"Do you think she will tell me as soon as either one becomes a company concern or will she act on her own to ameliorate the situation?" J. R. Gim-bal asks realizing that that could be indicative of where her loyalties are placed.

"I think she would rather tell you directly and not tolerate using the company's editors to telegraph her covert concerns regarding her father and Roy Abbit," advises J-R One.

"Even though the circumstances could be construed as strictly company business?" J. R. Gim-bal asks. "She has to shadow Roy Abbit as an assignment regardless of her loyalties," he adds.

"I suppose, at the least, you could tell her how to spin the company approach," J-R One suggests. "That way she could work through her editor with greater confidence in your knowing more of what is going on besides just information processing."

"If she wants the benefit of the Company, with a capital 'C', she has to cater to the charade," says J. R. Gim-bal, "I need the formality of this ruse to keep her or her father from tying my hands. I cannot guess what I will be told to do but I can be sure I cannot be vetoed by anyone, least of all the Mim-seys."

"And what or how the charade costs Roy Abbit is not her choice," says J-R One.

"At the very least I should be the one to tell her of these considerations," J. R. Gim-bal predicates. "She does not have any trouble telling the three of us apart and it would be a very different message coming from anyone but me."

MIDDLE CHAPTERS

1

1 1

1 2 1

1 3 3 1

1 4 6 4 1

1 5 10 10 5 1

1 6 15 20 15 6 1

Chapter 7

Chapter 111000....

Chapter ABC

Chapter $(1 + 3x + 3x^2 + x^3)$

Chapter $(a^3b^0 + 3a^2b^1 + 3a^1b^2 + a^0b^3)$

Chapter $2^0 + 2^1 + 2^2$

Episode 19

Dr. Mim-sey's office is not a warehouse. But that is because offices aren't as large as a warehouse. This is the regret of the receptionist.

The receptionist's desk is in front of a wall of furniture. It is not furniture to those who know better. Those are artifacts. If this were a museum each item would have a cabinet and a label.

Each item has a place in history and in some instances its place in a world. Some things are rare but most are not. Rare or not Dr. Mim-sey calls them amusing. They have most to do with the worlds and parochial languages to which they are associated.

His office is a warren of papers and books. And he rarely has trouble finding what he needs. The papers, books and artifacts are his thumbs. His own are insufficient. With the two, he can mark only two books. With his artifacts, papers and books, he can assemble a heap that is representative of a particular focus.

Each item in the heap is a jumping-off point or an issue that he wants to address in his never ending research. A research that now embraces both his planet and Earth. An embrace that includes more than mere research.

His daughter gave him a copy of William Shakespeare's *Coriolanus* and he does not know where to put it. He did not ask her for that and she knows better than to send him things he does not ask for. If he had asked for it, he would have a place to put it. But now, he could lose it because it is not in the right heap. For the time being it sits on the receptionist's otherwise uncluttered desk.

The room around his desk, on the other hand, is like the room inside his skull and there is no place for *Coriolanus*. Each time he puts it down, he might not get back to it before it gets lost. If it gets lost, it is lost for good. Then, if he does need it, his only hope is to ask Ali-ce for another copy. She will know he lost it. This is why she does this. She does this to drive him crazy. He cannot talk to Dr. W-abe and look for a place to put Coriolanus.

Now, he only sees the book when he picks up something from the receptionist's desk like an intern's matrix card. Dr. W-abe knows Dr. Mim-sey will go with him to a restaurant or to his house to eat. One way or another, when he does not want to eat alone, he does not have to eat alone. In any case, neither ever eats at the office. Money is of no concern. The real price is a little superficial conversation along the way. All along the way.

"As early as 2^1 the reasoning is that only 66% of the subsets on the number line comes before that power of two," grants Mim-sey.

Dr. W-abe walks past the nearby matrix elevators. Mim-sey is not surprised given the hour.

"That, by itself, is counter intuitive," advises W-abe.

"Intuition fails again," Mim-sey cautions, "since 2^{32} does not have 66% of its subsets before that power of two."

The homeward bound crowd is rapidly diminishing. But that crowd is still too dense for Dr. W-abe.

"In that case 50% of the subsets occur before 2^{32} and 50% of the subsets occur after," affirms W-abe.

"Not exactly," protests Mim-sey whose objective is the course of the conversation not the course of the crowd.

"We both have the same calculator and we both get the same results," discounts W-abe.

"I think it's the calculator," Mim-sey quibbles, "not the arithmetic."

"But the calculator is not rounding off 16%," disputes W-abe.

They choose to take the R. A. T. (Reticulated Atrium Transport) that circumnavigates the atrium in order cross to the other side.

"If you divide any power of two by the next higher power of two," Mim-sey says, "you will get 0.5 which is the 50% we expect."

As hoped and expected, it is a short wait for an atrium transport and it is unoccupied.

"Okay, so $2^1 / 2^2 = 0.5$ not 0.6666667," carps W-abe who feels vindicated by the vacancy of the R. A. T.

"But the real calculation is $2^n / (2^{n+1} - 1) = N$," says Mim-sey while taking some pride in getting a seat with the best view.

"Because the next power of two is also the next row on the Pascal or Chu triangle," W-abe clarifies. "Three is not a subset of four in row two of the triangle."

"We have not counted to four yet. Three cannot be a subset of four until we count to four," explains Mim-sey.

Both men are satisfied with the choice of transport. Each for his own reasons. Dr. W-abe is not there for the view and out of habit always lets Mim-sey choose his seat in order avoid the agitation that comes with Mim-sey not having the best choice of seats.

"I never was any good at percentages," says W-abe.

The glare of environmental lights confirm that it is after sunset.

"Our calculators are correct for any power of two when the numbers are powers of two," Mim-sey imparts. "But the subsets involve a number just one short of the power of two because of the null set."

"So the result is never really 50%," allows W-abe aware that the daily exodus is quite large. Again he has successfully insulated himself from the heave of the homeward-bound tide.

"Not with any number no matter how large," continues Dr. Mim-sey.

"But zero is a subset," says W-abe considering that the daily departure of 650,000 individuals is no small tide.

"And whether the math is correct or not pivots on that point - the zeroth subset," asserts Mim-sey.

"For any row on the triangle the sum of the subsets is correct," W-abe professes. "The difficulty is that one subset is the empty or null set."

"Row two of the triangle includes 0, 1, 2, 3 of the number line. The following rows include 0, ..., again, up to the value of 2^n where 'n' is the row number," declares Mim-sey reminding Dr. W-abe of something he had forgotten.

"So counting begins with zero for each row," W-abe says. "And if you want to count 2^n things you have to go to the row that includes $2^n + 1$."

Leaving work at the end of the day means leaving the building for almost 650,000 of its clients. In about three hours, the burden of the egress no longer taxes the building. Perhaps the exodus is more like a flood that begins with a rush or 250,000 for the first two hours and ends with a trickle of a little over 100,000 for the last hour.

"Four dimensions will not fit on row two of the triangle unless one of the dimensions is a non-dimension, not a dimension," contends Mim-sey as the R.A.T. nears its next stop.

"So a Theory of Everything with 1, 3, 6, 1 dimensions must include a dimensions for one no-dimension, one time dimension, the conventional three dimensions of space," W-abe maintains. "That leaves, the set of six dimensions to make the Theory of Everything work."

"But even then," Mim-sey says, "the theory is not finished until its sets of dimensions are a power of two."

"That is including the zeroth dimension," warrants W-abe exiting the transport.

Episode 20

Spending three to eight months rummaging through a library like the Library of Congress is not unusual for Ali-ce. Libraries are just beginning to convert to computer based media, so she works with the real thing: books, journals, microfilm and so on. She researches whatever her father needs and sends what she finds to the publisher she works for. They get it to him.

When he wants to learn a new language, she makes sure he gets the training material and eventually the classic literature. That includes the math and history in the native language. Sometimes that includes philosophy, politics and religion. It depends on what he wants.

Mim-sey gets what he wants completely with a minimum of extra material. Because Ali-ce knows how he works, he gets what he wants. When there are gaps he can tell Ali-ce what he is missing. The publisher, she works for, can convert it all to electronic media and transmit it, without creating suspicion or breaking any laws. Anyone with any authority can examine the media, if they want to, if they can understand it. The proliferation of computers just makes it easier. She does not have to worry about money so what she really earns is time off.

It takes him time to set new research goals and her time is hers. It pleases her to spend it with Roy Abbit. The same is so with Roy.

"What are your weddings like?" Roy asks while gazing at the Super-size Center on the street below.

"In some ways like yours," Ali-ce chats. "Both families are there especially the children."

"Why especially the children?" Roy asks with a view of only one person on the street below.

"It helps the children commit to a gender," says Ali-ce.

Roy and Ali-ce know this pedestrian as persistently homeless. He is homeless in the conventional sense but this is his neighborhood for many years. His neighborhood is his home. The streets are his hallways. The lights and benches are his furniture.

"Do weddings emphasize having children?" Roy asks.

"Not principally. That part is conspicuously understood," Ali-ce says. "What it does is to show how vast the support for gender choice is."

"The children see lots of people who have made the choice and they take it for granted that everyone chooses," interprets Roy.

"They see it as natural, inevitable, catholic...," renders Ali-ce.

"No exceptions?" Roy asks.

Some people know this homeless person as Willie but most do not know even that much. Roy calls Willie 'Bladderbox' because he always carries a box that is made to hold a five-gallon bag of milk.

"Everyone is on their best behavior," Ali-ce states. "Some of the older children might be a little suspicious of someone's commitment to his or her gender choice."

"But the youngest see only a huge public affirmation of gender choices and the acceptance of commitment," relates Roy who recognizes Willie's tote as one that is quite sturdy.

"Yes, because they choose their gender like they choose their words 'with habituation and deliberation', imposes Ali-ce who supposes the box as not only strong but easily replaceable.

"They choose to behave like a boy or a girl like their parents did and the longer they go between changes the more they habituate to a gender," guesses Roy.

"By the time sexuality begins," Ali-ce stresses, "they are almost completely committed to one gender or another."

Willie's box is his 'keep' and the neighborhood is his castle. No one can get into his keep except through the one hole which also serves as the handle.

"The years after childhood and before marriage must be very difficult," Roy marks, "given the cultural inertia to choose."

"For lives that are not punctuated by disaster, disease, death or some other devastation, it is the most difficult," underlines Ali-ce.

"I do not suppose," Roy postulates, "that they suspect they are at such a difficult stage, until it is behind them."

"The parents do but they do not tell the children that," affirms Ali-ce.

"Why not?" Roy asks.

"On the one hand, there are things that are worse than growing pains even if they are rare. On the other hand, it is not very effective to say that this is the toughest part of your life," prompts Ali-ce.

"Do the young adults figure that the beginning is not supposed to be the toughest, the end is?" Roy asks.

"That, and it is hard to think past the agony of the moment," Ali-ce explains, "and relieve it with the remote consolation of a future of bliss."

Roy sees the Super-size Center as one extreme of an engine. An engine of change for the neighborhood. Ali-ce sees the poor neighborhood as the other pole of the engine of change.

The two extremes of a homogenous and impoverished neighborhood and a center of retail wealth focus the difference of economic potential. A difference of potential creates a flow but, here, it is not one of electricity or falling water. This economic engine changes the neighborhood.

"So the children see a few weddings by the time they get married themselves?" Roy asks.

"As many as possible," Ali-ce informs. "Parents make every sacrifice to attend the weddings and bring the kids of all ages."

The neighborhood is drifting away from a homogenous society, its parochial identity. This change changes everything.

"But all those weddings must be confusing," says Roy.

"They miss things," Ali-ce says, "but the not the principal events."

"Like what?" Roy asks.

"Men and boy-hopefuls are one side of the room socializing. They wear some form of black, or a dark color," Ali-ce says. "The women and girl-hopefuls are on the other side of the room. They wear some form of white, or a pastel or even a very light grey."

"This way the children see people who have made their choices and how it fits them," allows Roy.

He has seen neighborhoods change in Coconut Grove and South Beach. As the neighborhood culture changes its social ethic changes mostly by unlinking. Social conventions fend for themselves more.

"Children see their side and the other side and decide which one they like best. There is no rope or wall or window between the two groups," Ali-ce says, "It is only their choice that puts them on a side."

Cultural and social linkages of the neighborhood are invisible until they start unlinking. This unlinking is marked by an inability of a culture to avoid random changes in social ethics and neighborhood conventions.

"The children see their parents separated. Does that make them unhappy?" Roy asks.

"They feel a price for the choice," Ali-ce remarks, "and wonder about it."

"But the separation is just for the wedding, so it is temporary anyway?" Roy supposes.

"And it is a strong contrast to the ceremony that joins the couple who are getting married," suggests Ali-ce.

"I see. Do they realize the contrast?" Roy asks.

"Most of the time, the youngest ones just feel the difference between separation and reunion as a change for the better," Ali-ce bids. "The ones that are slightly older recognize the contrast rationally and think they have discovered something."

"Do the grown-ups go along with this and let them think they have plumbed a secret?" Roy asks.

"Yes, of course. And at the end of the ceremony the two sides converge, mix and find the other half of the family," fills in Ali-ce.

Neighborhoods begin to integrate socially and sexually. People start marrying outside of their group. Absenting a gay ethic, there are still various heterosexual ethics. Differences among sexual ethics are not obvious until groups from outside of the neighborhood integrate.

These differences translate into stress and strife. Public conflict is the ultimate expression of these tensions.

"The separation ends and everyone enjoys the day," says Roy.

"Some of the youngest are frightened by the rush of everyone to get back to their other half," Ali-ce judges, "but that sorts out."

"I like that," lauds Roy.

The emergence of a gay ethic is indicative of integration. Such groups complicate and integrate a heterosexual ethic not just a gay ethic.

"It is quite wonderful and if it is done well, mother, father and the children will have a family color," Ali-ce extols. "A Father can be dark green, the mother a light green and the children a green in-between."

"And the children of each family group see their family colors and other family colors," pictures Roy.

Social change is not recognized by the things you cannot see no matter how many. So lots of social change goes unnoticed or at least unacknowledged until after the fact. By the time there is some ostensible indication of change it is after the fact and it is too late to resist.

"When parents and children go to lots of weddings, of even the remotest relatives," Ali-ce defends, "the children clue into the social inertia."

So change is usually only marked by the visible changes which are few or foreign. The emergence of a gay neighborhood is just one of the visible changes. A gay ethic is not exclusively a sexual ethic. It is a social ethic that is as extensive and complex as any.

"None of the weddings that I have seen are like that," claims Roy.

"I imagine the sentiments are apparent enough for the older family members," says Ali-ce.

"Yes, but I do not think the children get it," Roy affirms, "And weddings are often expensive and can last many hours."

"Our private ceremonies are that way," Ali-ce admits. "But those are just for the bride and groom and a few family and friends."

"So most of the family does not participate?" Roy asks.

"The private ceremony is optional," Ali-ce relates. "Costs are shared by the guests."

"So, a marriage can include both ceremonies?" Roy asks.

"Yes, but the public ceremony is the only legal marriage," Ali-ce describes. "Guests buy tickets that pay to rent the hall and that is it."

"What if something happens to a guest and they cannot attend the ceremony?" Asks Roy.

"The guest list determines the size of the hall and the cost of the ticket," Ali-ce says. "If you agree to be on the guest list you must pay the cost of the ticket whether you attend or not."

"Is that fair?" Roy asks.

"It keeps the cost of the wedding as low as possible," Ali-ce says, "small guest list, small hall."

The more extensive that an ethic is the more invisible it is. These invisible aspects of a gay ethic are as obvious as privacy. The emergence of a gay ethic is a visible indication of complex change.

"No small guest list with a large hall for ego's sake?" Roy asks.

"Extravagance is as vulgar as switching genders. The wedding is so much for the socialization of the children that parents should go to as many weddings as possible," Ali-ce says. "Your ability to do this is compromised if you spend a lot of money on one or a few weddings."

But Willie did not have to become gay, to exchanged his bladder box for an abandoned luggage bag with wheels. The neighborhood tide raises his boat along with all the others.

Episode 21

Officer Gim-bal lives and works out of a small house on a residential street that ends in a cull-de-sac. But that is only recently.

For many years before Glasscastle Publishing buys the five acre plot at the end of Gates Street, it has an even smaller house. There is a 3/4 acre truck garden and a fruit orchard with an unobstructed view of the railroad tracks that defines the plot's eastern border.

Glasscastle built a new house as close to the street as zoning would allow. Then they filled the plot with the tallest pine trees and fencing zoning would allow.

Everything about it pleases the neighborhood. The house is new and not too big or expensive looking. Trees are a joy all year long. Sunrise is from behind the trees, Ooooh. If there is a trace of rain, fog or snow, it is most prominent among the trees, Ooooh. You cannot see or hear the train if you want to, Ooooh. Hawks and owls have taken-up permanent residence there, Ooooh. The circumventing fence guarantees to eliminate any infestation of racoons and possums that the grove would otherwise promote, Aaaah.

Glasscastel does not do all this for the neighborhood. It is part of the hiring package for J. R. Gim-bal. That and a modest annual salary make the deal. The trees, fence and neighborhood make a perfect screen for Rus-ty's shuttle and there is room for a second one. The other one is usually Ali-ce's. At the moment there are two shuttles in Rus-ty's backyard.

"Is the jury still out on Roy Abbit's manuscript?" Asks Ali-ce closing her hatch.

"No one has heard the final word, yet," objects Gim-bal moving along the path with Ali-ce. "But already I am looking at a father and daughter who are poles apart over this author."

"I do not see him as significant or a menace, so my father must see him in opposite terms," infers Ali-ce.

"That, I think, is getting ahead of the situation," tenders Gim-bal as he turns left.

The proximity lights illuminate the path, just at their feet, with a light blue glow.

"So it is better put," Ali-ce considers, "that he expects either Roy Abbit or his manuscript to be significant or a menace."

"I think that is fair but he has not convinced anyone of that, yet," Gim-bal judges. "Perhaps not even himself."

Darkness swallows up the path immediately behind them as the receding pale blue lights blink off. The shallow pools of light serve them along the way but do not delineate the path ahead. The lights say 'you are here' and nothing of where you have been or where you are going.

"We become opposites only after he succeeds," says Ali-ce.

"But neither of you can contravene the outcome," Gim-bal says.

"If Mr. Abbit or his manuscript is deemed a threat, I can see how I would want to oppose its consequences and my father would not. But If Mr. Abbit and his work are conceded to be benign, why would my father object to that?" Asks Ali-ce.

"That is the surprise, even to me. He seems to need Mr. Abbit or at least his work to be a threat. When that is established, he turns the corner on vocational obscurity. Roy Abbit is his ticket to the big leagues. What is going to come along and put your father on a new vocational plateau besides Roy Abbit?" Gim-bal asks.

A trespasser could get no help from lights or landscaping. Ali-ce and Rus-ty turn left.

"So far, there has not been any glory for anyone. Only one of the worlds knows the other exists," presses Ali-ce turning right. "And all we have done is setup a bureaucracy to monitor the 'Earthlians' and promote fictions that prepare our culture to accept the possibility of another world."

"This could and must go on for many lifetimes before the reality surfaces," says Gim-bal.

"That works to our advantage. No one wants the shock of discovering another world with billions of inhabitants," Ali-ce says, turning left. "This could only fragment our culture."

"Nor do we want something that unites a significant segment of our world in a misguided or unsustainable campaign that would waste our resources," advances Gim-bal, following Ali-ce.

"So if Mr. Abbit and his manuscript are held to be benign, not just I and my father must support the status quo," Ali-ce says as she steps out of the trees and onto the grass behind the house. "Everyone must."

"Certainly," Gim-bal agrees.

The dim yellow porch light deepens the somberness of their faces.

"And if the opinion is against Roy Abbit, my father is on the winning side by default and I have to make a choice," affirms Ali-ce as she and Gim-bal cross the porch and reach for the door.

"Exactly," declares Gim-bal as he and Ali-ce enter the house.

$$1$$
$$1\ 1$$
$$1\ 2\ 1$$
$$1\ 3\ 3\ 1$$
$$1\ 4\ 6\ 4\ 1$$
$$1\ 5\ 10\ 10\ 5\ 1$$
$$1\ 6\ 15\ 20\ 15\ 6\ 1$$

Chapter 8

Chapter 0001000....

Chapter D

Chapter $(1 + 4x + 3x^2 + x^3)$

Chapter $(a^3b^0 + 4a^2b^1 + 3a^1b^2 + a^0b^3)$

Chapter 2^3

Episode 22

    Being almost directly below, it only took seconds to move the conversation to the restaurant sixty floors down. They both like this one so it does not take much to agree on it. Dr. W-abe likes the size, light and service. Mim-sey likes the size and weight of the drink glasses as well as seating with a view of the front door.

He always likes to see who comes and goes. He and Dr. W-abe have been to this restaurant a few times before and always order the same thing. If they wanted to eat something else, they would go to a different restaurant.

They both like that the service does not linger around the table. Dr. W-abe wants to be left alone especially during meals. Mim-sey does not want to craft the conversation for more than one listener. Conversation with Dr. Mim-sey is a manipulation.

When he chats you up, he wants something. He gets it and you never know it. This is difficult to do with more than one person at a time. It is almost always done through a series of conversations.

Dr. W-abe knows Dr. Mim-sey's compulsive subterfuge and could not care less. He accommodates him by keeping confidential information out of his working life. If it is restricted material, someone else deals with it.

Project Bump was the beginning and end of his cache of secret information. That was so long ago and so superficial that the details of the project are now common and public at least in the form of urban-fantasy. If his innocence bores Dr. Mim-sey that is fine.

"Besides being counter intuitive, What is it to say half of the subsets for any power of two occur after the number?" W-abe asks as he raises his glass to encourage Dr. Mim-sey.

While the drink glasses are large and heavy, the drinks are not. A blatant restaurant ruse that is a given forgiven. Mim-sey flatters himself that he sees through the trick but that at least some of that heft must be the generosity of the drink.

"So Mr. Abbit's idea is just an artifact?" Mim-sey asks.

"A good analogy. The amateur arithmetic archeologist digs the the fragments out of the layers of arithmetic rock, puts them together and has something to put on the shelf. Then spends the rest of his life with the the single task of trying to get someone to look at it," says W-abe watching Mim-sey weigh his glass with a distracted satisfaction.

"The game becomes, 'Guess what it is and tell me what good it is.' If he or someone finds a use for it, he can claim peerage," says Mim-sey who takes a sip like each one was a week's pay.

"Not in my lifetime," says W-abe watching Mim-sey luxuriate just incase his drink is not free.

"There it is again - time. Einstein had to add that dimension because it takes time for a change in gravity to get from here to there," huffs Mim-sey setting down his glass slowly almost reverently.

"There is nothing new in that anymore," says W-abe taking a drink that reminds him of the unnecessary weight of the glass and the insultingly small serving.

"What if the moment of time is 2^n? Then the present is situated in the middle of row-n of the Pascal-Chu triangle," says Mim-sey who recognizes the waiter this time as one who knows them.

"That is where the largest number of subsets is bunched. There the probabilities are highest. So, at least, more different things can happen," says W-abe contemplating what he will order for dinner.

"At the extreme ends of the row the 'future and past' are found. Entry-zero is the past before the first subsets. The last entry is the limit of the future for that row," Mim-sey relates. "For a given row, it is one subset - the only subset that includes all of the coefficients, all binary ones, all the black marbles."

"On the one hand that is as far as you can get into the future given the coefficients of the moment," responds W-abe who hands the waiter both menus on his approach and requests the usual meal and more drinks to the relief and satisfaction of Dr. Mim-sey.

"Time begins with the zeroth moment. The moment before the first constituent of past, present and future," says Mim-sey who can drink freely now that W-abe has established by ordering the meal for both of them that he will pay.

"And the future ends just before 2^{n+1}," says W-abe relaxing with the purge of ceremony.

"The next row is the next power of two. There are twice as many possibilities in the next future," enthuses Mim-sey as the waiter begins arranging their tableware and spices on the table space between them.

"More precisely, twice as many subsets. The possibilities derive from the subsets," says W-abe comfortable as ever with courses shared from the center of the table and delivered on his right since he gets the check.

Episode 23

For reasons that they never question Roy and Ali-ce each have their own apartments. When she is in Miami, they spend time in both apartments. He has a balcony facing north. She has a patio facing west.

One day Roy insists on going to bed early because he maintains that they have to get up early. Just before sunrise Ali-ce can smell coffee and hears Roy moving around the apartment. When she seems a little awake, he tells her she has to move to the chaise on the patio along with her pillow and covers.

He makes sure she is just as comfortable on the chaise as she was moments before in bed. When she dozes, she does not know for how long. Once she is awake enough, he brings her coffee and sugar. She likes wheat toast with salt instead of butter. Both the coffee and toast are warm.

As the sun comes up, the patio is on the shady side of the building. Dogs bark now and then. Just enough, according to doggy calculus, to earn their food. The parrots make more noise but they never stop in any nearby trees.

Roy waters the plants with a hose to raise the smell of water and earth. When the breeze is right, you can smell a frangipani tree. Being on the shady side of the apartment, the morning begins a little before six and lasts until almost nine.

That is Roy's gem. A quiet, comfortable, morning that lasts more than a minute or the few minutes between the front door and the rail station. The bliss is just across the threshold and it is accessible for hours. In a town known for heat, light, hustle and noise, all day every day, he shows her a nexus of peace and fragrance that can be had for more than a moment and more than once a year. Now, it is Ali-ce's treasure too. Eventually, conversation replaces the muse.

"If socialization of weddings is so important," Roy says, "you must go to a lot of them."

"Once a week is not unusual," says Ali-ce as she looks across the yard at the neighbor's patio.

"How can anyone afford a wedding every week?" Roy asks noticing her occasional focus on the neighboring house.

"It is just the price of the ticket, like a movie," Ali-ce says. "Only, you participate in the movie."

But the reality of ordinary Miami life tempers Roy's gift to Ali-ce. He knows it is not just the neighbor's patio that is the source of her distraction.

"What about the cost of clothes?" Roy quizzes Ali-ce.

"You can wear what you would wear anywhere. It is nice if you can share a color though," says Ali-ce who knows that the neighbors are functional in English but prefer not to speak it when she is around.

"Does a family color cost more?" Roy asks.

"A family color is not like a tartan," Ali-ce says. "It is just the color you share at the time. Just enough to clue the children in on the membership."

Ali-ce is predictably fluent in Spanish and is neither dilettantish nor illiterate.

"So, the important thing is the combination of sorting and mixing that is part of the ceremony. What about at home?" Roy questions her.

"We do not need clothes at home," says Ali-ce who feels obliged by her neighbor's hostility not to impose her language skills on them.

"What I mean is, how do you manage to get the children to grow into a gender choice before they become sexually active?" Roy asks. He, for his part, has no choice but to speak English.

"When they are youngest," Ali-ce says, "we pay attention to their biases for mother or father."

Before she met her neighbors, she noticed that they would stop speaking English when they saw her and change to Spanish.

"You do not suppose they are gender conscious at the earliest age?" Roy asks.

"No, but we do not want to reverse suppositions that we were not aware of at first," says Ali-ce.

She hears them speaking English from inside her apartment and derives that they prefer it.

"So their earliest choices are not based on morphology or artifacts?" Roy probes.

"Dad has the male behaviors and mom has the female personality," says Ali-ce who becomes 'the gringa' and invisible when she steps out of her apartment.

"How do they choose between them?" Roy asks.

"They do not at first," says Ali-ce who is presumed to speak only English by her neighbors while they pay close attention to her conversations with Roy.

"How do things sort out?" Roy asks unaware of the neighbor's Spanish pejoratives.

"For the children, mother is food and rest. The father is activity and water," says Ali-ce who always hears their casual contempt and pointed derision.

"What happens as more complicated behaviors and needs develop?" Roy canvasses.

"We steer the child to the parent that best supports the needs and behaviors," says Ali-ce.

"But one of the parents may end-up with most of the attention and most of the work," says Roy who presumes the neighbors to be literate in Spanish and just uncomfortable with a comparative insufficiency with English.

"You mean," Ali-ce asks, "like with four girl-hopefuls and no boy-hopefuls?"

"I was just wondering about a single boy- or girl-hopeful, but yes, what happens with four girl-hopefuls?" Asks Roy who would be surprised to learn that his neighbors were barely literate in Spanish too.

"The girl-hopefuls still learn how dads and moms behave with daughters," says Ali-ce. And as far as the extra work, dad does more of the things that could be done by either gender."

She had a glimpse of this illiteracy when they complained among themselves about the requirements for temporary work. The wife could not get work as a legal secretary. The agency requires filing skills that she could not show without knowing the 'nicknames' of the clients.

"Do the girl-hopefuls sit back and watch dad work?" Roy asks.

"The girl-hopefuls that are old enough," Ali-ce says, "help dad and learn too." So Ali-ce keeps to herself and keeps her Spanish to herself too.

"Only a few things become specific to gender?" Roy asks.

"And those things develop later and last. Early on, gender differentiation is more like mom and dad differentiation," says Ali-ce who hears her neighbors talk endlessly around a given topic until a consensus is reached on what the conversation itself is actually about.

"What about when they get older?" Roy deliberates.

"The roles are pretty well sorted out," Ali-ce says, "by the time physiology starts asserting itself."

"So mom and dad, have something of a head start?" Roy asks.

"Then clothes and ceremonies are used to emphasize the differences," says Ali-ce.

"I was wondering when you put your clothes on," says Roy with a smile.

The language barrier becomes a tool for segregation - a kind of illiterate elitism. The Anglo not only does not know the language he does not know how it is used. He is not supposed to know. As long as he is in the dark, he can be treated with condescension.

"The children are still quite young by the time they sort out the differences between moms and dads," Ali-ce says. "We use clothes and ceremonies to reinforce the differences."

"So they notice that you put on clothes before you leave the house?" Roy asks.

"Yes," Ali-ce says, "and they notice that everyone puts clothes on before someone comes to the house."

"No surprise visits?" Roy asks.

The Anglo is lead to believe that not being bilingual is indicative of his ignorance. The Anglo never knows who is literate and who is not on the other side of the language divide. He is, as often as possible, lead to believe that he is the ignorant one.

"If visitors do not know mom and dad have children, they find out," Ali-ce says, "and it is their responsibility to accommodate the parents by waiting until the parents and kids are ready to answer the door or video phone."

"That is the priority. Must one wait even if its bad weather?" Roy quizzes her.

"Except for emergencies, yes," says Ali-ce.

"What do you do about those years after childhood and before marriage about compromised gender choices?" Roy asks.

A second language is disadvantaged by a lot of work with no economic return. Native skills do not pay. A policeman does not get paid more if he is bilingual.

"I think I know what you mean," Ali-ce says, "but I am not sure."

"Suppose you have a son and he has a boy friend," says Roy.

"As long as neither one of them wants to reverse his personal gender choice," Ali-ce says, "that is fine."

Formal education raises the bar. A degree in a second language is a difference with distinction. Ordinary skills are differences with no distinction. They do not set you apart. There is no need to pay extra for the common skills.

"Growing up to be a boy or a girl is a life long commitment and changing is the taboo?" Roy questions Ali-ce.

"Remember our ancient history of mutilation to select gender?" Ali-ce asks.

"Yes," says Roy.

"We do not ever," Ali-ce says, "want to return to that."

So the translator gets paid for one translation not two languages. The motivation to learn one language or another is a consequence of social inertia. Economic incentives are a distant second and do not guarantee the life of the language. If a language dies, it is from lack of use.

Social inertia in this case means the atrophy of an alternative languages. It is the same thing that kills any language in isolation. What is not useful is not used. Language evolves. So words, even entire languages come and go.

"What has history got to do with the taboo of willfully changing gender?" Roy asks.

"We are afraid that the trend for switching will support a trend for mutilation," Ali-ce says, "either as punishment or as a social convention."

"In order to prevent changing gender in your lifetime, you can select it, say soon after birth and guarantee it with the help of a surgeon," says Roy.

"No one knows," Ali-ce says, "how to make the right choice at such an early age."

"What if your son wanted to spend his life with his boyfriend?" Roy inquires.

"That would not risk our social fabric as much," Ali-ce says, "as a change in choice of gender."

"But what if one of them got pregnant?" Roy asks.

"They could both be pregnant at one time or another and raise their children without any social stigma. But it would require a surrogate mother's role in order to assure that the children had a chance to develop a lifelong gender choice," says Ali-ce.

"Who chooses the surrogate-mother-role parent?" Roy canvases.

"The biological parents," says Ali-ce who chooses not to confront the neighbor's ruse.

"The same for a daughter with a girlfriend?" Roy asks.

"The same," says Ali-ce recalling the cattleya Roy gave her and that it is now gone.

Roy asks. "What about marriage?" Unaware that the Ginger Ficus and Phalaenopsis are gone too.

"All marriages are equally public and legal," says Ali-ce who never replaced the oncidium either.

"What I was wondering was, are there as many identical gender-role marriages as mixed gender-role marriages?" Roy asks as the morning and the chasm between houses dawn on him.

"Oh," says Ali-ce. "There are fewer identical gender-role marriages."

"Does that say stigma to you?" Roy deliberates knowing she does not need a picture window to see the light of day.

"No. Unfit parents lose their children. So everything depends on having a committed male and a committed female to raise the children," says Ali-ce who will not accuse her neighbors.

"So if two people of identical gender roles want to have children, and keep them, they know they have to agree on a surrogate gender role parent. If they cannot then they must plan on a marriage without children," says Roy knowing she will be satisfied with leaving her neighbors to a fate of their own making.

"So most marriages," Ali-ce says, "are mixed gender role marriages."

Episode 24

If you live in Lynchburg you are going to live by the rules. At the stroke of midnight, on Saturday it is not just Sunday it is the moment the Blue Laws apply and the zeal to conform is inspired by the zeal to enforce. If you recklessly find yourself on the dance floor, do not interfere with the rush to put chairs on the dance floor. It seems that it is illegal to, uh, dance on Sunday. Do not try to buy a padlock or a camera. They are not for sale.

Any reasonably, intelligent, ethical, alien, hermaphrodite who seeks to remain inconspicuous, will observe the law. At the same time, should you feel the need, urgent or casual, for beer and adult pornography - not to worry. Whoever wrote the Blue Laws had you in mind.

Not five blocks from the little house at the end of Gates Street stands the Food Zone ready to serve from 7 A. M. To 11 P. M. Well, the beer is self-serve to those who are 21 and over. Porn, for those 18 and over, is under the counter and you have to ask for it.

Ali-ce needed something to get her to Washington, D.C. and Gim-bal needed something to get him to Monday. This walk up the sloping street includes conversation about the rigors of life in Lynchburg. The walk back is unavoidably serious.

"So I must prepare myself for bitter consequences no matter what?" Ali-ce asks.

"If Roy Abbit and his manuscript pose a threat to our world," Gim-bal says, "you have to accept the circumstances that are imposed on him as a prophylactic or join him and pay the price with him."

The unhurried walk is slow in spite of the downhill grade from the highway.

"And If my father cannot accept that Roy Abbit is benign, he will be forced to?" Ali-ce asks.

"With a force," Gim-bal says, "that is equal to his resistance."

"The threat is a little short of naked or personal," says Ali-ce, "but I do not doubt it in the least."

"Then we understand each other completely," says Gim-bal.

The neighborhood is as quiet as it is dark. There are porch lights but no livingroom lights. Not even the dogs are up. It is after 9 P. M., and in a couple of hours there will not even be porch lights, just the dim and dated street lights.

"Is my father as clear in his comprehension?" Ali-ce asks.

Gim-bal says, "Unfortunately, no."

"He is at a disadvantage. Where there is a real threat, he must expose it," Ali-ce says. "In the event there is no threat, he must conform to that. Failing to do either, makes him a traitor."

"That is why I must give him some more room," says Gim-bal.

Dogs are 'dawgs' here. The 'w' becomes 'oo' even for women speakers. It is a shock hear a woman say 'da-oog'. And they say it unselfconsciously.

"Is it too early to hang a sword over his head?" Ali-ce asks.

"Yes," says Gim-bal, "but it is not too early to tell him he could be making a choice that could affect him personally and seriously."

"If he does not betray his integrity as a researcher," Ali-ce says, "he is safe."

"Yes, But if Roy is no threat and if Roy Abbit is dishonestly promoted as one," Gim-bal says, "your father becomes a loose cannon and everything that must be done to stop him will be done."

"Then you will not object to my saying the same thing to him?" Ali-ce asks.

"I am afraid that coming from his daughter is the same as coming from me or even his superiors. The danger is that it will artificially harden is resolve," Gim-bal says. "It will become a point of pride, personal pride."

"That is pitiful," says Ali-ce.

The Virginia accent is pleasant except for that one word - 'Da-oog'.

Southern accents are pleasant and natural. But perhaps the only exception is Virginia 'dog' spoken by a woman.

"I am sorry. No one wants to think of your father this way. But if I am not careful, I could force his hand the wrong way," says Gim-bal, "If he wants to hang himself with his own rope he is going to have to go through me or whoever is deputized to enforce whatever position is adopted."

"If the worst happens," Ali-ce says, "the least part of my struggle will be shifting blame."

"And it will be the least of my consolations," says Gim-bal.

'Da-oog' is not a nice word on Clidim. Here, on Earth, they have names for their 'da-oogs'.

1

1 1

1 2 1

1 3 3 1

1 4 6 4 1

1 5 10 10 5 1

1 6 15 20 15 6 1

Chapter 9

Chapter 1001000....

Chapter AD

Chapter $(1 + 4x + 4x^2 + x^3)$

Chapter $(a^3b^0 + 4a^2b^1 + 4a^1b^2 + a^0b^3)$

Chapter $2^0 + 2^3$

Episode 25

They are friends but Dr. Mim-sey does not like to go to Dr. W-abe's office. And Dr. W-abe does not like Dr. Mim-sey to visit his office. Again, friendship not withstanding. What Dr. W-abe notices is that this is the only place Dr. Mim-sey is not comfortable. He does not say why. Dr. W-abe does not ask. He focuses on making Dr. Mim-sey welcome and comfortable but never with much hope or success.

With the focus on Dr. Mim-sey and the futile effort to make him relax, he never gets around to asking what is wrong. (Dr. W-abe even changed his cologne once. When that did not seem to work, he went back to his old fragrance.) Dr. W-abe settles for any convenient excuse or lie that gets them both out of the office. And they get on with conversation along the way to someplace else.

If there is anything Dr. Mim-sey does not like about Dr. W-abe's office it is that it is so perfect. It has good furniture, good light and no distractions. It has no dust, no plants and no noise. He walks by the receptionist who always says," Dr. W-abe is in." while continuing to work. Dr. Mim-sey wonders if Dr. W-abe is ever out or busy.

Dr. Mim-sey resolves to visit once when he knows Dr. W-abe is not there just to hear the receptionist say something different. Then he considers the possibility of the receptionist saying," Dr. W-abe is in." anyway and then the existence of multiple Dr. W-abes.

This possibility has so many complications that it raises his anxiety quotient and mitigates Mim-sey's ability to concentrate on his purpose for being there. One way or another, Dr. W-abe wonders why Dr. Mim-sey is there. Dr. W-abe gets tense and this makes Dr. Mim-sey tense. For relief, Dr. W-abe presses a conversation that is sure to engage Dr. Mim-sey. Lately, this is Abbit's math.

"That not a very realistic time line. It progresses up to (2^n-1), pops to (2^{n+1}) and then," W-abe derides, "it is suddenly twice as long as before with twice as many subsets."

"As algebra or arithmetic, maybe," Mim-sey sighs, "but as calculus things smooth out."

"So, I win," W-abe gloats. "The carnival cashier cannot carry the challenge."

"There you go. Getting ahead of yourself again," Mim-sey says.

"When I am right, I do not mind saying so," says W-abe reflecting on a very productive week.

"Well if you do not mind doing most of the listening, it might open your eyes," suggests Mim-sey suspecting that Dr. W-abe is more receptive than usual to his visit in spite of the deflecting interruptions.

"As long as what you have to say is not as pedestrian as that metaphor, fine," says W-abe while taking a seat beside Dr. Mim-sey in chairs normally used by clients.

"Imagine a time-line, like a number line. Time is moving from zero into the future," utters Mim-sey who prides himself in his excellent guess at Dr. W-abe's openness to his visit.

"Um," W-abe taunts as he studies a small multicolored glass sculpture between them.

"I will take that as 'go on'," says Mim-sey resisting the temptation to fiddle with the sculpture. "On this time-line someone is born at the turn of the century and dies at the end of his 64th year."

"That is convenient - a time line from 2^0 to 2^6," says W-abe looking up from the artifact.

"In a moment we will let go of the conveniences. But while we are here, let us say his prospects improve evenly from birth to age 32," says Mim-sey settling for running his finger along the edge of the table. "Then his prospects decline, just as evenly, after that age."

"That is funny 2^5 is right in the middle," says W-abe wondering if Mim-sey found any dust.

"Anyway, let a dot on the time line represent the man at age 32. All the events that can contribute to his benefit are at a maximum then, they are at a level of 100%. Draw a line up from the dot. It does not matter how long. At the base of that line, on the time line is the subset for his lifetime," says Mim-sey who is disappointed but not surprised to find no dust in a prime location.

"I see where this is going," coaxes W-abe to engage Mim-sey.

"All the other contributions to his life are shorter and can be arranged as horizontal lines up and down the vertical line," says Mim-sey picking up a pamphlet without any attempt to read it. "The ends of all those horizontal lines can be used to draw a curve. A hill shaped curve that starts at birth and goes up until age 32 and then goes down to 64."

"It could be a bell curve," W-abe utters, "but that depends."

"If his ancestors, including his parents, contributed to his life-curve, it would not touch the horizontal time line at zero. And the values of an afterlife would prevent the curve from touching the line after 64. Even if his afterlife has little resemblance to a corporeal one, just so many atoms circulating randomly through the universe," ventures Mim-sey while he flexes the pamphlet without reading it.

"If his life was less than ideal," W-abe claims, "the bell curve would be lopsided."

"For the time being, let us stick to the ideal. What would his 'life curve' look like at age eight?" Mim-sey asks returning the pamphlet to the table without attention to its orientation.

"It would have to be the same bell curve but smaller and closer to zero (his zero) on the time line," says W-abe picking up the pamphlet with a plan to put it back where he wants it to be.

"Both curves would have that zero in common. But the sides of either curve would not match," Mim-sey says. "Any curve for any part of his life would not match the one at age 32."

"The peak of the curve would not match at any other age. At any time on the time line he could only expect to live 50% longer than he had lived so far," allows W-abe still holding the company brochure.

"At age eight," Mim-sey grants, "the curve would project him to die at age twelve."

"And most of the time the curve would not be conveniently located at 0, 1, 2, 3...," W-abe says. "It would take a calculus to describe the curve at most points on the time line. I win."

"I must tell you that you are not done listening yet," declares Mim-sey checking his pockets.

"What?" W-abe asks while reconsidering returning the pamphlet.

"For the arithmetician 'n' would be rational number most of the time, those times when it was not a natural number," adds Mim-sey who now has something to write with.

"How convenient is a number like $2^{3.6667}$?" Asks W-abe reaching for notepaper on his desk.

"It is not. But the limitations of no calculus are bypassed, albeit inconveniently," says Mim-sey while W-abe places the notepaper on the table where the pamphlet used to be.

"And so are the inconveniences of the quantum leaps from row to row of the Pascal or Chu triangle," says W-abe relieved to see Mim-sey pickup the notepaper and begin to doodle.

"Our subject can live any number of years. And our cashier can keep his calculations," says Mim-sey who seems to have forgotten the W-abe's company brochure.

"And he keeps his theory of 50% of the subsets following 2^n," conconcedes W-abe.

"Almost," says Mim-sey who does not notice W-abe returning the pamphlet to the table.

"Oh, yes. 2^n-1 depending on whether you are counting the null subset, or not," concedes W-abe as he watches Dr. Mim-sey pocket his notes.

"So if Blaise Pascal and Chu Shih-Chieh lived the same number of years," Mim-sey offers, "they would not have the same number of birthdays."

"How is that?" W-abe asks.

"The Chinese count the day you are born as a birthday. So after one year, in China, it is your second birthday," Mim-sey asserts. "One year after birth, Pascal is only celebrating his first birthday."

"Um," concedes W-abe.

Episode 26

It is too early for lunch, so Roy and Ali-ce settle on a mall-crawl before choosing a restaurant. Some of the malls are interesting experiences. If you like people, Lincoln Road is out doors and has a constant stream of tourists and natives. Everyone there is there to see people and of course eat and shop.

Other malls are completely enclosed and air-conditioned. These malls presumably do well for the purpose but make poor promenades. Somehow in the roofed malls while everyone is unique, everyone seems the same. They all seem to have the same attitude, purpose and destination. This kills the chances for the charm and variety you are likely to find at Lincoln Road.

For Roy and Ali-ce, they want to focus on each other and the two extremes for a mall do not suit them. At one mall you could have two drinks and a slice of pizza for $50.00 and the wait service would not leave you alone for a second. The corporate model and the science of business have pushed the business model to the last detail in optimizing cash flow and customer volume for what is supposed to be casual dining.

At the other extreme, the outdoor mall wait service is more natural but the flow of people make conversation difficult whether you are at an inside or outside restaurant. People stream among the restaurant tables.

Fortunately, there are malls with landscaping that permit walking and dining without corporate population streaming. In a mall like this, air-conditioning is reserved to interiors but it is not required outside. A mall like this permits comfortable touring on foot and restaurants without assembly line cuisine, rubber stamp clientele and clockwork wait service. This is where Roy and Ali-ce prefer to spend some time together and talk.

"What about mom and dad?" Roy asks watching water rush over some big round rocks.

"You mean us?" Ali-ce asks who always studies the plants, even lonely landscape plants.

Roy's gaze is fixed forward with some of his thoughts supposing that those rocks just might be fiberglass. "As far as I am concerned, everything is about us," says Roy.

"I like that. What about mom and dad?" Ali-ce asks studying one plant and then another.

"Can we have children?" Roy asks with the turbulent water providing constant change and attraction.

"I am sure we are not biologically compatible," says Ali-ce.

"So am I," Roy says, "but we could legally raise children once we were legally married."

"In my world they could be raised to make their own gender choices," says Ali-ce moving to change her view of the plants. "So the weight of society and law would be on our side."

"I am not so sure about my world. There would always be someone who thought he had something to say about our marriage and our children," says Roy moving to keep close to Ali-ce.

"We could ignore that," says Ali-ce who finds a new vantage point to study the landscape.

"I am afraid the carping would be endless," Roy says. "No one could guarantee our safety."

"What about the law," says Ali-ce who seems to have finished her careful plant study.

"The law could punish transgressions only after the fact. Then it would be too late," says Roy turning to walk along the waterway.

"Even if we could live independently of society, the children could not be expected to," says Ali-ce who is unimpressed by the obviously artificial recirculation of the water.

"Your father would not acknowledge your marriage," Roy says. "We could not support a marriage or children under the circumstances even if it was tolerated by society at large."

"They would not have to disapprove. They would just watch our lives succumb to hardship," says Ali-ce whose indifference to the waterworks is prompted by its tiny scale.

"That would be the lesson for their children and none of them would want to repeat our mistakes," says Roy who marvels at the movement of thousands gallons of water a minute.

"Our children would be taken away once our lack of resources begin to impact on the quality of their lives," says Ali-ce with some appreciation for the absence of whining pipes and growling motors at least.

"It would not matter whether we had an entire ship or a whole planet to ourselves," Roy says. "The children would have to be raised for a life in society. One that they were native to."

"And they could not do that," Ali-ce says, "unless we lived in the society they would join."

Marveling at the heaving, roiling water erupting among huge rocks, Roy says, "We would have to live separate lives in separate worlds as well as live our lives together."

"Does that sound familiar?" Ali-ce asks amused that the terminating pond never overfills.

"Not in the least," says Roy who is uncritical of receiving pond.

"You are innocent. Have you ever heard of cheating?" Ali-ce asks.

"Oh, I was thinking of a complicated marriage," Roy says, "but not that complication."

"I suppose," Ali-ce says, "I could marry and have children and not tell them about you."

"But I could not marry and have children and keep you," Roy admits, "even in secret."

"Why? Who would know?" Ali-ce asks stopping with Roy to look in a store window.

"I would. I know it is possible to meet and fall in love with someone after you are married," grants Roy who is pleased to see some opal jewelry. One of his favorite stones.

"There must be those who manage such a thing successfully," allows Ali-ce.

"But we only hear about the failures. They are tragic, especially for the children. Who could do that knowingly? I could not marry anyone as long as I love you, even if it means not having children," avers Roy recalling that Ali-ce's world uses opal for handles, knobs and buttons.

"And I could not ask you to forfeit marriage and children for me," declares Ali-ce.

"If the two of us cannot find a way, how are two civilizations going to manage?" Roy asks.

"My father says such a futile task is only be avoided," says Ali-ce.

"And that includes us?" Roy asks giving up on buying her an opal pendant or ring.

"He need not know about us," says Ali-ce resolving to bring Roy something from home.

"What about us?" Roy asks yearning to impress Ali-ce somehow.

"I do not know," says Ali-ce wondering if Roy could use an opal cutting board.

Episode 27

Leading, more or less, in a northeasterly direction from the house at Gates Street is a meandering path. For a quiet house in a quiet neighborhood it is a still quieter path. Even its maintenance is unobtrusive. The path is layered with pine needles and lit with very low wattage proximity lights that are only inches above ground. The meander is intended to disguise the very idea of a path because this path leads to the shuttle site used by Rus-ty, Ali-ce and others.

Along the way there are seats, benches, and small pools to serve meditation. A trespasser would suppose a poor connection among the few artifacts and the clearing that from time to time includes one or two glass pavilions. Least obvious of all is that anyone who knows the 'path' can move quickly without impediments between the house to the 'pavilion'. This time, Rus-ty Gim-bal's egress is measured by a conversation with his first assistant.

"Why are the Robes calling you back?" J-R One asks.

"You are never told why. You are just called," says J. R. Gim-bal eyeing the small stretch of back lawn separating the house from the trees.

"Yes, but what do you think?" J-R One asks as the porch light blinks on for the evening.

"If you and J-R Two were not intended to be identical clones of me, this conversation would never happen. But I must prepare you to do whatever I would do as if you both were me," clues J. R. Gim-bal wondering if he has enough time for this conversation before he leaves.

"The Robes have the last word on whatever is brought to them and they can choose which issues to consider," grants J-R One wishing for a way out but for which there is none.

"This project is not a public issue but it is the only significant thing I am connected with," allows J. R. Gim-bal steering his assistant away from his understandable but vain hope.

"Why do you suppose anyone sought to bring this to them?" J-R One asks with resignation.

"I do not think it was W-abe," J. R. Gim-bal says. "He does not have an agenda here."

"Ali-ce Mim-sey might be trying to protect Roy Abbit's life," says J-R One still in denial.

"She knows almost as well as we do that Roy Abbit or his manuscript may not be significant. It is too soon to trash the project or anyone's life with this kind of power play," says J. R.Gim-bal turning his assistant away from his misplaced hopes and toward a more likely truth.

"If she wants to use the Robes to spare Roy Abbit's life," J-R One asks, "could she?"

"The Robes could issue the death warrant in this case. They would not do that without listening to the arguments from the principals: She, I, Dr. Mim-sey and even Dr. W-abe as a dedicated mathematician," concedes J. R. Gim-bal redirecting his assistant to the significant parties.

"Have any of them been before the Robes, yet?" J-R One asks.

"If any of us went before the Robes, at this point, it would have to be to argue for the extermination of Roy Abbit," says J. R. Gim-bal looking at the wall made by trees and the deepening darkness.

"Even if Dr. W-abe thinks the death of Roy Abbit is justified he would not try to inject himself into such a secret project for the purpose," says J-R One hopefully but without much conviction.

"He could not do that without implying that he is more involved than he in fact is," grunts J. R. Gim-bal descending the steps and easing across the small space of lawn toward the tree line.

"He cannot advocate the fate of Roy Abbit, either now or in the future and insist he knows nothing about the project at the same time," says J-R One keeping himself close and talking low.

"Ali-ce Mim-sey's expertise and sympathies do not justify her going over everyone's head to advocate Roy Abbit's death," says J. R. Gim-bal leaving room on his left side for his assistant.

"I am in a position to know that you will not argue for the death of Roy Abbit under any circumstances," says J-R One watching the low watt lights ease on as they move among the trees.

"For a lot of reasons - I am not a mathematician; I am not the one who decides his fate. If an official death warrant is issued, I am the only one on this planet to enforce it," worms J. R. Gim-bal broaching his real trepidation to his call from the Robe's Chamber.

"So your summons to the Robes must be because of machinations of Dr. Mim-sey," sighs J-R One with the disappointment of approaching the shuttle and guessing that he has uncovered the truth.

"This could only mean that he was successful in getting a death warrant for Roy Abbit and I am going to be told to execute it," says J. R. Gim-bal while he reaches out to touch the shuttle.

"Then Dr. Mim-sey could represent himself as a world-saving hero only among project members," probes J-R One desperately hoping not to end the conversation that is only just beginning.

"That would move him up in the project organization to say the least," says J. R. Gim-bal.

"And, in time, the inevitable leaks will become a testament to his abilities and worthiness to things that are not even connected to the project," judges J-R One with a sense of the real focus.

"The project could end on the day of Roy Abbit's death and Mim-sey could reap benefits from then to the end of his career," concludes J. R. Gim-bal wishing that the conversation could have ended sooner.

1

1 1

1 2 1

1 3 3 1

1 4 6 4 1

1 5 10 10 5 1

1 6 15 20 15 6 1

Chapter 10

Chapter 0101000....

Chapter BD

Chapter $(1 + 4x + 5x^2 + x^3)$

Chapter $(a^3b^0 + 4a^2b^1 + 5a^1b^2 + a^0b^3)$

Chapter $2^1 + 2^3$

Episode 28

    Dr. W-abe is beginning to give up on spending any more productive time at the office. Dr. Mim-sey is the only one he knows that looks comfortable and uncomfortable at the same time. He settles into a chair like he lives there. There he shows the signs of commodious accommodation and then without any apparent reason, he gets up.

He insists on examining things on the desk even though he has done it many times before. Sometimes Dr. W-abe thinks Dr. Mim-sey forgets anyone is there. The seascape which is viewed best at a distance, gets the once over from only inches away. Mim-sey looks at different parts of the painting, so he must be appreciating it at some level.

He never goes near the bookcase without repositioning something. But the conversation never lags. Dr. W-abe knows after all these years not to challenge his focus. Mim-sey hears everything and even seems to control the conversation.

Dr. W-abe is reminded with his remark, "A time line that is just a number line is not very original or representative of the notion" in a vain attempt to divert Mim-sey from the bookshelf that everybody else seems to know not to touch.

"But that, along with the bell curve, provides for the inclusion of information that is more than counting units of time," says Mim-sey matching a short book with the nearest one of the same size.

"Particles, people and planets," W-abe says, "all have to be defined by the same phenomena."

"But planets have more to do with inertia and gravity and particles have more to do with electromagnetism," Mim-sey says. "The inertia and magnetism of people are more metaphorical."

"What good is this time line as a bell curve?" W-abe asks as his books assume that all too familiar shape.

"It will take a little more arithmetic to substantiate this conception of time, but the analogy it provides is a good beginning," says Mim-sey wishing for more short books. "Its initial assumptions will not have to be abandoned in favor of a better metaphor or a more sublime mathematics."

"Particles and planets have a history in physics and something in common. That is why the people got left behind with the arrival of alchemy," argues W-abe trying to change the focus.

"But they returned with probability. The present, which is so ably represented by the bell curve and the time line, is consistent with the inertia of the present and the higher probability of a present. One that is not likely to change because of the waning influence of things past and the as yet effects of the things of the future," says Mim-sey eyeing his bookbinder's version of a bell curve.

"Things?" W-abe asks steering from the figurative to the literal.

"A more specific term will not serve the analogy any better. As for the physical state of particles and planets," Mim-sey rallies, "I only have to remind you of Galileo's spin on inertia."

"It requires an outside force to make an object change what it is doing," clues W-abe grateful for the turn of the conversation. "An object that is not moving will not move until it is forced."

"And if the object is moving?" Mim-sey asks forcing the point.

"A moving object will continue that movement until it is forced to change its movement," apprises W-abe wondering what Mim-sey plans to do with these tandem principles.

"The bell curve and the time line reiterate and represent either the remoteness or the urgency of past, present or future forces that effect change," declares Mim-sey pressing the new alignment of books together as if to physically assert his conclusion.

"Past influences on inertia change inertia," W-abe objects. "They do not maintain the motion."

"Future influences of inertia have not happened yet but there is a probability of some effect," says Mim-sey occupied with the admiration of his bookshelf bell curve where there was none before.

"The present effects of inertia are the meat and potatoes of pedestrian physics," protests W-abe stalling for the clue to the point of this conversation.

"The science culminates in quantum theory," declaims Mim-sey suddenly coming to the point and casually turning to face a perturbed Dr. W-abe.

"You planned that" W-abe protests walking away from Mim-sey and the bookshelf as if to walk away from the conclusion.

"I am on your side," rejoins Mim-sey to a spot on the back of W-abe's head.

Episode 29

Roy and Ali-ce take the Metrorail train north to catch the Tri-rail train north that she needs to get the northbound Amtrak train. While a mere fence and 100 feet come between the Metrorail and Amtrak stations in Miami, money and bureaucracy cannot surmount it. True, anyone willing to carry or trundle his bags night or day from either of the nearby stations can get to the Amtrak gate just five or six blocks along the fence without a taxi.

And true, the muggings and murders are infrequent. And true, most of the perpetrators are ultimately brought to justice. And true, there must be a taxi that is willing to make a $4.00 one-way four minute trip. Somehow, the inclination to make these easy choices is almost universally declined.

The Metrorail does connect to a commuter train which is almost as close as the Amtrack station. For less than the cost of a taxi to the Amtrak station you can twirl through the turnstile to get to the Tri-rail train ticket kiosk and station. The Tri-rail travels north on the Amtrak rails to Hollywood, Florida and beyond.

Hollywood is just twenty miles north of Miami. But unlike Miami, metropolitan Hollywood has done the seemingly impossible. The two train systems not only share the tracks but they also share the station. There anyone who wants to take the Amtrak train to greater North America can do so bags and baggage without the impediments of panhandlers or put-upon taxi drivers targeting you as both redeeming them from and insulting to their life of subsistence.

This circuitous circuit provides ample time for Roy and Ali-ce enjoy each other's company as a kind of long, albeit temporary, good-by. They are in the mood to talk and grow closer. When seclusion and circumstances permit, they talk. Discretion is a habit but it is not a rigorous custom in Miami. Any conversation that is overheard or understood does not necessarily raise eyebrows until it is recounted as gossip. And then it is intended to reinforce a stereotype.

"So even though you are a hermaphrodite," Roy says, "you are a confirmed woman."

"Socially and psychologically a woman, and anyone that I marry will be a confirmed man," answers Ali-ce as the train stops at a northbound station and a bicyclist exits with his bike.

"This is not mere curiosity. What do you do with the gonads you are not using?" Roy asks observing that the car is theirs until the next stop.

"Have you not noticed," Ali-ce asks watching Roy's eyes, "my second bellybutton?"

"I cannot believe that little dot could be a normal penis," says Roy without looking down.

"It is all part of the big lie," says Ali-ce steering the conversation above the belt.

"So you have been lying to me?" Roy asks pretending exaggerated shock and surprise.

"Do not be frightened. I have not lied to you and I will never," chides Ali-ce amused at Roy's histrionics. "The big lie is what we call those years when we first become sexually active."

"You mean you are not committed to your gender by then?" Roy asks with curiosity.

"Actually over committed," answers Ali-ce reflecting back to those years.

"Does that mean you are promiscuous?" Roy asks digging for secrets.

"No. Social integration works against that. Were you hoping?" Ali-ce asks.

"Kind-of, sort-of," says Roy feeling a little resentful of some of his mental reflexes.

"You do not have to worry about my appetite. What I mean is that the girls are deliberately focused on being girls and the boys are focused on being boys," articulates Ali-ce.

"So they are denying their alternative capacity?" Roy asks through the announcement of the next stop.

"Yes and it is emerging at the same time," says Ali-ce watching the nearby platform-side doors. "So in order to repress the competing capacity in themselves they emphasize their chosen gender and discourage any satisfaction of their own alternative gender."

"But they realize it in the person they pursue - sexually and socially," probes Roy.

"This works against masturbation," mutters Ali-ce as a younger man enters the train car.

"Because it diminishes the pleasure or satisfaction that can be achieved with the 'opposite sex'?" Roy asks watching the passenger walk to the end of the car and lay down on the last seat.

"And the lack of tension between potential mates is conspicuous and embarrassing," stipulates Ali-ce undistracted by the self-absorbed passenger with his own train routine.

"So the boys do not appear to be the men they want to be and the girls do not appear to be the women they want to be. So where is the lie in that?" Roy asks with some interest.

"It is a tacit lie, I suppose. The parents do not tell the emerging adults that they do not have to segregate their physical sexuality," details Ali-ce now with the perspective of an adult.

"You mean they can throw all of their gonads into the game, so to speak?" Roy asks.

"After they are married," particularizes Ali-ce a little suspicious of his enthusiasm.

"What is the difference?" Roy asks trying to appear circumspect.

"They know each other well enough that they can protect each other's social gender commitment," Ali-ce offers. "The male can tell the female when she is slipping out of her socially chosen gender and vice versa."

"Does the male tolerate being told he is acting like a girl?" Roy asks with trepidation.

"They depend on it," says Ali-ce wondering what Roy's feminine side would include and whether she would recognize it. "No one wants to pay too high a price for their guilty pleasures."

"So couples find sometime after marriage that they do not need to segregate their sexual activity," says Roy wondering what Ali-ce's masculine side includes besides her bellybutton.

"I imagine there are some couples," Ali-ce theorizes absently, "who never find that out."

"People would know that. Would they not?" Roy asks with no one in particular in mind.

"That is the other part of the big lie." Ali-ce says realizing scope of the truth. "Everyone says that his sexual activity is consistent with his chosen social gender, whether it is true or not."

"So the men have to practice birth control?" Roy asks with a new sense of practicality.

"Yes. And the women have to make sure that their bellybutton does not tumesce," says Ali-ce with a new perspective on a social convention.

"So by the time the children become sexually active the big lie is firmly established everywhere," says Roy as the stop ahead is announced.

"Almost everywhere," Ali-ce prompts. "There are the 'experts' and other advocates that say it is possible to be hermaphroditically active without compromising your chosen social gender."

"But nobody wants to risk eroding his commitment to his chosen gender?" Roy asks.

"The commitment is not so much at risk," Ali-ce relates as the train stops to exchange passengers. "Nobody wants anyone else to know that they are hermaphroditically active."

"Because of the stigma?" Roy asks as a jowly man on a cell phone enters the car.

"It might break up the marriage. There would be competition from other adults for the skilled partner," conveys Ali-ce while the man heaves into a seat a few steps from the door.

"So the big lie is maintained," says Roy unperturbed by the portly passenger's shouting into his cell phone.

"And the idea of being hermaphroditically active is a fiction that only has a place in fantasy and scandal," says Ali-ce considering the benefit of a telepathically induced debilitating headache on this loud phone caller.

"Do I have to worry about your belly button?" Roy asks as the passenger focuses on Roy without compromising his conversation or volume.

"No. Do I have to worry about your belly button?" Ali-ce asks deciding to spare the cell phone caller a punishment that he cannot conveniently connect to an awareness of his trespass.

Episode 30

The Robe's Chamber is where the Robes study the issues and issue their opinions. It is, in fact, four buildings. Three of the buildings are square and surround the center building. It goes without saying that the center building must be a triangle.

In its early centuries the triangle shaped building is the only edifice. Each face of the triangular building fronts a square. Two of the squares are identical in size. The third square is equal in area to the sum of the two other squares. Over the centuries the Robe's Chamber's need for building space tends to increase. Ancillary buildings grow up around the three squares which are called the Robe's quadrangles.

Interested parties assemble in a 'quad' on the days that the Robes hand down an opinion or hear arguments. They are there with the hope of seeing, at least, one third of the Robe's Procession move to the Robe's Chamber. That centuries long tradition ends with the misguided assassination of one of the Robes. The assassin does not understand or care that the assassination would make little difference in the Robe's opinions. The opinions are written and printed for distribution before they are announced to the public.

For murder to affect the behind the scene debate, research and council of the Robes, murder is best done early and often. That, of course, does not happen. After the assassination, the quadrangles become the footprint of the ancillary buildings. By then the triangular Robe's Chamber building is three floors high.

Since then, those three floors are always raised above the quad-buildings. The roof of each of the quad-buildings preserve the quadrangles on their roofs. Interested crowds still gather in the acreage outside the Robe's Chamber complex. But now, for them, it is limited to discussing issues on the way to forming their own opinions.

"Dr. Mim-sey is advancing two arguments. The putative one is that Roy Abbit is a threat to his world and ours," conveys 1st-Robe trying to decide if he wants a hot or a cold drink from the kiosk.

"He is smart enough not to favor one or the other," says 111th-Robe deciding, as usual, on his usual hot drink.

"That will fragment the debate," says 1st-Robe while considering whether or not to have a sweetener. "The issues will be decided by who is on which side rather than the nature or substance of those issues."

"The argument will divide along two lines. One camp will argue that we will survive to be superior," imparts 111th-Robe. "The rest will be those who think we will be subjugated by the Earthlians."

"Mim-sey will avoid taking sides and rise above both," advises 1st-Robe settling for a hot drink.

"And he uses the Robe's Chamber to do it," says 111th-Robe turning toward his usual table.

"There is another argument, the unspoken one, of whether or not we should even take up the issue," collogues 1st-Robe wondering why this table suits 111th-Robe.

"Not taking up that issue will raise questions about our competence when the issue surfaces as a public one," considers 111th-Robe waiting for his colleague to choose a seat.

"Either way, we are advancing his agenda," says 1st-Robe sitting down without hesitation or deliberation.

"What becomes of us," says 111th-Robe, "when we are represented as Mim-sey's or anyone's agent."

"We cannot alibi in advance of the allegation to deflect the stigma," says 1st-Robe stirring his drink.

"Can we enlist a mathematician to disarm the issue?" 111th-Robe asks while looking for a spoon.

"Not without Dr. Mim-sey producing an equal and opposing expert," weighs 1st-Robe.

"Would that be Dr. W-abe?" 111th-Robe asks reaching to a nearby table for a spoon.

"Dr. W-abe might have his private opinions but he will not take a public position either way," deliberates 1st-Robe testing his drink for a drinkable temperature.

"Is that a setback for Dr. Mim-sey?" 111th-Robe asks wishing that his colleague had not taken the chair he wanted.

"Not in the least," relates 1st-Robe privately ignoring the other's distraction over the choice of chairs.

"And Dr. W-abe can avoid the issue," says 111th-Robe, "without tarnishing his reputation."

"I do not doubt that Dr. Mim-sey can recruit as many mathematicians of Dr. W-abe's caliber as he wants," says 1st-Robe.

"Dr. Mim-sey has enlisted his daughter in this effort of his. Is she like him?" 111th-Robe asks while adjusting his chair to get out of the glare and the heat of the sunlight enveloping his space.

"Do you mean, is she a proxy who will do what it takes to gain and maintain power, in this case, for her father?" 1st-Robe asks.

"She has spent enough time with Roy Abbit to know what kind of threat he is. Her casual and voluntary association with him suggests that she does not consider him a threat," reasons 111th-Robe now moving his chair, more than a little, to avoid the sunlight.

"I suppose she is a disappointment to her father. She is not a mathematician and cannot be used at Roy Abbit's expense. She has not murdered him or facilitated that, so far," estimates 1st-Robe convinced now that the other Robe has moved his chair the wrong way for his purpose.

"Can her choice be used against her?" 111th-Robe asks.

"As a professional, her association and even affection for Roy Abbit cannot be held against her," says 1st-Robe who is pleased to find that his drink is at a good temperature.

"She and we will be long gone before anyone can hold anyone personally responsible for the course of history," voices 111th-Robe who settles for moving to a chair beside 1st-Robe.

"She has a personal integrity that more resembles Dr. W-abe's than her father's," utters 1st-Robe who moves 111th-Robe's drink more into his convenient reach.

"Does she have any relevant opinions?" 111th-Robe asks.

"She maintains that Roy Abbit is neither the megalomaniac nor the genius that it takes to launch a war between the two worlds, Earth and ours, uh, assuming they can find us and get here," affirms 1st-Robe.

"Yes, prevailing pedestrian opinion on our planet and Earth is that any number of alien worlds probably exist but are merely yet to be discovered. Is that Roy Abbit's opinion?" 111th-Robe asks seeing that his spoon in now out of reach.

"Abbit and Mim-sey are among the few, besides us, who know without speculation that their respective planets are not unique in supportting an advanced civilization," says 1st-Robe following with another comfortable sip.

"Can this be held against Ali-ce'?" 111th-Robe asks looking for another spoon in easy reach.

"Not at this point. If she campaigns, like her father, for the revelation of Earth and our world to each other, that is another issue," ventures 1st-Robe who finds a handy spoon and gives it to 111th-Robe. "But on a one to one basis, like hers with Roy Abbit, no one can call her a traitor; even by association."

"But a formal opinion that Roy Abbit is not a potential threat pulls back the curtain on the legend. It reveals Earth as a real world and an independent civilization," vents 111th-Robe who resumes stirring his drink without taking a sip.

"Dr. Mim-sey uses us to do that. And with Roy Abbit as the putative villain the presumption is in Dr. Mim-sey's favor," wagers 1st-Robe as he finishes his drink with satisfaction.

"Dr. Mim-sey cannot lose," complains 111th-Robe who finds his drink a little too cold.

1

1 1

1 2 1

1 3 3 1

1 4 6 4 1

1 5 10 10 5 1

1 6 15 20 15 6 1

Chapter 11

Chapter 1101000....

Chapter ABD

Chapter $(1 + 4x + 5x^2 + 2x^3)$

Chapter $(a^3b^0 + 4a^2b^1 + 5a^1b^2 + 2a^0b^3)$

Chapter $2^0 + 2^1 + 2^3$

Episode 31

Dr. Mim-sey's apartment is not far from Dr. W-abe's if you cross the archipelago of green spaces that separate them. The interior of his apartment is not far from his office in terms of clutter and disorganization. Fortunately, his apartment is not as overburdened as his office. But W-abe usually has to move something to put anything down.

Every chair and shelf has pillows and puppets most of them given to him by friends who know his penchant for them. Dr. Mim-sey can tell you who gave him gave him each one and when. W-abe has heard most of the stories many times and cannot remember any of them except when Dr. Mim-sey reminds him with endless retelling.

Most of his pictures are framed but few are hung on the walls. They lean against each other until one comes up in conversation. That one gets propped-up or hung somewhere. The displaced picture or painting becomes a standee among the others.

Mim-sey does not fret for a place to put anything. Dr. W-abe will never know the logic. The clutter, mild as it is, would wear on him if Dr. Mim-sey were not so diligent in making way for his friend. W-abe can go anywhere and use anything in the apartment without inconvenience. Mim-sey is always cheerfully making-way by moving something, telling its history and making Dr. W-abe welcome and guiltless of imposition.

Dr. W-abe, unknowingly it, accommodates Dr. Mim-sey perfectly. He always sits in the same chair. He uses the same table and follows Dr. Mim-sey's lead when it comes to doing something else. His short memory, for Dr. Mim-sey's stories, allows Dr. Mim-sey to tell them over and over. With each recounting Dr. W-abe can honestly say, "Oh, yes. I remember now." which never fails to please Dr. Mim-sey and drive his enthusiasm.

To the extent that Dr. W-abe grows distracted or impatient, they can both honestly attribute that to anything else other than the environment, the company or the hospitality. His otherwise ambiguous discontent is personified in the perturbation of Roy Abbit of Earth. Thus, Mr. Abbit becomes the topic of conversation which implies his position as enemy number one.

"So time is a hump," W-abe quarrels while moving some pillows to make more room to sit.

"In a perfect universe it is. Every subset representing every probability and actuality is relevant," allows Mim-sey while he looks for a new 'home' for the offending blue-black pillow.

"But is the universe ever perfect?" W-abe asks following Mim-sey with his voice.

"I am not falling for that trap. We do not know when and if the universe is perfect. We are usually some part of the universe," says Mim-sey squeezing the orphan pillow. "We are not the whole universe."

"Then, as a rule," W-abe says, "we must deal with some subsets in some kind of association, with its universe."

"But we do not deal with just the sum of some subsets. We deal with 'Some Universe Minus Subsets'," declares Mim-sey putting the transient pillow on top of a slightly lighter, dark-blue pillow.

"S. U. M. S.?" W-abe asks while repositioning an antique puppet so it is less likely to fall behind the chair.

"For the life span of someone born at the turn of the century and who lives to the last day of his 2^6-year," Mim-sey deems, "any 2^n chunk of the universe is a 'S. U. M. S. Universe.'"

"So a 'S. U. M. S. Universe' is a 2^n subset of the whole universe," accepts W-abe who allows Mim-sey to construe his gesture to be one of appreciation for the puppet.

"That is the beginning of the math. Now the irrelevant subsets must be subtracted," advises Mim-sey who is pleased to see W-abe's remoteness to his puppets is eroding.

"Are you saying your register savant can do that?" W-abe asks.

"It is easier than you think, if you remember that any row of the Pascal or Chu triangle is equal to 2^n and 'n' is the row number," says Mim-sey considering where to get another antique puppet.

"So our sixty-three year old is row six and the sixty-four is the sum of $1 + 6 + 15 + 20 + 15 + 6 + 1$," W-abe attests. "The first number is zero the rest of the numbers represent his sixty-three years."

"Now suppose he lives more or less than sixty-three years," offers Mim-sey trying to imagine where to get a cheap antique puppet for his friend. Never mind a display case. W-abe can get that himself.

"Living more than sixty-three years the 'S. U. M. S. Universe' becomes one hundred twenty-eight or row seven of the triangle," W-abe grants. "But our specimen is not going to live much more than 100 years."

"In either case, some subsets have to go," says Mim-sey deciding to give his wife the task of finding a puppet for the purpose.

"Okay, lets say he lived fifty-six years. He died seven years too soon. I am waiting for this to be difficult," remarks W-abe as he watches Mim-sey prop up his wife's picture next to the phone.

"You may want to write this down. Year fifty-seven in binary numbers is 100111. Fifty-eight is 010111. Fifty-nine is 110111. Sixty is 001111. Sixty-one is 101111. Sixty-two is 011111. Sixty-three is 111111," says Mim-sey, smiling. "The binary number with all ones, 'all the black marbles' is six binary ones. Now let us call it 'X^6.'"

"Maybe it should be 'X_6' if anything or even 'X_{abcdef},'" says W-abe.

"You will recognize it as soon as I tell you. But first notice that there are three numbers with four binary ones and three numbers with five binary ones. Let us call those $3x^4$ and $3x^5$," says Mim-sey.

"Are we there yet?" W-abe asks still smiling and wondering why Mim-sey's smile is lingering.

"As soon as you remember that the numbers in a row of the Pascal or Chu triangle can be expanded into a polynomial. In this case: $1 + 6x + 15x^2 + 20x^3 + 15x^4 + 6x^5 + x^6$. If you count from zero to sixty-three with binary numbers, one of them will be all zeros, six of them will have one binary one, fifteen will have two binary ones and so on," clues Mim-sey resuming his smile.

"Oh, it is the expansion of $(1 + x)^n$," assents W-abe smiling in return. "You subtract the polynomial for the seven years of his shortened life as $(1 + 6x + 15x^2 + 20x^3 + 15x^4 + 6x^5 + x^6) - (3x^4 + 3x^5 + x^6)$."

"The polynomial for our prematurely dead soul is $1 + 6x + 15x^2 + 20x^3 + 12x^4 + 3x^5$," says Mim-sey now considering an anatomically correct, amusingly exaggerated, antique puppet as the perfect gift.

32.

Ali-ce resonates with Roy's choice of a diversion. A good restaurant experience is not to be underestimated for its worth. And the opportunity to stroll the green space of a mall in a stridently urban environment is a distinct pleasure. Water, stone, light, shade, space, breeze, and encroachments interplay constantly. Every step through the green space changes everything, but, somehow, one change is never disconnected from the next.

Each new experience is defined by its proximity to the old ones. There are no inorganic lines, fences, nothing that would count as a delineator with the exception of perimeters. How can things be distinct and continuous at the same time? It is more of an observation than a question for Ali-ce.

But the contrast of opposites in close association is not the conscious motive or focus of Ali-ce or Roy. It is one more chance to enjoy life and each other. Conversation punctuates, competes with and complements their purposes.

"If your father is not going to murder me, who is and why?" Roy asks as they turn away from a jewelry store and its opal trinkets.

"I do not know," Ali-ce says eyeing a particularly dangerous looking cactus. "He will not tolerate any prying on this topic."

"So how do you know it is not me?" Roy asks stopping to let Ali-ce study the plant.

"I told you he is not going to get me to murder you," asserts Ali-ce wondering if this plant is edible and how to prepare it. "And what have you done to justify his fears?"

"Lately?" Roy asks as he resumes the walk on a cue from Ali-ce.

"No offense but it is my impression that a cashier is not going to promote a clash of civilizations," proclaims Ali-ce now looking through the door of a candle shop.

"I could forget to redeem your coupons or forget your discount on a jumbo pack of toilet paper," jests Roy hoping Ali-ce is thinking of a candle for herself but not for him.

"Is that cause for murder?" Ali-ce asks considering a candle for Roy but not for herself.

"I suppose anything is possible," says Roy turning to look at some Sansevieria Cylindrica. "But it is righteous to complain to the employer and insist that the money is incidental."

"Then why complain?" Ali-ce asks looking at the pike-like plant.

"The offended customer will tell the employer that the error was deliberate and part of a personal campaign to treat customers of their ilk with prejudice and contempt," belabors Roy.

"So if the employer wants to protect his business and livelihood, he must remove you from your position of power and obfuscation?" Ali-ce asks recalling Sansevieria as flat not cylindrical.

Roy remembers the flat ones with yellow stripes as 'mother-in-law tongues' and says, "The customer is satisfied that you are deservingly removed from your means of subsistence."

"So the complaint is justified on a lie?" Ali-ce asks.

"The lie is inflated to fit the desired punishment - being fired," beefs Roy as they stop to take in a new waterfall with its boulders, flowering plants and primeval ferns.

"No one can expect the employer to fire someone for a small mistake that can be fixed on the spot," says Ali-ce pleased to see the water emerge, almost believably, from under a large rock.

"So the customer cannot exploit petty motives. It has to be a big lie and believable no matter how hackneyed," says Roy attracted by the volume of the flowing water and the riot of the fall.

"Does it work?" Ali-ce asks thinking moss and mollusks would not hurt the believability of the landscape.

"In a sloppy sort of way," says Roy moving to a better view. "The employer expects that some of the truth is just as the offended customer asserts and raises in pay are not justified."

"You cannot afford to work for that employer forever," says Ali-ce moving with him.

"So you change employers," says Roy settling in for more contemplation and appreciation.

"But if anyone can manufacture the lie," Ali-ce carps, "then almost anyone will."

"So you change jobs until a kind of equilibrium is achieved. The opportunity to launch the lie is minimized by making the discount process bullet proof," says Roy who is getting hungry for lunch. "And the wages for a cashier are so low that there is no advantage in believing the lie."

"There are not many people waiting to take the job?" Ali-ce asks thinking more of lunch too.

"Usually, and the cost of training a new cashier becomes relevant. "That makes it inconvenient," says Roy crossing a bridge that shortens the path to the collection of restaurants.

"But can you live very well on that income?" Ali-ce asks passing a fast food restaurant.

"I can lower my expectations and look for another job," says Roy with a quick look back.

"So your math book is a source of income?" Ali-ce asks with pseudo-innocence.

"In theory," Roy says. "But math books do not sell very well. They are not entertaining."

"So why did you write a math book?" Ali-ce asks with interest.

"It supports my interest in language and philosophy," says Roy trying to justify himself.

"That does not sound like something that will threaten civilizations," says Ali-ce pointedly.

"The topic almost guarantees that it will not," says Roy reminded of the unintended effect.

"Because no one will publish or read the book?" Ali-ce asks trying to follow his thinking.

"That and elementary math and amateur philosophy do not prop up or tear down civilizations," chaffs Roy hoping there is more truth here than he or Dr. Mim-sey suspects.

"So neither one of us has anything to worry about," affirms Ali-ce with similar hopes.

"Just be sure you do not pay for your father's crimes," says Roy flinching at the remark.

"He shares your feelings but for different reasons," says Ali-ce with complex sympathies.

"Not knowing the victim must make your job difficult," says Roy.

"I do not need to know the victim personally," coaxes Ali-ce trying to deliver herself from self-indictment. "I just send my father books on which he can model his mission or justification."

"Like what?" Roy asks with the idea of supporting her defenses.

"Have you read Shakespeare's *Julius Caesar*?" Ali-ce asks.

"No, I have read about it," responds Roy acknowledging his sparse academic experience. "It is easier, I expect, and faster than reading *Plutarch's Lives of the Noble Greeks and Romans*."

"A head of state is assassinated with knives by senators and peers," explains Ali-ce.

"Do you suppose our heads of state are planning a war against your world?" Roy asks.

"I do not think things are that far along," imparts Ali-ce with some self-consolation.

"An assassination or the threat of one could divert attention from things that would lead to our discovery of your world," allows Roy feeling something less of a target.

"Or like the cashiers, reduce the resources for any more significant unfavorable actions," frets Ali-ce with the suspicion that reality could involve just the opposite.

33.

The two identical rooftop quads front the offices for eight of the twelve Robes. As a consequence, those offices are equal in size and similar in floor plan. The remaining four Robes' offices front the remaining quad which is equal to the sum of the other two quads. These offices, consequentially, are larger than the other eight. These offices are allocated to the four senior Robes of Robe's Chamber. 1st-Robe is just returning to his office and finds 001th-Robe waiting for him.

"I just left 111th-Robe having. Please allow me to apologize," says 1st-Robe observing that his protegee was not in her preferred chair.

"There is no need. My work is unusually current so I do not mind or feel constrained to hurry," grants 001th-Robe without referring to her preference for chairs. "In fact, I am suspicious of my confidence in my opinions when they are insulated from consultation. I welcome other opinions."

"We twelve are alike in that respect. We come together, each with a lifetime of experience, But we all know we need to test our certainties," says 1st-Robe with a clue to 001th-Robe's unstated turn of habit. "The gentle effrontery of an unsympathetic peer is welcomed no matter how it may sting our ego."

"But what of the insult of a pretender?" 001th-Robe asks referring to a third party.

"That can only be Dr. Mim-sey," says 1st-Robe unperturbed by his colleague's rush to the point.

"Indeed," predicates 001th-Robe taking 1st-Robes unequivocating response as his accession to the theory and permission to proceed at more than one step at a time. "It is one thing to petition the opinion of issues that have their roots in our culture and circumstances. But it is quite another to use the Robes for political insurrection. That is the single most likely consequence of our consideration of a death warrant for an 'Earthling who poses a threat to our civilization'."

"Dr. Mim-sey uses us to establish, officially, not only that there is another inhabited planet in the universe. He also proposes that at least one of its members represents a threat to our existence," contends 1st-Robe who cannot bring himself to take a seat to talk even though it is his own office.

"It is a shock as naked as an unannounced alien attack," continues 001th-Robe.

"But now when anyone asks, "Who will save us?" he is positioned to promote himself as the champion," says 1st-Robe pushing the theory without a sense of exaggeration. "As the putative discoverer of the threat, he is a response to the threat, in advance of the threat at the same time."

"Each and every citizen can say for centuries, "See how wise Dr. Mim-sey is. Before we even knew of another world, he plumbed the depths and singled out our enemy in the person of Roy Abbit," maintains 001th-Robe with resignation rather than hype.

"Our descendant citizenry will acknowledge that we have not been attacked in all those subsequent years," says 1st-Robe acknowledging the mundane that can replace terror with time. "It must be that Dr. Mim-sey knew exactly what was needed to disarm the blood thirsty Earthlians."

"If we cannot find a majority of Robes to issue the warrant or if we simply dismiss Dr. Mim-sey's petition as unworthy of consideration, our civilization, newly surprised by the existence of another world, will be convinced that we are unable to appreciate the urgency of the threat and the wisdom of Dr. Mim-sey," says 1st-Robe with a diamond hard focus.

"Every Robe that is appointed from now on will be appointed with the priority for the Robe's Chamber to be more dedicated to the defense of our world and the subjugation the enemy world," says 001th-Robe with another taste of the mundane that sullies pivotal significance.

"And it is no fault of Dr. Mim-sey's that we might not do our duty before we were utterly destroyed by who knows what kind of enemy Earth is," says 1st-Robe returning to the villain.

"The absence of proof is proof beyond contradiction of how dangerously subtle an enemy Earth is and how urgent action on our world's part is needed," deduces 001th-Robe aware of how the absence of proof also declares the absence of jurisdiction for the Robe's Chamber.

"Simple silence roars of Dr. Mim-sey's righteousness," concurs 1st-Robe finally taking a seat with heavy resignation.

1

1 1

1 2 1

1 3 3 1

1 4 6 4 1

1 5 10 10 5 1

1 6 15 20 15 6 1

Chapter 12

Chapter 0011000....

Chapter CD

Chapter $(1 + 4x + 6x^2 + 2x^3)$

Chapter $(a^3b^0 + 4a^2b^1 + 6a^1b^2 + 2a^0b^3)$

Chapter $2^2 + 2^3$

Episode 34

Dining for Dr. W-abe, at Dr. Mim-sey's house, is a rare pleasure for him. Both the informality and the style of food take him back to his earliest years. It is more familiar than reminiscent. Dr. W-abe does not decide what is going to be cooked or how but he can count on its being good.

Though he does not do the cooking, he is not barred from the kitchen or the process. He can do a few things along the way to participate and even contribute. He feels included and unselfconscious.

Regional tastes do not intrude. By some irony of history, cooking styles transcend regional differences. Those differences are shaped by economics, politics and history. But everybody has to eat.

Hundreds of years ago and for scores of hundreds of years before, cooking and housework was a class activity. Those that did the cooking and labor intensive work raised their children to do the same with pride. The classes that depend on them simply do not know how to do the work and were stigmatized accordingly.

In time, class distinctions tend to shift and disappear. But most of the food and cooking styles persist. Everyone likes the food they grow up with. Regional cooking styles evolve but no one thinks of their food as anything other than typically of family or familiar.

Economics and technology, in the food arena, have to complement its antecedent. An improved stove or oven is no improvement if you cannot cook the food you have had all your life whether you are young or old. Faster, cheaper and easier is fine as long as you can have your food the way you like it - know it.

Dr. Mim-sey has a dinning table and a full complement of chairs. But it is usually too much trouble to clear the clutter from the whole table for just two people. So the doctors, usually put a small table in front of the chairs they are using in a room handy to the kitchen or dinning room. It is there that they eat and talk.

"So what if any row of the Pascal or Chu triangle can be a polynomial? So what if those polynomials can fall between the rows?" W-abe asks. "What does that tell us about one's life or time in general?"

"If you did not ask the question, I would have," declares Mim-sey passing the sweetener.

"Because of what you are going to tell me next?" W-abe asks with a taste of his entree.

"Or what your cash register roustabout can tell you about quantum time," says Mim-sey.

"Do we have enough years left to get there from here?" W-abe asks picking up his glass.

"We have already been there. Suppose our specimen dies before the last day of his sixty-third year," Mim-sey says. "Let us suppose he dies at fifty-seven. Before our fifty-seven year old dies, he has a past 1 - 56 (A to DEF), present 57 (ADEF) and a future 58 59, 60, 61, 62 and 63 (BDEF, ABDEF, CDEF, ACDEF, BCDEF and ABCDEF)."

"Soon he will just have a past," says W-abe focusing on his next bite.

"Do you remember that his sixty-three years in binary numbers is six ones?" Mim-sey asks.

"Yes. It is just one short of a power of two," discounts W-abe.

"If you label the binary digits A through F, you have an alphabetical combination for every year of his life," designates Mim-sey taking a bite and passing the conversation to his friend.

"His first year was A. The second year was B. The fourth year was C. The eighth year was D...." describes W-abe with a sip of warm tea.

"Now the years in between are represented by some combination of those letters. His third year was AB. His fifth year was AC," Mim-sey delineates. "The sixth year was BC."

"His seventh year was ABC. Do you think I have it?" W-abe asks with a smile and an effort to lighten the conversation.

"Not like you are going to get it. What is his sixty-third year?" Mim-sey asks with a sip of his own drink and ignoring his friend's ploy.

"The six binary ones are the six letters ABCDEF," renders W-abe.

"All of the subsets between 32 and 63 include the letter F even if he does not live out 63 years. The future he never sees after age 57 is included in subsets between age 32 and 63. Is 'E' a subset of his past?" Asks Mim-sey cutting a portion for his next savory bite.

"Yes, from age 16, along with A, B, C and D. The coefficients A through D are factors of any number larger than 32," chimes in W-abe figuring he has time for a satisfying bite of his own.

"You will agree that he cannot get to his 57th birthday without the first thirty-one. But at the same time the subsets of his past are waiting for him in his future. He is ADEF years old at 57. Factors of his first, eighth, sixteenth, and thirty-second years contributed to his life to that point." Mim-sey says forgetting his plate for a moment. "If he lives to 58 (BDEF), factors from his second year, in this case factor B come in to play."

"Just like any quantum, the various elements of his life include aspects of his past, present or future," narrates W-abe.

"Not just his life but time itself. Any moment in time consists of its past, present and future," states Mim-sey getting back to his plate and giving his friend a moment to anticipate his point.

"If all that is true," posits W-abe just before Mim-sey takes a bite, then time, like matter and energy, can be interpreted as either a wave or a particle."

"The one-dimensional time line is the first hint of time as a partiicle," Mim-sey says. "Each increment of the time line is essentially a point of time and so time is a particle or a series of particles."

"And the two-dimensional bell curve or the polynomial interpretation of that line is a primitive version of time as a wave?" W-abe asks forgetting his plate for the moment.

"At least. And our heuristic fifty-seven-year-old is an analogous wave of time until his death, at which time, his past, present and future transform into a particle of the past," Mim-sey discourses poised for his next bite. "His end, like a photon or an electron, etc. is a scintillation.

"The analogy suggests that his ongoing life is momentum and his death fixes that momentum, like a particle, of fifty-seven years," agitates W-abe focusing on his next bite.

"The scientific definition of time can be distinguished from the philosophical version as long as you limit the definition of the dearly departed's universe," says Mim-sey looking at W-abe.

"This is done by limiting him to a universe with no existence of his own before his birth or after his death," says W-abe. With this thought he seems to end the conversation and turn his attention to enjoying his meal.

"A universe he cannot leave and return to except by the convention of death and birth in that universe," says Mim-sey who in spite of his apparent indulgence forgets his meal to digest a conclusion that not only seems to end the conversation but threatens to end his quest as well.

Episode 35

Thirty years ago the largest community concern in Greene County, Ohio was the County Fair. Then, as now, there are individual concerns, family and school events but the single event that involves most people is the annual county fair.

Ali-ce is just returned from a trip home and a visit with her mother and father. Roy is just graduated from high school. Ali-ce is not overwhelmed by the scale of the event but she is completely charmed by its content. The prize livestock are beautiful and she sees the devotion and work each person commits to their roles.

This is not just a show this year or once a year. It is a view of life itself and an all too brief one at that. Pies and beans are important in a personal way and the fair, by making them important, in a public way acknowledges just a few of the things that make everyday life so valuable.

Here, both Roy and Ali-ce adore the sight of the mother pig soundly sleeping with all her piglets on top of her just as completely asleep. All the while, her regular and rapid breathing motors them up and down like little pink pistons. They are totally enveloped in mutual peace and warmth in contrast with the chilly autumn air. Roy and Ali-ce enjoy the sight without drawing any parallels to their own emotional peace and warmth.

"Ali-ce," hums Roy at the sight of the pigs asleep and oblivious.

"Yes," returns Ali-ce marveling at their uncompromised comfort.

"Today is my birthday," reveals Roy with pride.

"This is an important day. Is it okay to ask how old you are?" Ali-ce asks.

"Of course, I am eighteen," Roy illuminates with certainty. "Today is August thirteenth."

"I thought you were not supposed to ask how old people are," presses Ali-ce dubiously.

"I think that is for women and older people," Roy asserts with the full confidence of all his years. "Women are always trying to look as young as possible."

"I suppose older people shun the stigma of old age," says Ali-ce with a similar confidence.

"How old was I on your last visit?" Roy asks without a clue.

"You were in your fifties," stirs Ali-ce with a smile that comes from pleasant memories.

"How was I?" Roy asks a little uncomprehending.

"You were fine. Are you trying to pry into your future?" Ali-ce asks with a firm voice.

"Sort of," responds Roy mindful of the warning that her tone signals.

"I have told you not to do that," urges Ali-ce in a softer tone.

"But I am curious," Roy defends. "I want things to be better than they have been so far."

"I do not think anyone can promise that. Maybe you will have more of an answer with some more years to look back on," advises Ali-ce trying not to sound like her parents and possibly Roy's.

"You told me, or maybe it was a dream, that you cannot change history. So why not tell me a little about my future?" Roy asks looking for a way around her resolve.

"A lot has to change just to change a minor part of someone's history. But that is not why your future is safe with me," grants Ali-ce without any sense of risk to her commitment.

"Then what is it?" Roy asks as they walk through the horse stables.

"You would discount the bad experiences, which have their value," Ali-ce says, "and you would try to promote the good ones; perhaps beyond their merit."

"People do that anyway," says Roy wishing more of the horses were sticking their noses out of their stables for Ali-ce's benefit.

"That is a simplification," says Ali-ce visibly impressed with the few horses she does see. "There is the risk that you might be overwhelmed by setbacks that do not otherwise have enough future significance. And the good parts will not be as good if you knew too much about them before they happen."

"It is like knowing what a gift is before unwrapping it," says Roy.

"Yes, even if you guess what it is," says Ali-ce slowing and hoping to see more horses. "There is the confirmation, and the times that follow it, that are better for each of the surprises."

"Well, I am surprised and very happy to have you here," says Roy.

"As long as my father does not find a way to object, there is no compromise in my happiness either," says Ali-ce who is rewarded for her stroll with the sight a horse that seems to want to see her as well.

"How are he and things at home?" Roy asks feeling like he never asks often enough.

"He is fine but I do not get home as often as he or I would like. It is easier to 'phone home'," poses Ali-ce weighing two different truths that do not always qualify as true.

"Oh. Where do you stay?" Roy asks hoping for an answer he will comprehend.

"When I am not here doing research for my father," Ali-ce offers without any attempt to describe home itself, "I can also stay on my ship and park it the other side of the sun."

"You are not going to be discovered by any accident there," says Roy with chagrin that replaces the vanity of wishing to visit her on her world.

"And I can take advantage of the sun as a power source when I need to," continues Ali-ce.

"So, when I see you, is that good or bad?" Roy asks as they leave the stable.

"How can it be anything but good?" Ali-ce asks as they cross a patch of darkness outside the horse barn.

"I mean, is your dad unhappy because he does not have what he is looking for?" Asks Roy.

"He does not blame me," says Ali-ce as they enter a barn that is surprising both for it interior space and the animals within. "I give him what he wants. But with a little study and work he manages to eliminate the potential that he thought he had and needs to try another approach."

"What did you send him the before?" Roy asks with some tension.

"William Shakespeare's *Julius Caesar*," tenders Ali-ce with surprise at the spectacle arrayed before them.

"So it was a waste of time?" Roy asks with no help from Ali-ce on what to say or feel.

"In one sense, yes. He thinks a well-placed murder will prevent a clash of your world and ours. As long as I am here, he is still looking for his solution," says Ali-ce with an emerging smile.

"He cannot be happy about that," says Roy aching for the chance to explain the obvious that this is the Bull Barn.

"He has spent his life as a researcher," rambles Ali-ce knowing that she is not looking at cows and that they were in their stalls for the night.

"It is too soon to throw away his knife collection?" Roy asks watching her smile shrink.

"I suppose that was supposed to be funny?" Ali-ce asks avoiding a glare with some effort.

"Sorry," says Roy. "I just do not see our world as a threat."

"Now that you mention it," says Ali-ce discounting the reflex fear of extraterrestrials, "you seem to be the only one, on this world, who knows that there is another world."

"But theory and belief populate the universe with many worlds. If I were to say that I know of another world the believers and doubters would disagree about whether I was crazy or keeping the faith," states Roy feeling that he ought to explain why the bulls are stabled this way.

"No one knows how many on our planet know about this one," says Ali-ce smiling again.

"That must be why he was interested in Caesar. If anyone else knows anything here, they are probably few and all high officials," says Roy mindful of a little known fact about most bulls.

"No doubt that is true here. In my case my father did not build my pav... ship with private money," recites Ali-ce without a clue to what Roy is thinking. "And the money for the research has to come from somewhere."

"And his solution, when he comes to it, has to be okay by his, your, law somehow," says Roy recalling that bulls are unwilling to walk backwards. Putting them in nose first helps to keep them in their stalls.

"I cannot imagine my father being executed for his efforts especially if he is successful," says Ali-ce wondering why openly independent and circumspect farmers uniformly stable the bulls so uniformly and immodestly.

"You are the one who is at risk," warns Roy grateful that they are nearing the exit to this barn. "No one could protect you from a fanatic once you are identified as an alien."

"I have certain advantages and I will be careful," rejoins Ali-ce thinking of looking back.

"That will give us more time together, on and off, for a little over thirty more years," concludes Roy hoping that she is not thinking of looking over her shoulder.

"You see," Ali-ce chides. "I should not have told you I knew you in your fifties."

"But I am glad to know that I do not have a bull's eye on my back," answers Roy.

"What is that?" Ali-ce asks wondering at his choice of words as they approach a large oak in an even larger pool of darkness

"It is what marksmen use to practice shooting accuracy," says Roy inhaling the night air.

"A *'baxin'* or a *'piwem'*. Even though you were born on the Ides of August?" Ali-ce asks with the intention of changing the subject.

"But that is the fifteenth. I was born on the thirteenth," ripostes Roy with some confusion while welcoming the comfort of a bail of hay.

"In Caesar's time the 'ides' is the fifteenth of March, May, July and October. It is the thirteenth in all the other months," says Ali-ce delighted by the dry, grassy smell of the hay.

"Today is the thirteenth and the ides of August?" Roy asks without suspicion.

"Happy birthday. Are you ready for your present?" Ali-ce asks without guile.

Episode 36

Now located 17 floors up from centuries before, the quads are deserted most of the time. Staff, on sixteen floors beneath, consider it an accomplishment just to get to the office on time each day and leave for home or dinner venues at the first opportunity and by the most direct route. Time off for lunch, however that is managed, is not ample enough for a trip to a quad and an appreciation of the dramatic space.

The same would seem to be true for the Robes themselves. But a twelve hour workday and the urgent need to think without distraction can, from time to time, drive a Robe to one of the quads. That is 101th-Robe's intention this evening and he almost succeeds. But the arrival of 010nd-Robe offers the opportunity to talk about an issue of mutual concern. The option of a dialogue depends on the the senior Robe to speaking first.

"Am I disturbing you?" asks 010nd-Robe as 101th-Robe stands and the environmental lights encroach on the fading day.

"Fortunately, no," divulges 101th-Robe taking a seat in order to assure his colleague of his openness to a mutual dialogue.

"Why do you say 'fortunately'?" 010nd-Robe asks noting the invitation to engage in a substantive conversation.

"Because I am able to respond with the truth. 'You are not disturbing me.'," rejoins 101th-Robe pointing to a seat inviting 010nd-Robe to some comfort and signaling the consent for unlimited dialogue.

"I would hope that you would speak truthfully in all ways," says 010nd-Robe settling into his chair and the conversation.

"I have something that I need to ask you, or at least some Robe, but to address your concern, I would lie and say 'Yes' or 'No' if I thought truth did not apply. By way of explanation, suppose I were wishing today's lunch were later rather than earlier when you ask your question. Certainly, the thought would be interrupted. But that thought could not be defended as disturbing something of substance," says 101th-Robe sometimes looking at his chair arm to relieve the sense of lecturing.

"I want to get to your greater concern but are you saying that an answer to a direct question deserves a response with or without a 'degree' of truth?" 010nd-Robe asks.

"For a 'Yes/No' question where the response is insufficient to the truth, yes," declares 101th-Robe rising to find a new perspective for his view and his concerns.

"Of course. It is too easy to think past the obvious and not do it justice. You make me want to give particular attention to your real and current concern," says 010nd-Robe rising too.

"How long has that other world been known to exist?" 101th-Robe asks getting to the point without anymore delay.

"I was still in school and you were not born yet when it was first detected but it was not located for many more years. Almost no one can estimate the chronology precisely because the earliest years were ultrasecret," says 010nd-Robe approaching a landscaped pond with most of its plants in the water.

"Almost everyone has heard legends of other worlds but always consider them a fiction. We have come to depend on that," shares 101th-Robe staring at the water that is made especially black by the dimness of the evening lights.

"One can imagine that some of the fiction is distilled from the facts that were supposed to be contained," says 010nd-Robe taking a seat with his back to the pond and facing the chairs he just left.

"Dissemination of the facts of what came to be called the planet 'Earth' among official circles eventually must have included the Robe's Chamber," says 101th-Robe gazing at the nearby pool of light that isolates the now empty chairs.

"Yes and unofficial legends were and are deliberately cultivated to soften the blow to our culture," says 010nd-Robe ignoring the blackness of their vantage point.

"So I suppose any official public disclosure of a second world, even now, would be premature and a significant shock," 101th-Robe says while feeling an increasing consciousness of the night's enveloping blackness.

"Such an attempt has to be considered irresponsible or worse," stipulates 010nd-Robe.

"I suppose the degree of 'irresponsibility', as you so gingerly put it, depends on two things: the nature of the threat that Earth presents and, of course, how you disclose the threat," says 101th-Robe feeling obliged by the blackness to speak even more quietly or move back to the chairs and their island of light.

"In this case, the irresponsibility is that Earth may never become a threat given the depth of technology it would take them to find us and get to us," says 010nd-Robe while turning away from the light.

"Both civilizations would have to survive long enough to get together in some mutual fashion," says 101th-Robe reluctantly turning to face the disorienting darkness.

"The knowledge that we exist is not the same as knowing where we are and then there is the getting here. It is a one-sided situation that could last millions of years," says 010nd-Robe.

"Suppose a biological weapon strikes our world killing millions of us," says 101th-Robe.

"If it were extraterrestrial, the capsule that delivers the agent has to survive to the surface, our biosphere. It could be located and its construction would determine whether it is native or not," says 010nd-Robe taking a few steps and in that way deliberately drawing the other Robe along too and deeper into the night's darkness.

"So the only real danger another world, or at least this other world, poses to our civilization is panic or a skewed social commitment to a misrepresented threat," says 101th-Robe.

"And we are being made the instigating agent of this panic by Dr. Mim-sey," concludes 010nd-Robe.

"Denying that Earth poses a threat to us confirms Earth's existence officially and publically, for the first time," says 101th-Robe following his colleague still deeper into darkness.

"For our denial to work against Dr. Mim-sey's exploitation of public fears, it would require a majority of the population to believe that Earth will never become a threat," submits 010nd-Robe.

"But given enough time of innocent coexistence, this inevitable population of our 'believers' will feel compelled to take over Earth to insure our own welfare," proposes 101th-Robe.

"No doubt, whether by haste or deliberation, subjugation will cost millions of lives on both sides," estimates 010nd-Robe.

"Billions," 101th-Robe testifies into a moonless night and uncounted white, red and blue stars flecking a jet black sky.

$$1$$
$$1\ 1$$
$$1\ 2\ 1$$
$$1\ 3\ 3\ 1$$
$$1\ 4\ 6\ 4\ 1$$
$$1\ 5\ 10\ 10\ 5\ 1$$
$$1\ 6\ 15\ 20\ 15\ 6\ 1$$

Chapter 13

Chapter 1011000....

Chapter ACD

Chapter $(1 + 4x + 6x^2 + 3x^3)$

Chapter $(a^3b^0 + 4a^2b^1 + 6a^1b^2 + 3a^0b^3)$

Chapter $2^0 + 2^2 + 2^3$

Episode 37

Who can explain the reciprocating symmetry that makes dining at Dr. W-abe's house an indulgent pleasure for Dr. Mim-sey. The food is gourmet and exotic. Each bite is better than the last. Mim-sey does not know from one serving to the next whether it is aquatic or terrestrial. He does not know whether it is the last week's or the last century's best.

His long association with Dr. W-abe assures Dr. Mim-sey, at least, that each meal is the best. Dr. W-abe does not notice Dr. Mim-sey's awe. W-abe buys, keeps, and eats what keeps the longest in his pantry, freezer and refrigerator.

He buys keeps, and eats what comes in single serving sizes or not much larger. Whatever can be eaten without preparation is better than something that has to be griddled, grilled, broiled or boiled. If an entree needs that, it rarely gets on the home menu.

Fresh is possible. But it has to be easy to get and easy (not too expensive) to throw away as a leftover. W-abe, as his fortunes improve, has less time and interest in cultivating the basic kitchen skills of long ago. It suits him all the more that he does not have to educate, cajole or impress Dr. Mim-sey into eating what they are eating. Dr. W-abe can enjoy his dinner without distraction which is the way he likes it. It is as natural as conversation.

"Birth is a fundamental requirement of existence in this universe, but is death a requirement for leaving and returning?" W-abe asks vaguely mindful of the previous conversation but clearly recalling the pleasure of the circumstances of that exchange with his friend, Mim-sey.

"Leaving and returning to this universe after birth and before death would have to be done within the time-line, or curve, of the individual," says Mim-sey as an investment in his dinner.

"Why cannot birth in this universe be the departure from some other universe and the entrance into this universe?" W-abe asks while rummaging around and gathering the components of his meal.

"I do not see why it cannot and death in this universe can just as validly be the departure from this universe and the entrance into some other universe," conveys Mim-sey.

"You may be able to leave your universe at any point in your lifetime but you cannot return to a significantly earlier point in it. Individuals would find themselves as adults returning to childhood circumstances," advises W-abe while pouring cold, pulpy, sky blue drinks into tall, opal tumblers.

"So leaving and returning to a universe between the conventions of birth and death must be done within a relatively narrow window of time for the individual's universe," says Mim-sey.

"The universe this individual visits in this window of time may not be hospitable," says W-abe plating the contents of some chilled containers and adding serving utensils to the plates after a bite.

"In which case," Mim-sey considers, "the visit is meaningless."

"Upon visiting a hospitable universe, the duration of the visit will depend on the time line of that universe," W-abe holds. "If it is similar to his, the visit will be too brief to be significant."

"But supposing a time line in some hospitable universe allows a significant visit. This universe would have to be comprehensible to the visitor," relates Mim-sey with a sip.

"The visitor would have to be comprehensible to his host as well," says W-abe assembling two crystal chargers with pates, fruits, and fungi.

"The host universe or elements of it might have to prepare a venue for the visitor," states Mim-sey contributing by gathering the spices, sauces and oils that finish each bite.

"That means that the visitor would have to communicate the requirements of a significant venue in advance of his visit," says W-abe while surveying the dishes on the counter for completeness.

"And the host would have to understand the requirements for creating the venue," chimes in Mim-sey who is surveying for his opportunity to begin the meal.

"This venue would have to be created for each visit or maintained for subsequent visits," delivers W-abe filling his foamed glass ice bucket with ice before he finally picks up his fork and signaling the opportunity to begin the meal.

"Then there are the issues of entropy and conservation," says Mim-sey picking up his fork.

Episode 38

Ali-ce does not eat breakfast. Lunch, dinner and fast food are fine but she has no interest in the breakfast menu. She likes coffee and wheat bread with salt on it. Roy has never heard of that but her coffee is good and she likes it with sugar.

Roy does not plan on giving up his butter but he is thinking of getting her a toaster. He likes to take a shower and brush his teeth as soon as he wakes up but he does not want to wake her. So he waits until he is sure she is waking before getting out of bed. She keeps extra toothbrushes so he feels free to use them. He usually gets back into bed while she takes her turn in the bathroom.

The blow dryer is her last obligation to herself. She starts the coffee maker and joins him in bed. He does not make the coffee because he wants her to make it the way she likes it. She is pleasant but does not start the conversation, this time, so Roy, eventually, takes the initiative.

"So, how did you find us?" Roy asks beginning his remark from where his thought ended.

"You are excellent as usual. I am pleased to say," replies Ali-ce.

"Oops, I mean how did you find, uh, locate our planet?" Roy asks with an effort to be clearer.

"You were found before I was born," responds Ali-ce with a glance at the coffee maker.

"You mean Earth?" Roy asks following her glance and checking the coffee maker too.

"Yes, but I do not know the technical details," vouches Ali-ce getting up to pour some coffee. "As you know our ships can emerge at any point in the universe and at any point in time."

"I am still getting used to that but yes," conveys Roy.

"This made navigation and communication very important. The navigation part was the simpler problem," imparts Ali-ce parking two cups of coffee on the table and taking a seat.

"The known landmark galaxies and clusters are numerous enough that we could tell our relative location at any time."

"So, what was the solution to being able to communicate?" Roy asks buttering some bread for himself.

"At first, we settled for debriefing whoever came back and downloading data packets that we call eggs," says Ali-ce sprinkling salt on her bread without concern for measurement. "In the meantime, we looked for ways to phone home."

"If you could count on the debriefings why was telecommunication necessary?" Roy asks.

"Mostly, it was cheaper," declares Ali-ce tasting her bread and adding some more salt. "We could get information and act on it without the cost of the trip back each time."

"But the time and distance must have been impossibly large sometimes," suggests Roy.

"I suppose so. But the solution to that was to use Q-particles. As long as one was between home and the ship, it could be used as a lens or amplifier or something like that," states Ali-ce.

"What is a Q-particle?" Roy asks chasing his bread and butter with a sip of coffee.

"It is big," Ali-ce explains "It takes light about twenty years to cross one particle."

"That is big for a particle," observes Roy measuring some sugar with the tip of his spoon.

"But it has the advantage of transmitting out whatever is transmitted in, presumably at speeds greater than normal light speeds," conveys Ali-ce taking her turn with the sugar.

"So you heard our radios or televisions along with some of your communications?" Roy asks finally satisfied with the taste of his coffee.

"Not exactly. History has it that there was a burst of gamma rays in the path between the ship and the Q-particle," imparts Ali-ce getting up and returning to the table with the coffee pot. "So the burst of gamma rays was included as part of the information that was received."

"So that was easy enough to fix," infers Roy while watching Ali-ce refill his cup.

"There was no fix," says Ali-ce setting the pot on some napkins. "It has not happened before or since."

"That must have made the solution impossible," says Roy reaching for the sugar again.

"And important," Ali-ce recounts. "Was it deliberate interference or a cosmic accident? If it was deliberate, it was something that could happen again and that could be dangerous."

"So what was the source?" Roy asks stirring his coffee and considering another piece of bread and butter.

"It turns out that the source of the gamma rays was something on this beautiful blue planet," relates Ali-ce turning sideways in her chair and picking up her half-finished piece of bread.

"And that is how you found us?" Roy asks pushing the sugar bowl closer to Ali-ce.

"Yes but it was a huge controversy. You were very hard to find since your orbit took you out of the equations we used to locate the source of the gamma rays. But your sun was not the source of the gamma rays," says Ali-ce with a sip of coffee and a quick look at the sugar beckoning bowl.

"So you kept looking?" Roy asks buttering a slice of bread.

"Well, whoever was responsible for the solution at the time," says Ali-ce adding sugar to her coffee. "One of the theories for the gamma ray source was a weapon or a weapon test."

"So they began looking for a planet near our sun?" Roy asks sipping his coffee.

"Yes, but the planet had to support life and technology and that life and technology might discover us while we were discovering them," says Ali-ce finally happy with her coffee.

Episode 39

110rd-Robe likes to use the redundant zero in his title. It intimidates everybody. It does not matter the potentate, they get it wrong every time and he reminds them at the most inopportune time, for them. The potentate has to blame it on his Secretary of This-or-That and usually realizes that he does not know the Secretary's name or at least better not use their true name in a lie. So he has to make up a name and he does not have time to think of a believable phoney name. The pause or stutter that defines the gap takes the potentate down a notch that they never seem to get back.

110rd-Robe likes to get to work before sunrise. He does not use it as an excuse to leave early. He goes to work early to enjoy the morning, a hot drink and sunrise at a decent temperature most of the time. There is a generous awning for the rainy days and he dresses for cold days.

It is rare weather that robs him of his cherished solitude and his ownership of the morning. Of course, the few and far between bad weather days guarantee his promised isolation and with that he is never disappointed. But, all the same, on those days he can be sure someone will be waiting inside for him.

Then, as a matter of course, he mentions to whoever is waiting that they could have joined him. He wants to see if they will lie and say they wanted to respect his solitude. So he makes sure to lock eyes with them while they are fabricating and iterating the lie and trying not to look like a liar. But today is a fair weather day and there are plenty of people who know his habit so he is planning on being imposed upon sometime after sunrise.

Surprisingly early, before the sun can extinguish the environmental lights, he sees a colleague crossing the quad and walking directly his way. Just as surprisingly it is 0011th-Robe, the most junior Robe of the twelve. Knowing the courage that took, 110rd-Robe's presuppositions of superiority abandon him. Fortunately, he is in his space and can honestly greet 0011th-Robe as a peer.

110rd-Robe greets his colleague beyond speaking distance to let her know she is welcome, not imposing, and can speak freely and informally.

"You are not this early every day, are you?" 110rd-Robe asks gesturing to an empty chair.

"To tell the truth I have trouble going to sleep at the end of the day and I do not want to wake-up in the morning," volunteers 0011th-Robe.

"So I can suppose without much chance of being wrong that your being here is not only important to you but must involve me as well," says 110rd-Robe admiring 0011th-Robe's lack of guile.

"You make it easy to get to the point. I want to ask you about the death warrant that is being sought by Dr. Mim-sey," admits 0011th-Robe while an intern brings a tray with two hot drinks.

"Since it is an original warrant, it requires a unanimous vote. It is not on appeal to reverse the finding of a lesser authority. Dr. Mim-sey has brought the request to us without any consideration of any other court. We have to choose to consider it or not. If we choose to consider it, it will take a unanimous vote to condemn the person, or persons, named in the petition," relates 110rd-Robe adjusting his cup to a comfortable distance.

"Before I get to my question, what keeps us from being flooded with millions of such petitions every year?" 0011th-Robe asks without touching her cup or any intention to drink.

"In fact we are but a petition of less than 100,000 pages is rarely given a first look. That is justice by the pound, I guess you would say," says 110rd-Robe supposing 0011th-Robe has not had breakfast.

"Well, we have the staff to review petitions that size. Any and every excuse we can find to return it to the originator is seized upon. That much everyone knows, even many outside the legal system," dismisses 0011th-Robe.

"So, on the face of it, a death warrant is infrequent," says 110rd-Robe taking a tentative sip.

"My staff tells me," says 0011th-Robe, "that a death warrant from a lower jurisdiction gets consideration every ten years or so."

"Has your staff told you how many 'original' death warrants have been considered?" 110rd-Robe asks pleased to find his drink is approaching an appealing temperature.

"No, not yet," says 0011th-Robe rolling and unrolling the corner of her napkin.

"When they get in today, tell them they do not have to bother. This is the first one, ever," reports 110rd-Robe hoping he can get to his drink before it is unpalatably cool.

"The first one in more than twenty-five hundred years?" 0011th-Robe asks letting go of her fretted napkin.

"In all those years the warrants originated from local circumstances there was no need to originate a death warrant in the Robes Chamber. But Roy Abbit has no local history. There is nowhere to spawn his warrant except here," says 110rd-Robe taking a gratifying sip of his still warm drink.

"So, on the one hand," acknowledges 0011th-Robe without getting back to toying with her napkin, "this is not a measure of the importance of Roy Abbit,"

"Indeed, but that is something Dr. Mim-sey is not likely to shout from the rooftops," charges 110rd-Robe looking at his drink when he is not looking at 0011th-Robe.

"And, on the other hand, it is a measure of the underhanded zeal of Dr. Mim-sey as well," says 0011th-Robe hoping this is not the first time 110rd-Robe has considered or heard this opinion.

"Underhanded?" 110rd-Robe asks before drinking nearly half of his morning drink.

"If Roy Abbit is not the threat Dr. Mim-sey says he is," affirms 0011th-Robe, "then yes."

"And you want to test your feelings for the substance, or lack of substance, of Roy Abbit's merit for an original death warrant from the Robe's Chamber?" 110rd-Robe asks.

"My perspective is not so confrontational, if I may presume. It is simply my inability to attribute a threat to our civilization to any single person, either here or on Earth, least of all this Roy Abbit," divulges 0011th-Robe watching 110rd-Robe push his unfinished cup away so that he can respond without distraction.

$$1$$

$$1\ 1$$

$$1\ 2\ 1$$

$$1\ 3\ 3\ 1$$

$$1\ 4\ 6\ 4\ 1$$

$$1\ 5\ 10\ 10\ 5\ 1$$

$$1\ 6\ 15\ 20\ 15\ 6\ 1$$

Chapter 14

Chapter 0111000....

Chapter BCD

Chapter $(1 + 4x + 6x^2 + 4x^3)$

Chapter $(a^3b^0 + 4a^4b^1 + 6a^1b^2 + 4a^0b^3)$

Chapter $2^1 + 2^2 + 2^3$

Episode 40

When he is not working, Dr. W-abe prefers to be intoxicated. It is an expensive habit, financially and socially, but Dr. W-abe thinks he can afford it. He knows it does not help his personality but his few friends do not think it hurts him. He is a little more approachable but drunk or sober, Dr.W-abe is always reluctant to condemn, criticize or commiserate.

W-abe's veil of intoxication is easily pierced by Dr. Mim-sey when it serves his purpose. Dr. Mim-sey drinks sparingly except when he is with Dr. W-abe. Then he drinks freely. He pays for his drinks as much as Dr. W-abe lets him but he can always count on Dr. W-abe picking up the tab after the first two or three.

Mim-sey does not keep any intoxicants at home. So Dr. W-abe always brings something with him on a visit and leaves the bottle no matter how much is left. Mim-sey is never disappointed by that generous courtesy.

When Dr. Mim-sey becomes drunk, he likes to tell jokes. But he has been a researcher for so long that he 'explains' jokes rather than tell them. Even if he tells it, word for word, like a joke, it sounds like a lecture. Dr. W-abe always smiles when he is confident that Dr. Mim-sey is telling a joke.

That way his light chuckle is a little more believable when it comes time to laugh. Dr. W-abe is not insincere. He likes Dr. Mim-sey's company when they are drinking and as long as he can carry on a conversation.

"The universes beyond our own go unnoticed simply because we cannot comprehend them," says Mim-sey poking at an ice cube. "We do not know them when we see them."

"Suppose we find a universe that is similar enough to ours that we can establish its existence apart from ours. Can we visit them or can they visit us?" W-abe asks before taking a sip.

"If it is a closed system it would be very difficult. The visit would mean adding or removing energy from the system. Energy is manifested by the coming and going of the visitors and their accouterments. Those parameters change the entire system. In this case the entire universe," says Mim-sey.

"That reduces visitations to an unbounded similar universe whose total amount of energy is constantly changing anyway. And if our universe is a closed system we would still have to leave the total amount of energy unchanged while we exit our universe and return or the total amount of energy must be unchanged while they visit ours and exit," says W-abe refilling his antique style foamed-glass ice bucket.

"We would have to supply the energy that is lost with our exit and eliminate the excess that results from our return," says Mim-sey helping himself to some ice and then replacing the deeply flanged lid that is more like a bowel with a handle in the middle.

"The possibility of anyone visiting another universe, no matter how similar, in his lifetime is almost out of the question. How do we prove whether the universes are open or closed?" W-abe asks putting ice in his drink. "How do we maintain the total energy for the ones that are closed?"

"The visitor has one possibility," says Mim-sey.

"Not visiting and no trespassing allowed?" W-abe asks refilling both glasses.

"Every universe no matter how dissimilar it is from all the other universes is identical to itself," says Mim-sey studying the ice bucket.

"The visitor can visit himself?" W-abe asks reaching for a piece of salty, white fungus.

"The visitor can visit any part of his own universe," says Mim-sey amused at W-abe's sarcasm.

"No matter where he goes in his universe the total amount of energy is always the same. It does not matter whether the system is open or closed," says W-abe chasing the fungus with an ice cold sip from his glass.

"If the system is closed, it is elastic to itself. If any of the parameters change the entire system changes," says Mim-sey recalling that the deep rimed lid was originally used to hold ice. This kept food cool and edible before technology extended the range of refrigeration to cabinets, drawers and vessels.

"Entropy will increase. Chaos or the number of choices will increase," says W-abe studying the aggregation of bubbles just visible under the crystalline glaze which he finds more charming than the obsolete style of the foamed glass vessel.

"Someone can leave his time line and spend some time in another time line as long as he does not leave his universe," says Mim-sey with a helping of some sweet red fungus.

"Or we can borrow someone from his time line as long as they are in our universe," says W-abe mindful that foamed glass is the one thing that cannot be made on any planet at any cost. Not even Clidim or Earth.

"In any case," Mim-sey says, "our juggernaut and its passenger have to get back before its absence is noticed."

"He has to be gone an insignificant amount of time?" W-abe asks.

"Insignificant for his time line," says Mim-sey marveling at the huge value of the of this rare foamed-glass object that W-abe uses as a household utensil. "Suppose he was being photographed at the time and the shutter was open for 1/1,000 of a second . If he was out and back in less than that, the camera would not record his brief but total absence.

"That is okay for the in-laws, say a visit of 1/10,000 of a second. A significant visit would have to last, here or there, a while longer," says W-abe charmed that foamed glass is manufactured in a weightless environment far beyond the gravitational influence of any planet.

"If his time line is incremented identically with ours, his visit would be 1/10,000 of a second," says Mim-sey while refiling his glass amply. "But that is true only if his time line intersected ours on the perpendicular. Any angle less than or greater than 90 degrees means an intersection of more duration."

"If the angle is steep enough he could visit for three seconds, three minutes, three days...," says W-abe

"Such a visit is significant," says Mim-sey.

"Is it sufficient of any use?" W-abe asks.

"Perhaps not," says Mim-sey, "but several visits can be sufficient."

"If we do not get him back in a short enough interval, his absence will be noticed," says W-abe considering himself lucky to own a piece of foamed-glass.

Mim-sey, wishing hopelessly to visit the single factory that makes foamed-glass once in his lifetime, says, "The other time lines that comprise his experience will close the gap if he is gone too long."

"What will that be like?" W-abe asks with no illusions of ever visiting the foamed glass factory.

"I do not know. I guess it depends on what is going on at the time. He will return as the dearly departed or nearly departed. His remains may be found, perhaps, uncorrupted and presumed naturally deceased. Or the circumstances that may resemble a lightning strike may fill the gap. The local speculation will be that the event vaporized him. The truth may be that pinching off all the lines of mass, energy and time at the moment produces the lightning or whatever the event comes to be called," says Mim-sey.

"Things are difficult enough when just working within our own universe," says W-abe.

"As long as we get him out and back in time," Mim-sey says, "his life line will appear normal."

"His fortunes and his circumstances will serve him and fail him as probability dictates," says W-abe.

"We might be able to shorten his life," Mim-sey says, "but never extend it."

Episode 41

Centuries before Ali-ce is born technology on Clidim focuses on taking the sensation of motion out of transportation. You are not supposed to see, feel or hear anything that shows you are moving. Windows are tinted or mirrored. Acceleration and deceleration are compensated for by pneumatics in the seats and floors. Surfaces are reticulated to reduce the sound of wind or water flowing by.

Any suspension that is mechanical is engineered to avoid transmitting vibration or changes in speed to the passenger. Video and music are supplied as a palliative to the real purpose. If there is a sense of movement it is attributed to the object in the video or it is masked with musical effects. This custom is not so pervasive on Earth.

Roy and Ali-ce decide to take-in some of the Greene County Fair before eating lunch. Even if it means eating a meal after four o'clock. That way they can sample the carnival food between rides. Ali-ce is completely thrilled with even the simplest rides. A lumbering Ferris wheel is totally scandalous, so huge and so slow. The very idea makes her laugh. She has to make sure no one thinks she is ridiculing the ride.

It is hard to say which is her favorite. A Whirlygig is high on the list for its blatant exploitation of centrifugal force. Ali-ce is delighted that anyone would think to build a machine for such a purpose and then put people on it for money.

She has mixed feelings about the roller coaster. The weightlessness of the downhill ride is not so novel to her and the acceleration downward is too reminiscent of a failure of technology to be any fun. For her sake and Roy's, she thinks of the Ferris wheel while she is on the roller coaster. Ali-ce laughs delightfully and Roy is sure she likes this ride as well as the others.

They take some time to sit under a tree and have some corn-on-the-cob. The melted butter helps keep the salt that she loves, on the corn. She thinks it is no wonder that the horses and pigs love corn-on-the-cob. They are comfortable enough to stay for a conversation.

"That gamma emission not only located you," Ali-ce says, "it made you very important."

"Because we could be a threat?" Roy asks as he settles on the grassiest spot he can find.

"Not that so much. High energy explosives are not as dangerous as a technology that delivers it," says Ali-ce choosing the most prominent and hopefully wide enough stretch of root.

"So your world," Roy says, "had to know if we could find you."

"Of course we had to do it without being discovered in the process," reveals Ali-ce.

"I suppose you settled for monitoring our radio and video transmissions," presumes Roy.

"At first," Ali-ce explains, "and this would also tell us how close we could get and so on."

"What were your first impressions?" Roy asks wondering if that tree root is really comfortable.

"This was before you and I were born," Ali-ce says, "but evidently it was quite a shock."

"We were already a threat then?" Roy asks.

"The shock was mostly biological," says Ali-ce while handing Roy his corn.

"What do you mean?" asks Roy unwrapping his corn-on-the-cob.

"We were not expecting such a huge population," says Ali-ce looking for a good starting place on the cob.

"We have a long history but only recently came to such a large world population," explains Roy deciding that more salt would do and looking among the napkins for salt packets.

"So do we but the surprise was that of the planets known to support life," says Ali-ce licking a finger, "this was the only one we have found with an anthropoid species."

"I suppose most of the planets that could support life had some form of animal or plant life but nothing like yours," construes Roy focusing on generously applying his salt to the buttered corn.

"We do not know what the triggers are," says Ali-ce picking up a salt packet. "We know that throughout the universe the atoms are the same, so life will be similar wherever it occurs."

"That is not the same as identical?" Roy asks.

"No, but usually, there is microscopic and macroscopic life, a food chain and those life forms depend on light or chemical energy for survival," says Ali-ce taking a turn at her corn.

"We have some bacteria that use light to make their food," inserts Roy setting his corn on the wrapper and picking up some napkins to wipe his chin and hands. "But that is rare here."

"On most planets, that we know, this is a rule not the exception," says Ali-ce handing Roy more napkins.

"So what did you learn from this huge human population?" Roy asks balling up his used napkin.

"We learned that we had to keep you under permanent and continuous observation," advises Ali-ce moving closer to Roy as if to change the subject without changing the subject.

"Do you have a station up there somewhere listening?" Roy asks a little self-consciously.

"That is too easy to detect," apprises Ali-ce focusing on her corn.

"You guys are down here?" Roy asks seeming to forget his corn-on-the-cob.

"Your own laws protect us from intrusion and therefore detection," clues Ali-ce.

"You cannot be a public institution or a dry cleaner without any place in the information stream then," says Roy moving to eliminate as much of the remaining distance from Ali-ce that he could.

"We could not remain anonymous as a public or semi public institution," says Ali-ce.

"A dry cleaner with a big antenna would raise questions," says Roy propping up a knee.

"The information stream is just part of the picture," says Ali-ce draping her arm on his knee. "We cannot be caught transmitting large amounts of information or restricted information."

"The spy business - I guess you need to screen the information so you do not send trivial or misleading information also," says Roy pulling his corn on its wrapper a little closer to himself.

"That is the toughest part," flinches Ali-ce taking another bite of corn.

"I suppose you need to support or justify your risks without precipitating a war or other counterproductive events," interposes Roy picking up his corn and taking a few more bites.

"That," grants Ali-ce setting her corn on its wrapper, "and we want to be able to replace old information, which is less valuable, with new information that could be more valuable."

"I never thought of that. You have some old stuff still hanging around?" Asks Roy.

"We started sampling your video stream before you could transmit in color," allows Ali-ce dabbing her fingers with a napkin, "and to this day there are a few bureaucrats who think Earth people are grey."

"Do they know that we have color transmissions now?" Roy asks studying her diligence.

"Yes, but they are out of the information loop and do not get access to current information," acknowledges Ali-ce putting her napkin in the bag that originally carried the corn.

"So they imagine we are grey in the color images too?" Roy asks then finishing his corn.

"I suppose. The popular legend for books and movies is that Earth people are grey. I guess it comes from information leaks that go back a long way," construes Ali-ce with her last bites of corn.

"Is this why you have so much trouble deciding what color you want to be?" Roy asks.

"In a way. When I was a little girl," confesses Ali-ce putting the empty cob in a bag, "I was sure that Earth was not a fiction and that I would go there, come here, on a secret mission."

"I guess you were belittled for that," says Roy putting his empty cob in the bag too.

"My mother came into my room and caught me practicing being grey," says Ali-ce smiling.

"Did she laugh?" Roy asks trying to imagine the scene.

"No," recalls Ali-ce becoming a little lost in the memory. "But I could tell she was shocked and made an excuse to leave without telling me why she came to my room in the first place."

"But your secret was out," says Roy bringing her back to the current moment.

"She must have said something to my father," says Ali-ce smiling a little. "He went to a lot of trouble to tell me that even if there was an Earth the people probably would not be grey."

"You saw through him?" Roy asks.

"He was trying so hard not to be technical that I figured mom must have said something," says Ali-ce with a distant memory of an all too brief moment shaping the expression on her face.

"I decided then that there must be an Earth with humans, grey or not."

"So you never gave up on going to Earth," says Roy giving up hearing more specifics.

"And I always wonder when my father looks at me if my current color is the right one for Earth," says Ali-ce looking directly at Roy as if he had the same question in the back of his mind.

"It is a good thing changing color is not easy," says Roy.

"Who knows what I would have become," teases Ali-ce.

Episode 42

For scores of thousands of years the construction material of choice is glass. And on elite buildings like the Robe's Chamber there is the added feature of a polychromatic glass frieze. On three sides of the three buildings that occupy the footprint of the original quads, at the very top of those buildings is a tableau of civilization and justice. It is a procession of heros and artifacts all interwoven and to a scale that is intended to be appropriate to the view of the mortals passing by seventeen floors below.

The mingling of shapes and colors becomes a fabric that finishes the edges of one of the most important buildings belonging to this civilization. While the frieze is not supposed to look too large from ground level, to a visitor on the quad-level the scale is titanic. For architectural reasons only the top eighteen feet of the frieze rise above the deck of the quads. This wall of glass provides privacy and protection from wind along with a climate that supports three enviable parks.

110rd-Robe's quad is on the sunrise side of the building and the significance of this is just dawning on 0011th-Robe. This normally featureless perimeter transforms the inside into a looming wall of blazing colors. The riot of colors sorts into giants and artifacts whose themes thunder the obligations of the Robe's Chamber.

For the two judges immersed in the sunrise-spawned kaleidoscope, the cascade of color inspires the conversation as much as it competes with it. The distraction comes from colors laid upon the conferees. Once the sun is above the horizon everyone and everything is flooded with color.

Every turn or lean anyone makes is amplified by a wild change in the color of their clothes and skin. Blues, golds, reds and greens overwhelm what used to be mundane features of faces, hands and clothes. If you want to keep your equilibrium this a good time not to remember you are seventeen floors above the safety of the ground.

Unfortunately, this is the moment and place 0011th-Robe must argue the insignificance of Roy Abbit. On that proof hangs Roy Abbit's life and 0011th-Robe's sense of justice.

"I would like to begin with a question, if I may," says 0011th-Robe.

"Of course. We are not likely to be more informal than this," remarks 110rd-Robe with a hand gesture that went unexplained, "at least until the ballot for this warrant is taken."

"Have you read any of Abbit's manuscript?" 0011th-Robe asks.

"It is a fair question, if not an intrepid one. A copy of the manuscript is part of Dr. Mim-sey's petition. I, frankly, tried. It is just 240 pages but I am not ready to focus on that yet. I have settled for a promise to myself to do a better job of reading it before the vote on the warrant."

"This is only an observation. I am not being defensive in this instance. The promise I have made to myself, is one that I made before this conversation. This is mostly to bookmark a point of thoroughness for me," explains 110rd-Robe putting off the substance of the issues for the moment in favor of leveling the playing field in order to facilitate the coming exchange.

"This is ordinarily an aggressive question and it intends to elicit a defensive response but this is the exception and you got it. It is, in this instance, a fair question not an aggressive question," says 0011th-Robe greatly relieved to know that her mentor is allowing so much room to come to the point. "The aggressive question, if I may wax pedagogical, assumes Mr. Abbit's manuscript is impeccable and it is your job to recognize it. But the fair question allows that the merits of Mr. Abbit's manuscript are initially unproven to you. Consequetly you cannot make more than a tentative judgement about the substance of his manuscript. Perhaps even after the most thorough reading."

"If I may anticipate you, Dr. Mim-sey does not want the 'fair question' to be asked. The question is not whether or not Mr. Abbit is a threat. Dr. Mim-sey has already decided. Dr. Mim-sey needs Roy Abbit to be a threat. For him, it would be ***unfair*** to ask 'whether or not Roy Abbit is a threat'. If any of us asked the question the challenge would be, 'That is an unfair question.' He wants each and every one of the twelve Robes to be on the defensive when it comes to the question of the threat that Roy Abbit poses," says 110rd-Robe as an intern enters with a pair of alexandrite cups and chrysolite carafes.

"He wants the judges to doubt privately and individually their ability to judge," 0011th-Robe reasons. Now she is guessing that the unexplained gesture was a signal for more refreshments. 110rd-Robe seems to be anticipating a conversation of some duration.

"He thinks, I suppose, that we will hand down a decision that we do not have to defend or perhaps one that is easier to defend," mirrors 110rd-Robe as the intern places the hot drinks on the table.

"If we vote for the warrant we will never have to defend the decision," 0011th-Robe says. "We will never have to admit or face our individual doubts about the issues and the death warrant."

"It is all right with Dr. Mim-sey if we spend the rest of our lives questioning whether we made the right decision privately, severally," imputes 110rd-Robe watching his colleague position her cup.

0011th-Robe says, without looking away from her cup, "None of us will wonder if any of us have been *used*."

"If we do not admit to ourselves that we do not know whether or not Abbit or his manuscript is dangerous," 110rd-Robe credits, "we will never suspect we have been used."

"'Being used' is arguably a euphemism for being violated ,extorted, ransomed...," says 0011th-Robe lightly touching her pale yellow cup.

"If we assume, that we do not know whether Roy Abbit is a threat then we can vote against him. That is a yes for the warrant," says 110rd-Robe testing his drink for taste and temperature with a sip.

"We do have to know with certainty that he is not a threat to vote 'no' on the warrant," induces 0011th-Robe watching 110rd-Robe add a spoonful of salt to his drink.

"But to vote 'yes' should be because we are, individually, certain that his death will serve to change the course of history to our advantage," prompts 110rd-Robe with a taste of his drink.

"The change that is intended is a substantial one. We can only justify this warrant if there is a significant effect on the course of history," apprises 0011th-Robe reaching for the sugar.

"It is unlikely that the history of mathematics will be deflected with or without Roy Abbit or his mathematics," cautions 110rd-Robe stirring his drink to a cooler temperature.

"Their mathematics will eventually be on a par with ours regardless. Their mathematics is like ours. It is based on a culture of mathematics not not a cult of mathematicians," instructs 0011th-Robe following with another taste of her drink.

"The murder of Roy Abbit might affect the latter but not the former. We have to live as if our world and Earth will meet no matter what we do," construes 110rd-Robe as comfortable with his current opinion as he is with his current refreshment.

"If it never happens, fine," grouses 0011th-Robe with a taste of bitterness.

1

1 1

1 2 1

1 3 3 1

1 4 6 4 1

1 5 10 10 5 1

1 6 15 20 15 6 1

Chapter 15

Chapter 1111000....

Chapter ABCD

Chapter $(1 + 4x + 6x^2 + 4x^3 + x^4)$

Chapter $(a^4b^0 + 4a^3b^1 + 6a^2b^2 + 4a^1b^3 + a^0b^4)$

Chapter $2^0 + 2^1 + 2^2 + 2^3$

Episode 43

The days are gone when passenger liners on the river are free floating. As air, land and intra-urban transportation advances in technology, the perceptions of transporting disappear. Passengers now expect to travel without feeling it. This is not easy or economical for the floating passenger liner.

For those who have an interest in transoceanic passenger ships there are a few in museums. The history and the relics of those behemoths are still a fascination to some. What saved the passenger liners on the large rivers is an accident of history. As the passenger liners died, the amusement parks worked to preserve their history. It makes money. One or two of the great liners ply the ersatz waterways of the largest amusement parks. Riders pay to cruse and have a meal. You can even stay overnight for a fee.

A passenger ship is the ticket to the big leagues if you want to be an amusement park magnate. In order to enhance the amusement park passenger liner experience and without announcing it to the world, the liner rides on an underwater rail. As an added incentive, the rail reduces the cost of the artificial waterway. As a necessity, the park owner must deliver a ride that does not nauseate the passengers. Now, with the rail, it is possible to control not only the speed of the ship but every move it makes.

Ridership on the river-going passenger liners is saved by this subterfuge. The ships in demand are the ones on the submerged rails. What used to be a secret is now a selling point. Tourists pay extra for the luxury of the glittering crystal ships and competition is discouraged to keep the profits high.

Dr. Mim-sey and Dr. W-abe are cruising down river. Dr. W-abe is returning from a consulting job upriver and Dr. Mim-sey made it his business to visit a conveniently nearby university. Now Dr. Mim-sey can travel back with Dr. W-abe and continue his favorite conversation.

"This conversation is getting expensive," pleads W-abe while looking over the side at the flowing water.

"Since you value time as much as money," Mim-sey quips focusing on the horizon, "it was only a matter of time until you pushed that button."

"That does not deflect the point of whether or not we should **end** this topic. This is not the only thing I do," W-abe says. "It is bad enough to neglect things I have to do. It is worse to neglect things I want to do."

"Then you should be asking if we should **continue** this conversation or not," says Mim-sey.

"I suppose there is a difference," W-abe scoffs, "and I do not doubt that you will tell me what that difference is."

"If we end this conversation its value is eclipsed by your needs and wants," says Mim-sey.

"And if we continue this discussion it is because it is more important than my needs and wants?" W-abe asks with more interest in the breezes and wave action created by the moving ship.

"Only as far as it could make the difference in whether or not you can attend to the things you have to and want to," reprises Mim-sey continuing to compete for W-abe's attention.

"What in time, space and all the known dimensions could make this conversation that important," fleers W-abe looking up from the churning water and turning his back to its attraction.

"Whether or not anyone in their time, space and dimensions know about us in our time, space and dimension is what is so important," judges Mim-sey turning his head and shoulders toward Dr. W-abe.

"That is a lot closer to home than any next universe," says W-abe.

"Exactly, time and space alone, comfortably and conveniently, locate entire galaxies billions of light-years away from us," Mim-sey says. "And this is without the inconvenience of leaving our own universe."

"But the insulation of time, space and universes are just a fabric of so many dimensions," says W-abe to the ship's illuminated lounge sign. "A fabric that the right person or people could pour through."

"Like water through a sieve," argues Mim-sey stepping a little toward the lounge door.

"He or they could 'surface' uncomfortably and inconveniently close to us," says W-abe to stress the point.

"We are not our own best company by ourselves. An alien world would be difficult to assimilate," says Mim-sey. "For these and other reasons our civilization can be overwhelmed if it is too easy to migrate here."

"We need the insulation of time and space - perhaps, even, the isolation," contends W-abe.

"The one who can cross the barriers of the dimensions dissolves the barriers of time and space," emphasize Mim-sey waiting for Dr. W-abe to move toward the lounge or enter it.

"That cannot or should not be made easy or convenient," says W-abe entering the dimness of the ship's lounge.

"That depends on what you call easy or convenient," says Mim-sey relieved to see by the empty chairs that most of the passengers are still indulging in their evening meal. "Suppose a civilization, naturally, through a long history of development, adopts the ease and convenience of exploiting ever more dimensions."

"Along the way the barriers of time and space become less of an impediment," says W-abe surveying the oxbow bar and the empty bar stools.

"Then they begin to explore and pursue the possibility of using a pyramid of dimensions to move from place to place in time and space," maintains Mim-sey studying Dr. W-abe closely. "There must be a tipping point where the incentives exceed the costs, difficulty and inconvenience."

"And then by some accident or method they find us in time and space," W-abe contends.

"The rest will be history," says Mim-sey moving toward a table and a little away from W-abe.

"Our civilization could be long gone by the time these polydimensionals pounce," says W-abe.

"Before you get too cozy with the probability of our two cosmic ships passing in the night consider this - Suppose an individual persistently considers the ease and convenience of exploiting ever more dimensions? What then?" Mim-sey asks sitting down on the chance that W-abe will follow his lead.

"Ultimately, he sees holes in the barriers of time and space," says W-abe settling into a chair too.

"He begins to explore and pursue the possibility of moving through time and space via an archipelago of dimensions," persists Mim-sey browsing through the drink list aimlessly.

"The costs and difficulties of a well-focused individual are a lot lower than for a team, an institution or a civilization," designates W-abe without picking up the drink list.

"This only serves his chances of success, he has the same probability of finding us as anyone, any team, institution or civilization," Mim-sey says. "It does not matter who is behind the telescope."

"Multiplying the individuals multiplies the number of telescopes that can detect our existence," says W-abe.

"All those individuals with all those telescopes, all those eyes, looking around the universe," deplores Mim-sey closing the drink list as a signal to the wait service to take their drink order.

"The 'telescopes' being the metaphor for the ways and means to find us and approach us. By ignoring this, it seems, I am not just risking the things I want or have to do but I could be risking what our civilization wants to do or has to do," W-abe objects before the waiter gets across the room and to their table.

"How expensive is that?" Mim-sey asks without referring to the cost of the drinks to come.

Episode 44

On her trip home, Ali-ce takes corn-on-the-cob and butter so she can share this exotic delicacy with her father. The butter, of course is easy to freeze and transport. Learning to thaw, shuck and boil the corn is easy as well. What surprises her is her father's lack of enthusiasm. After a few bites his apparent enthusiasm is eclipsed by his concerns for her work, her life, and the details of Roy Abbit's work and life.

Each time he prepares to take a bite after the first one he comments on something. He is busy talking and she is hoping by being attentive and eating too that he will begin to eat his corn-on-the-cob. It is not long before she figures things out and lets him off the hook by putting down her corn and turning the monologue into a dialogue. Along the way she says that the corn is not fresh enough and he had better not try to eat it. She tells him it will give him gas even though she knows better and this gives him the excuse he needs not to eat the corn.

On her way back to Earth, she vows to herself that if he ever tries caviar she will not tell him what it is until he says he likes it. It is similar to a rare fungus that they both love. So, given the chance to try it without knowing what it is, will give him the chance to like the caviar.

Ali-ce finds Roy living in a garage apartment that takes half his cashier wages. He tells her he is lucky to have it. He can afford it as a rental. If it was a condominium it would cost more than he makes each month. While the conversation drifts to other topics Ali-ce estimates that the apartment is only about 225 square feet and that not just his teens but even his forties are behind him.

"Is that why you made your work language and literature?" Roy asks picking up a hangar.

"Is *what* 'why I made language and literature my work'?" Ali-ce asks handing him a shirt.

Roy says, turning to the closet, "So you could realize your dream of coming to Earth."

"I do not suppose anything is that simple. Earth was a legend and I had more ordinary urgencies to attend to," relates Ali-ce making herself comfortable on the bed rather than the chair.

"You wanted to be like mom or dad?" Roy asks pleased at her lack of self-consciousness.

"Mom was math and anthropology and dad was math and language but I urgently wanted not to be like them," observes Ali-ce making room for Roy on the bed.

"Were they bad parents?" Roy asks taking the four steps to the sink to fill the teapot.

"Never," Ali-ce says, "but I thought independence hinged on being different from them."

"And you did not like math?" Roy asks reaching for the English Breakfast tea bags.

"Actually, I preferred it but my being completely ignorant of math meant that handle was gone for them," offers Ali-ce pleased with his choice of the slightly milder alternative to the regular black tea.

"So language was the only common experience," judges Roy hoping she likes the change from the ususal tea.

"And that could be further limited by the choice of languages," says Ali-ce while Roy places a large oak cutting board on the bed, "as long as we did not study the same languages."

"But that left family connections," views Roy placing cups on the improvised bed table.

"Those were on the weak side since their professional lives were so demanding," asserts Ali-ce as Roy puts the Reed and Barton spoons beside the cups, "but he finally took up English."

"So you forgot your dream of coming to Earth?" Roy asks without his usual reminder that his friend Bob gave him the flatware.

"I made sure my parents knew I had an interest in Earth languages and literature," offers Ali-ce who by now appreciates the flatware his friend gave him as much as he does.

"The need to gather and process that kind of information was not going to go away and along with it," poses Roy with a glance at the hotplate, "there would always be a need for talent."

"Even if I never went to Earth, I would be part of the information stream from there," says Ali-ce as Roy moves the celadon teapot from its shelf and places it conveniently close to the hotplate.

"But you did come here," submits Roy pouring some boiling water into the teapot.

Ali-ce says, adjusting the pillows for sitting, "I came here to work for a publisher."

"And they publish books on math, history, language and philosophy?" Roy asks.

"You sent your manuscript to them," says Ali-ce laying an arm across the stacked pillows.

"It was rejected," objects Roy pouring the rest of the hot water into the teapot.

"That is what got my attention," rallies Ali-ce while Roy adds the tea bags to the teapot.

"Because it had no merit?" Roy asks pushing the table and tea ensemble closer to Ali-ce.

"This manuscript that had 'no merit'," Ali-ce says, "kept moving upstream."

"It was not going to be published but it never seemed to go from the desk to the dump?" Roy asks as he completes the setting with the teapot and a pair of expensive kitchen towels.

"It always was on someone's desk," says Ali-ce as Roy positions his chair beside the bed.

"So you took the trouble to remember the author's name?" Roy asks pouring Ali-ce's tea.

"That is not all," reflects Ali-ce while Roy pours some tea for himself. "On one of my rare trips home, I was asked to take a large envelope with me."

"Was that unusual?" Roy asks getting up and stepping over to the cupboards.

"In a couple of ways. If it was your manuscript and if it was without merit, why was it getting this attention?" Ali-ce asks as Roy returns with some Chinese almond cookies.

"Your information is relayed home anyway, is it not?" Roy asks with a warm smile.

"Exactly, and if this is the only copy," speculates Ali-ce equally charmed by the company and tea, "it was being removed so no one else in the company was ever going to see it or copy it."

"Was it my manuscript?" Roy asks returning to his place elbow to elbow with Ali-ce.

"Yes, it was," confirms Ali-ce picking up her cup for a sip.

While dunking a cookie in his tea, Roy asks, "Did you open the envelope?"

"No," says Ali-ce taking particular notice of Roy's maneuver. "That would be detected."

"How did you learn you took back my manuscript?" Roy asks handing a cookie to Ali-ce.

"Your manuscript was about English and mathematics," says Ali-ce dunking her cookie and wondering about the anthropological significance of the practice.

"And English is one of your father's languages," Roy says.

"The publishing house did not dare translate the manuscript to Bemyem," states Ali-ce with a bite of her cookie. "Father had to see it in the original language to validate the math and philosophy."

"So my manuscript has something," says Roy pleased to see the smile that fills her face.

"Math and philosophy do not circulate like a romance novel from the public library," says Ali-ce distracted by the flow and flavor of warm tea from the cookie.

"How did you find out he had it?" Roy asks with half a cookie and half a dunk.

"I made it my mission to see what was going on," charges Ali-ce wondering how many years he has enjoyed this cookie, this tea, this dunking. "He is not going to have me executed."

"I hope," says Roy wondering if the tea or cookies are new to Ali-ce.

"As long as my knowledge is unofficial," says Ali-ce with another cookie, "we are both safe."

"Both you and me?" Roy asks warming Ali-ce's cup with a little more tea.

"Both my father and I," answers Ali-ce studying her teacup.

"Well. What did he say?" Roy asks stirring his tea. "I am always ready to be flattered."

"And I am always ready to flatter you," says Ali-ce, "but he said he had not read it, yet."

"After he reads it, ask him again," says Roy as Ali-ce puts a piece of cookie on her spoon.

"I do not like to say this," says Ali-ce dunking the piece of cookie, "but my father lied."

"Did you see him reading it?" Roy asks watching Ali-ce sip the warm tea from her spoon.

"No, maybe that is why he thought the lie was sufficient," rejoins Ali-ce finishing the softened cookie. "But he had marked some of the pages and made some separate notes too."

"Did he ever say he finally read it?" Roy asks placing a cookie next to her cup.

"He had to," Ali-ce replies. "He did not want me wondering why he had not read it."

"So, what did he say?" Roy asks with a cookie for himself and a dunk.

"He was as kind as he could be when he said the manuscript has no merit, nothing new," relates Ali-ce picking up her cookie and looking to make sure there are plenty more.

"But that made you suspicious?" Roy asks making the four-step trek to the kitchenette.

"Yes, but it was a suspicion with no merit, nothing new. Suspicion is always an option," says Ali-ce while Roy returns with two small bowls and some chilled Mandarin orange sections.

"The publisher has not returned my manuscript. Did you bring it back with you?" Roy asks placing a bowl of chilled fruit next to each teacup and taking his seat again.

"No and that is suspicious," says Ali-ce dunking her cookie in the light orange syrup without being prompted. "I offered to take the manuscript back to Earth with me and he said that he wanted to think about it."

"He made sure you left without it?" Roy asks surprised that he had not thought of dunking his cookie in the oranges.

"And he asked me to be the one to tell you that this publisher would not publish your book," says Ali-ce pleased to find the syrup was not too sweet for the cookie.

"But you were supposed to do it," says Roy wondering if her dunking is a lifelong custom, "without telling me about the existence of your world and its secrets."

"That is why I took advantage of the time shift that is normally a problem," says Ali-ce.

"We were well known to each other by the time I wrote that manuscript," says Roy.

"On and off for more than forty years," reflects Ali-ce.

"So you can count on me not running to the tabloids with this story?" Roy asks.

"And can you count on me when I say your manuscript must have some merit that deserves a defense," says Ali-ce wondering at so much pleasure coming from the simplicity of the tea, cookies and oranges.

"It would be enough just to publish it," says Roy pleased that Ali-ce seems so pleased.

"But that is asking too much now," says Ali-ce.

"Why is that?" Roy asks with a piece of cookie and some orange on his spoon.

"I have not told you everything, yet," says Ali-ce with the suspicion her father knows Roy Abbit better than she, "but any effort by you to published could put you at really serious risk."

"And after I am out of the way nothing stands between you and your father," says Roy. "Then you would pay for your affection for me and betrayal of your father and your world."

"To say the least," sighs Ali-ce but more afraid for Roy than herself.

"I cannot throw away your life," says Roy with more thought for Ali-ce than himself.

"And I cannot let you trade yours for whatever is in your manuscript," says Ali-ce.

Episode 45

Imperious is not easy and anyone who says it is has never really been. Dr. W-abe's rare encounters with the Robes has taught him that.

His appointment with 011th-Robe is in what appears to be a small library. Less than a hundred of the nearly four thousand volumes are off the shelf and stacked on the desks of the Robe's clerks. Each of the eighteen clerks is removing the bindings and fasteners for each volume so that every page can be digitized and indexed.

Only the digi-book version of these pages is indexed. Every significant word is tagged so that at any time a Robe highlights a word or a phrase, he can get a standard definition and the number of times the word or phrase is used in the petition. A link is listed for each occurrence of the word or phrase so that the sentence and paragraph can be called up.

All citations that are native to the petition are linked to the historic library for cross referencing precedents. As each volume is completed, it circulates among the twelve Robes who add relevant citations. These citations are indexed and linked to the historic library and the petition.

This process will add 450,000 pages of indexing to the million or so pages of the petition. This library room holds one single reference. It is Dr. Mim-sey's petition. With indexing it will comprise 4,780 volumes, most of which will be over 300 pages each. It is 011th-Robe's responsibility to make sure the petition is ready, in a timely manner, for all the Robes.

No errors or omissions can occur that do not reflect on his personal and professional skills. This is not the only thing 011th-Robe has to do. He has to contribute to the substance of the review as well. So he has made the time to talk to Dr. W-abe.

"Thank you for coming in," says 011th-Robe centering a piece of paper on his desktop.

"My respect for the Robe's Chamber makes it easy," says Dr. W-abe. "But a rather complete lack of any idea of what service I can be makes me wonder why I am here or what I can do."

"I can almost say that includes both of us," says 011th-Robe with a nearly imperceptible flash coming from the desktop. "But since you know Dr. Mim-sey a little more as a former co-worker, perhaps there is something you are not even aware of that could help us."

"You might tell me what the conundrum is. Maybe we could start there," says Dr. W-abe watching the image of a page being cropped from the footprint of the desktop on a monitor.

"Believe it or not it would help to level the playing field between Dr. Mim-sey and Roy Abbit," stipulates 011th-Robe glancing at the image of the page now double in size with annotations. "I know you are familiar with them both - on a circumstantial or anecdotal level, for the most part."

"I infer that you have to choose between them and that Dr. Mim-sey has all the marbles," says Dr. W-abe while 011th -Robe turns over the page on the center of his desk without looking at it.

"Right now there is no choice," cites 011th-Robe ignoring a flash of light. "It would help put things in a better perspective if the two individuals were more equally deserving of merit."

"Let us reverse the perspective and ask, "What do we owe Roy Abbit, vis-a-vis Earth?" Dr. W-abe asks as the image of a page is cropped from the footprint of the desktop on the monitor.

"All right," allows 011th-Robe as the image of the page telescopes in size with annotations.

"Truly, we are the more advanced world," states Dr. W-abe, "but what Dr. Mim-sey will not tell you is that Roy Abbit is just another example of our plunder of Earth."

"Another? Has he left out a few things?" asks 011th-Robe removing the page from the center of his desk and replacing it with a new one without acknowledging a subsequent flash.

"The necessity of secrecy has worked in his favor all these years," vents Dr. W-abe.

"Then we must owe Earth and Roy Abbit more than we think," underlines 011th-Robe.

"And, perhaps," submits Dr. W-abe, "we owe Dr. Mim-sey less than he thinks."

"The details must be endless," proposes 011th-Robe as the image of the page is cropped from the footprint of his desk. "But can you give me just a hint of what we could be looking at?"

"I can give you two. At first you cannot see the connection but with a little, uh, history you will. First, have you noticed that the footprint of this building is similar to the geometry of the batball field?" Dr. W-abe asks as the image of the page expands with electronic annotations.

"Yes, It is a very organic geometry," says 011th-Robe as he turns over the page centered on his desk. "It is a template or framework for many of our cultural concepts. Like fairness in competition, justice in controversy, mechanical and moral strength, just to name a few elements that come to mind."

"We have endless versions of the ideas, even now, but the geometry is a geometry we borrowed from Earth," says Dr. W-abe now less distracted by the almost imperceptible flash.

"But Project Bump does not go back as far as batball," says 011th-Robe as the page image grows.

"Project Bump became part of the larger scheme to introduce Earth to our world without a cultural catastrophe," says Dr. W-abe, "It is just the tip of the iceberg that prompted romanizing our alphabet and language. This is done by taking advantage of the uncertainty of time and space that comes with high speed travel."

"And so English came to be promoted as a professional asset. I have a feeling for that," proposes 011th-Robe removing the page from the center of his desk to the pile of scanned pages.

"It is one thing to find planets with fantastic life forms and climates. But a technologically advanced world could cause complex polarizations we could not anticipate or survive," says 011th-Robe.

"We might survive the polarization but we would say good-by to a global economy, global justice, all the homogeneity that provides us with food, shelter, peace, safety, justice and technology across the planet," says Dr. W-abe as 011th-Robe places a new page on the center of his desk.

"Those things would divide along the lines of advantages and disadvantages and compete among themselves?" 011th-Robe asks ignoring the flash but glancing at the image on his monitor.

"Groups with surpluses would trade or barter to compensate for deficiencies," says Dr. W-abe with no notice of the mushrooming image. "One group would maintain peace at the expense of justice. One group would maintain food supplies in exchange for technology and so on."

"I suppose the benefits of our currently complex integration would be lost - possibly forever," says 011th-Robe reaching for the page on the center of his desk without looking and turning it over.

"We digress," says Dr. W-abe. "The other thing we owe Earth besides an integrated social geometry is the ability to get there and back."

"I must be anticipating too much. How can it be that we get there, to Earth, but cannot return without their help?" 011th-Robe asks removing the page from the center of his desk but without placing it on the stack of scanned pages.

"Irony is not the same as inconsistency," alludes W-abe as 011th-Robe deposits the page on its stack of scanned pages. "The solution for this irony is mathematical. Like our social geometry and Abbit's categories."

"It looks like we owe a lot more to Earth than anyone knows," says 011th-Robe, "and Dr. Mim-sey, without acknowledging the details, accrues prestige to himself."

"That is our Dr. Mim-sey," says Dr. W-abe as 011th-Robe centers a new page on his desk.

"It appears that this will take more time to unpack. I doubt that we can do it all in this session, given our existing commitments," says 011th-Robe without watching the new image crop to the page and expand with annotations. "Let us arrange to meet again soon."

"The essentials are more than a few, so forgive me, it will take some time to get it all on the table," says Dr. W-abe as 011th-Robe turns over the page at the center of this desk.

"If I may impose on you in one more way," says 011th-Robe uninterrupted by the blink of light from his desk.

"Of course," endorses Dr. W-abe.

"My notes will be available to the other Robes," enjoins 011th-Robe. "But sometime after we finish meeting, will you write a summary of these things for the archives?"

"I am afraid the written summary will be longer on details and will include things we will not get to discuss," pleads Dr. W-abe without any conscious acknowledgment of the silent process of cropping and annotating of the page centered on 011th-Robe's desk.

"That will not be a problem," maintains 011th-Robe without any concern for the possible number of pages Dr. W-abe's summary might include. "Just do not include so much that it is never finished."

"I can do that," promises Dr. W-abe wondering how he is going to do it.

1

1 1

1 2 1

1 3 3 1

1 4 6 4 1

1 5 10 10 5 1

1 6 15 20 15 6 1

Chapter 16

Chapter 00001000....

Chapter E

Chapter $(1 + 5x + 6x^2 + 4x^3 + x^4)$

Chapter $(a^4b^0 + 5a^3b^1 + 6a^2b^2 + 4a^1b^3 + a^0b^4)$

Chapter 2^4

Episode 46

Dr. W-abe likes the technology of the matrix elevator. You insert your ID card at the elevator. Usually, the wait is tolerably short because, among other reasons, there are so many cars.

Before the elevator arrives, press 'Office/*Pimnumncli*, Exit/*Diebemnjam*, Home/*Gei*, Airport/*Vaegeslimn*, R.A.T./*Rlete Az Teimsla* or Emergency/ *Gemge Kemnqoimn*'. This elevator will take you up or down and horizontally to your floor and the nearest egress in terms of your circumstances except for 'Exit.'

The elevator will select the egress according to your personal information. The system is optimized to the shortest path that includes your trip along with anyone in the car that is going that way. If you press 'Exit' you have to indicate which face of the building you want and you get the ground floor for that face. If you press 'Emergency' that gets priority and the car goes to the nearest medical/security station.

Your path is optimized before you remove your card. Your destination is listed among any others included. If your destination is close enough to the top of the list, you can press 'confirmed.' If not, you can press 'cancel' or 'next car'.

There is no talking. Dr. W-abe likes that. All eyes are on the path indicator anticipating each exit. Dr. W-abe does not like that. He does not want strangers making judgments about his home or any other speculation that his choice raises. So he takes the R. A. T. when he is going home.

The R. A. T. is the 'Reticulated Atrium Transport'. It is a horizontal transport with a textured hull to reduce the noise of airflow around the car. There is a R. A. T. for every three floors of the atrium and it circumnavigates the archipelago of green spaces of the building-burb.

Unfortunately for Dr. W-abe the convoluted ride with the ever changing and spectacular views promotes conversation. He rarely hears anything new. This time, however, only Dr. Mim-sey is with him, so the conversation is not so predictable. But then again....

"So the danger is that Roy, the arithmetic archetype, will point to holes in the fabric of space and time. And that will lead to finding a compatible sub-universe and exchanging visits," says W-abe as he watches Mim-sey changing his line of sight to take in the best possible views.

"And that," rants Mim-sey, "will precipitate the decline of one of the interfacing sub-universes."

"We, our universe, can justify a prematurely short life for such a dangerous individual," says W-abe while Mim-sey takes in the vaulted roof and its intersection with the atrium wall.

"Do you want the job?" Mim-sey asks focusing his fascination on the architectural ornamentations.

"No," W-abe says. "I could not do it, even to save my civilization."

"I can spare you the conundrum," says Mim-sey wishing the R. A. T. was slower. "Or at least, I can give you an excuse. Rubbing him out could lead to our detection."

"That solves our problem but it does not solve the problem. By the way, Earth people are about 75% water. Rubbing him to death would make an awful mess," snickers W-abe.

"You give me the chance to call you facetious," says Mim-sey.

"You are the one that wants to use this impossible language, this English," W-abe taunts. "Even though I can infer what 'rubbing someone out' is I wonder where you find such a strange fragment.

"If I remember, I will tell you," says Mim-sey resolving to learn more of the English terms that are used in architecture. "But the language is another part of the clue to the solution of the problem."

"The problem is more than personal vanity now, so I cannot hide behind my superiority anymore," scorns W-abe with less interest in ornamentation and more in function. "So clue me."

"The basic unit of information in English," Mim-sey lectures, "is the sentence."

"I hope you will be kind and get back to quantum time real soon," W-abe mocks.

"I will get there as fast as the interruptions permit. Words fall into three or four categories depending on how you define them. The main categories are nouns and verbs. All the rest can be called modifiers," apprises Mim-sey. "We can limit the categories to two, if we lump the noun modifiers with the nouns and the verb modifiers with the verbs."

"Even if English were my natural language, I do not like the abstractions of nouns and verbs," confesses W-abe watching the environmental lights blink on at ground level.

"It is a quick and dirty way to quantum time" Mim-sey says becoming aware of the fading daylight.

"Okay but quicker, please," says W-abe this time to the path indicator and the next stop.

"Most English speakers do not concern themselves with the categories either," says Mim-sey taking in the emerging brightness of the water features scattered below. "But if you represent the nouns with binary ones and the verbs with binary zeros, any sentence becomes a string of ones and zeros."

"The noun modifiers and nouns get binary ones and the verb modifiers get binary zeros, just like the verbs," says W-abe.

"Different sentences will have different strings of ones and zeros," advises Mim-sey.

"By the way," conveys W-abe conceding for a moment, that all that water is composed of particles and sub-particles, "if you substitute one noun for another, it will not change the string of ones and zeros."

"Those sentences will be similar so what we have to say will still apply," imparts Mim-sey. "Most of what is true for a string is true for similar strings - just like similar triangles."

"Oops. I am the one slowing things down now," confesses W-abe reminded of his personal rule to resist his reflex for underestimating almost everyone.

"Anyway, a seven word sentence will have some combination of ones and zeros," Mim-sey states. "It can have a binary value up to 128. That is the same as row seven of the Pascal/Chu triangle and the polynomial $1 + 7x + 21x^2 + 35x^3 + 35x^4 + 21x^5 + 7x^6 + x^7$."

"That polynomial is the combination of the subsets representing the noun and the verb. Subtracting the values of the noun from this numbers will result in the values of the verb," vouches W-abe in a friendly effort to show interest and support for Mim-sey's monomania.

"What does the curve look like for the three polynomials?" Mim-sey asks.

"They will be pretty flat because all the coefficients are separated with plus signs. Who does not know that?" W-abe asks pretty sure that there are a great many who do not know that.

"If anyone knows, it is you," concedes Mim-sey. "I asked because I will need your attention for what comes next."

"Is there any hope for some minus signs? That will put some wiggle in the curve," says W-abe.

"Exactly. The zeros in the binary number for the noun represented the opposite category," Mim-sey warrants. "The opposite number in arithmetic is the number with the minus sign."

"So the zeros in the noun binary number indicate which numbers in the verb polynomial get the minus sign," says W-abe a little surprised at how subtle the obvious can be sometimes.

"The verb version of the polynomial will have its minus signs where the nouns occur. If you ignore the plus and minus signs the two sub-polynomials will add up to the original polynomial in this case, row seven of the triangle: $1 + 7 + 21 + 35 + 35 + 21 + 7 + 1 = 128$," summarizes Mim-sey hoping to keep W-abe's flagging attention for just a few seconds more.

"Any combination of nouns and verbs can be rendered as a pair of curves. One for the nouns and one for the verbs. A third curve represents their sum," says W-abe to summerize. "Their sum cancels the zeros and minus signs and results in a flat curve that is consistent with a string of binary ones."

"This does not tell you how to draw the curves but you do not need to use Newton's method to derive them," says Mim-sey satisfied to have W-abe cornered like a R.A.T. in a loop.

"That arithmetic acrobat sorted this out?" W-abe asks. "I think I am going to miss him."

"That is not the saddest part. It is this method, his method, that sows the seeds of his own destruction," says Mim-sey. "Even his death will have to be attributed to his pedestrian circumstances."

"The hand of heinous genius can never and will never be seen," says W-abe.

Episode 47

Ali-ce likes coffee with sugar but she does not like her tea with it. Making a good cup of tea is that much harder for her. She likes the way Roy makes his tea. So, if he does not offer to make it, she can ask for some without imposing on him. His mind does not wander, so if the conversation is not in the neighborhood of tea, he will not think of it. This time, Roy thinks of the tea first.

He has a hotplate next to the sink and microwave. His teapot is made of clear glass. He likes to see the water boil. It gives him some latitude. The teapots that whistle demand to be removed from the heat and will not be deferred. You cannot see how much water you have in them and for Roy that is important. With his glass teapot he can always use the same amount of water.

This way he always uses the same amount of tea leaves. By allowing the boiled water to stand a few minutes before adding the tea, he can be sure that the water is not so hot that it steams off the aromatics - an important part of the flavor.

Brewing a few minutes is better than brewing a few seconds even though the water changes quickly from clear to its translucent mahogany. The color is as important as the taste and the color is the first thing he misses when he has to change brands.

When Ali-ce is there, he uses his celadon teapot and gives her a bright stainless steel, Reed and Barton, teaspoon. He usually does not expect her to use it. It just looks nice with the cup and saucer. He always tells her his friend Bob gave him the celadon and the flatware.

He knows it makes him seem absentminded but it is important to acknowledge his friend Bob's gift when he uses it. Ali-ce honors his care as a ceremony and holds any conversation until after the predictable comment.

"This publisher, you came here to work for, is it your father's company?" Roy quizzes.

"Yes," Ali-ce confides on the way to his door with her tea. "This is one of the ways we monitor your world."

"And pays for it?" Roy asks following her outside with his cup and the celadon teapot.

"Even if the company paid for itself, it would not pay the costs of all we have to do here," allows Ali-ce who, like Roy, intends to enjoy some of the dry, cool November air in Miami.

As he hears someone in the kitchen, Roy asks, "What do you have to do here?"

"I do not know about everything but the information does not just come to us," concedes Ali-ce. "We have to cultivate native talent and use them to find and collect the best information that can be used back on our planet."

"So most of the employees are from here?" Roy asks listening to the ersatz poltergeist in his tiny apartment.

"Yes," says Ali-ce. "Most of them are building careers in math, philosophy, languages and history."

"So what do you do? Print dollars?" Roy asks enjoying her talent for mimicking sounds.

"Crimes and accidents risk exposing the few of us we have on your planet. We import diamonds from our planet," admits Ali-ce who maintains that the mimicry is done with the throat muscles not the vocal cords.

Roy asks, with the microwave beeping away as if it were being used, "Your father has a diamond company?"

"Semantics," holds Ali-ce as she adds the sound of the bell and door of the microwave opening. "These companies exist to interface our information gathering and personnel transfers."

"You do not use volunteers for life?" Roy asks almost believing that the sounds were real.

"Many of them are," says Ali-ce with some gravity. "What we cannot do is risk an autopsy."

"So some of your personnel transfers are in a box," says Roy regretting the turn of the topic.

"No one would volunteer if it meant never returning - especially after death," says Ali-ce.

Hoping her mood has not turned sour, Roy guesses, "I suppose this is pretty important for the family."

"For the time being," vouches Ali-ce following with a sip of tea and anticipation of lighter thoughts. "If the immediate family lived here then the deceased volunteer could stay here."

"So you have been cultivating me all these years to get me to write my little math book?" Roy asks hoping to turn the corner on the topic of conversation.

"No, You did that on your own, I did not know you until you sent your manuscript to this publisher," declares Ali-ce adding the sounds of an old mechanical typewriter working furiously.

"I am surprised they took the time to read it," responds Roy on the brink of laughing.

"It is a small company and it is more important to publish than to make money," declares Ali-ce delighted to see Roy so genuinely amused.

"As long as you are publishing," concedes Roy, "you are situated to collect information."

"And it is not illegal to lose money," claims Ali-ce to the bell and grind of a mechanical cash register.

"As long as you pay your bills," advances Roy to the flutter and crinkle of counting paper money.

"Exactly," says Ali-ce watching Roy's neighbor busy herself with laundry and whatever she can hear of their conversation, "but becoming famous for making a lot of money would complicate things."

"So when I turned up you said that you want to be the one to monitor Mr. Abbit?" Roy asks.

"No. I was assigned. They did not want the mathematicians to know you existed. Besides, my language career interfaces with your theme," says Ali-ce pleased to see the neighbor's busybody toy dog working his way over to them.

"So you were supposed to pepper my life from beginning to end with support for my career?" Roy asks reminded by the sight of the dog why he never gets much use of his table and chairs or the yard.

"I am supposed to insinuate myself into your life immediately," cites Ali-ce watching the furtive little dog zig zag closer; ingenuinely sniffing at plants and constantly eyeing her and Roy. "It is my job to frustrate your efforts to publish for the rest of your life."

"As far as anyone knows you have been involved with me for only a year?" Roy asks.

"And only in a disguised professional way," confirms Ali-ce choosing her moment to confront the dog.

"So what made you throw the baby out with the bath water and change the mission?" Roy asks now listening to the sound of the neighbor's doorbell coming from Ali-ce.

"The dishonesty of insisting that your work was meaningless and the implied threat that if I did not keep it that way you would have to be murdered," charges Ali-ce focusing on the encroaching dog.

"You mean, if I was a 'real' danger you would not mind my murder?" Roy asks over the barely audible imitation of a doorbell.

"I would not murder anyone but I could not prevent it," says Ali-ce still watching the dog and punctuation every opportunity with the mimicry of the door chime.

As he hears the door bell again, Roy says, "That does not sound like a simple remark."

"If you were a real danger, I would have done things the way they were planned. I would have stayed out of your life before you were discovered. You would have remained a stranger to me until the end," says Ali-ce as the little dog suddenly races back to her master's house in a barking rage.

"Even if that meant my murder?" Asks Roy watching the neighbor race inside after her dog.

"I would have worked to avoid that simply by making sure you never published," says Ali-ce as the neighbor adds her curses to the yapping din. "Even if that meant many decades of shadowing you."

"You would have made it a life long mission?" Roy asks with some concern for the dog's safety.

"With little to regret," says Ali-ce and with no doubt in the dog's predictable and undistracted ability to avoid harm while barking relentlessly at the intruder's signaling bell with all the fury of its breed.

"You would not regret that I lived out my life without the benefit of my talents or you would not regret my murder?" Roy asks gratefully for the return of privacy to the yard.

"If you were an open threat to my world there would be little regret for either," says Ali-ce.

"How can you say that and mean it?" Roy asks

"Because that truth has become hypothetical," deposes Ali-ce with quick glance across the yard when the neighbor turns off the yard light. "You are not the villain. It is my father."

"How can you be so sure?" Roy asks recognizing Ali-ce's reflexes for light cues.

"The punishment does not fit the crime. He is proposing murder for the potential threat of your theme. You are not proposing to uncover unknown worlds to be looted and destroyed," says Ali-ce who comes from a world where anyone and everyone can vocalize any sound.

"Things could go that way," says Roy recalling Ali-ce's indifference to bells, buzzers and aural cues.

"So many things are so far from that that your murder is not justified here and now," Ali-ce says. "There is the crime."

"Is that why you broke with the plan and became involved with my earlier life?" Roy asks considering the chaos on a planet of people responding to miscuing bells from uncounted mischief makers. (Ersatz flatulence is passe on Clidim's elevators as every school child soon learns. But Ali-ce is delighted by the unflagging effects on the lumbering elevators of Earth.)

"Even then I did not find a crime equal to the punishment," protests Ali-ce.

"'My only love has sprung from my only hate...'," professes Roy with reflection.

"You are a poet," banters Ali-ce weighing the force and economy of Roy's remark.

"I was quoting Shakespeare's *Romeo and Juliet*. I think you have sent that to your father," Roy says and beginning to realize how we use sound to anticipate and prompt action.

"Who gets murdered in that one?" Ali-ce asks aping sounds of a sword unscabbarding.

"Do not be bitter," Roy appeals. "You said, yourself, that he is a researcher first."

"So you think he will give up on any plan to murder you?" Ali-ce asks returning the virtual sword to its scabbard smartly.

"With our help. The plot discovered is the plot foiled," says Roy lifting the teapot and pouring some tea into Ali-ce's cup. "You can convince him his plan has no merit."

"He will not be convinced," says Ali-ce accepting more hot tea.

"He will be convinced by our cooperation and the time that follows without publishing my manuscript," says Roy while serving himself tea to the sounds of marbles pouring into his cup.

"How can you give that up?" Ali-ce asks lifting up her cup.

"Long or short, I want to spend the rest of my life with you. Am I going to die soon?" Roy asks without any expectation of a real answer while her smile disappears.

"You know," says Ali-ce looking into his eyes. "I do not like to answer questions about your future."

"If I live a long time, it could be because I never published the book," says Roy pressing his point.

"How long you live may have nothing to do with whether or not you publish that book," says Ali-ce to the susurration of the previously unknown Florida winter crickets and cicadas.

"So allow me to do whatever I can to live with you," says Roy amused at the new species.

"And allow me the same," says Ali-ce grateful for the relief of tension brought by these phantom creatures.

Episode 48

Perhaps, it is because all the Robes share their information, that Dr. W-abe (or anyone else for that matter) rarely sees the same Robe twice in a row. Or perhaps, it is because none of the Robes rarely see anyone twice in a row that they all share their information.

In any case, Dr. W-abe is neither surprised nor concerned that he is waiting in the Robes dining room for 0001th-Robe. He has every confidence that he will meet 0001th-Robe and right on time without regard for the informalities.

No, Dr. W-abe does not know who 0001th-Robe is or what he or she looks like. He will not be in a robe because they do not wear the robes except for official functions. But this much Dr. W-abe can be sure of, 0001th-Robe will find him.

Dr. W-abe has never been in the Robe's dining room before so he will be easy to pick out, presumably. To a lesser degree he cannot be sure how the conversation will begin but has no doubts about where it will end.

"Thank you for your time. Where do we begin in the quest of what we owe Earth/Roy Abbit?" 0001th-Robe invokes approaching the table with two servings of a rare jellied fungus and without introductions or other formalities.

"In order to be as brief and direct as possible, I would like to begin by reminding you that we have a process that reduces our organic waste to oils, salts and water by heat and pressure," urges Dr. W-abe shocked to see a Robe carrying his own tray and serving his guest the refreshments.

"Yes I like that scheme. "It generates some of its own heat and the products are very useful," collogues 0001th-Robe. "I would rather be 'recovered' that way than by cremation or burial."

"There is a trend to that," advises Dr. W-abe, "but I mention it because deep ocean exploration found similarly high pressure and temperatures in the deepest parts of our oceans."

But they also found fish, crabs, and various life forms including bacteria, did they not?" 0001th-Robe asks without waiting for Dr. W-abe to begin his drink or the prized jellied fungus.

"At that pressure, surface evolved organic material decomposes. Salt does not mix with water. Oil does," says Dr. W-abe. "Steam does not form even at those incredible temperatures."

"But these creatures have legs and digestive systems. They are mobile and compete in a food chain," says 0001th-Robe pleased to see that the conversation and refreshment are not imposed upon by fawning or formality. "If light and sound got that far down they would have eyes and ears."

"It is conceivable that given enough time and no continents, the creatures would evolve a civilization," says Dr. W-abe trying to remember how long it has been since he had this particular fungus.

"Why no continents?" 0001th-Robe asks adding some salt to his drink before tasting it.

"The evolution of life in either case is just as likely. The problem is that the two are so different that they would form civilizations at different rates," construes Dr. W-abe tasting his drink before deciding how much salt to add. "One of them would get a competitive advantage."

"Eventually, one civilization would penetrate the other to no mutual advantage but rather the exploitation of one of them," comments 0001th-Robe still stirring his drink to a cooler temperature and without tasting it.

"Do you suppose, one would become food and space for the other?" Dr. W-abe asks.

"I suppose that is so. It is difficult to say. We do not have any examples like that on any of the known worlds that I know of," acknowledges 0001th-Robe finally tasting his drink and smiling with satisfaction.

"But two things come from this before we even leave the planet, conveys Dr. W-abe. "Life is possible under extreme conditions and we can build ships to get there."

"I presume getting back has its own set of problems," concedes 0001th-Robe.

"Exactly, our integration of metals and glass allow us to build ships strong enough to reach the bottom of the oceans," acknowledges Dr. W-abe wishing he had a little more salt in his drink but does not want to put a wet spoon in the salt and does not see another salt service anywhere around.

"However, getting back is an energy problem," says Dr. W-abe.

"I should think the energy requirements would be the same both ways," says 0001th-Robe.

"You are correct," acknowledges Dr. W-abe deciding that he is very close to the right amount of salt for his drink anyway, "but the problem has proven to be that the **type** of energy required is not the same both ways."

"You can use gravity to sink to the bottom but how do you get back?" 0001th-Robe asks while collecting a spoonful of chilled, jellied fungus.

"Exactly," says Dr. W-abe hoping Mim-sey does not have a taste for this costly fungus. "Part of the solution is to return in a much smaller ship, by abandoning some of the weight of the support equipment."

"Obviously, the smaller ship brings back the explorers and the equipment left behind relays data," declares 0001th-Robe with confidence in his supposition.

"With respect, not as obviously as you may think. When we began exploring space, the smaller and returning ship brought data and the explorer were left behind to gather and return more data," confesses Dr. W-abe with a mix of regrets from the turn of the conversation and seeing the end of his portion of jellied fungus.

"That is terrible. No one knows this?" 0001th-Robe asks.

"A few," says Dr. W-abe with a little tension. "The returns were 'staged' to earn support to continue exploration. Everyone assumed exploring space was a lot like exploring the oceans."

"The astronauts did this knowingly and willingly?" 0001th-Robe asks with incredulity.

"There were conditions. Primarily, the hope was that return trips would ultimately become possible because of their efforts and no one must ever know about the sham," says Dr. W-abe wishing he did not know this.

"You mean the astronauts did not want to look like martyrs?" 0001th-Robe asks.

"They did not want to put that burden on their descendants. They did not want the controversy to imped the pace of progress," says Dr. W-abe who understood and shared in the conviction. "The sacrifice had a better chance of solving technological problems faster."

"The practice might uncover resources and solutions that could never be found any other way," predicates 0001th-Robe without pretending to understand or share in the rational.

"Exactly, the solutions and resources were found," affirms Dr. W-abe relieved that he could bring the conversation to this point. "Our astronauts come and go freely and productively, now."

"Does hindsight justify the cost in lives?" 0001th-Robe asks more than rhetorically.

"Not until it involved Earth" braves Dr. W-abe with a sense of shock that is the same every time he is brought back to this realization.

"We are in that kind of debt to Earth?" Asks 0001th-Robe haunted this time for the first time.

1

1 1

1 2 1

1 3 3 1

1 4 6 4 1

1 5 10 10 5 1

1 6 15 20 15 6 1

Chapter 17

Chapter 10001000....

Chapter AE

Chapter $(1 + 5x + 7x^2 + 4x^3 + x^4)$

Chapter $(a^4b^0 + 5a^3b^1 + 7a^2b^2 + 4a^1b^3 + a^0b^4)$

Chapter $2^0 + 2^4$

Episode 49

    If it were not so late Dr. Mim-sey would never have talked Dr. W-abe into the water club. At this hour, he has the consolation of almost no one being there. Besides, Dr. W-abe can count on going home from here, which is where he ultimately prefers to be. Here they have the incentive of plenty of service and drinks.

At the water club you can move from lounging areas in the hot pools to lounging areas in the robe rooms. Pools are clothing optional. Robe rooms require robes. The stairs and ramps that are submerged are well lit and comfortable to use.

Every time you get out of one of the pools, heated robes are immediately available so you can sit on chairs and benches that have deep upholstery - wait-service when and where you want it and none if you prefer. If you do not mind signing-in, you do not have to bring a means to pay as a building tenant.

Dr. W-abe cares nothing about the cost but he does not like to sign in so he always goes as Dr. Mim-sey's guest. Which is fine with Dr. Mim-sey because he does not prefer to entertain at home and this is a way to corner his friend for his self-serving conversation.

"With binary numbers, any number you mention is accessible to any schoolchild and many adults as a polynomial. These polynomials can include positive and negative values that affect the shape of the curve that the polynomials represent. So what is the danger of that?" W-abe asks almost absently.

"First, this kind of elementary math is accessible to millions of Earth people," scorns Mim-sey sitting neck deep on a submerged, illuminated chaise lounge. "Not to mention the endless generations that do not have the reach of calculus."

"I will not have to fear for my job. Will I?" W-abe asks stirring the warm clear water with his hands.

"Not in our lifetime and you know it," asserts Mim-sey measuring the distance to his glass on the stabilized floating table with his eyes before reaching for his glass. "The second danger is the sunset of point particle physics.

"That is the end of describing subatomic particles as a collection of an even smaller collection of even smaller particles," says W-abe reaching for his glass without looking.

"We still use that concept but only as a convenience," affirms Mim-sey feeling free to take a drink now as well. "The math is simpler and predictive in the cases where we still use it."

"It is tried and true," declares W-abe getting back to agitating his crystalline neighborhood. "Wave based descriptions of the universe, on the other hand, are difficult even for me, sometimes."

"But wave based physics can be applied from the smallest artifacts to the largest," Mim-sey predicates while watching Dr. W-abe distract himself with vortices and eddies. "It can tell us which universes have atoms and photons like ours because they come from the smaller pieces that we also know."

"So, by opening the door of wave mechanics to the largest objects means that eventually someone will be knocking on the door of the next most accessible part of the universe. Ours?" W-abe asks.

"In actual practice, probably, both the largest and smallest artifacts will get attention along the way. Theories will test into facts. Facts will test into universes. The large and small, windows and doors to our part of the universe will be discovered," says Mim-sey feeling like he deserves another sip and taking it.

"Windows and doors do not open by themselves," W-abe says. "We can count on that."

"Finding and penetrating to a meaningful sub-universe is not easy even if the math and physics pave the way," Mim-sey maintains as he sets his glass on the table with care.

"The math and physics are so difficult that, for Earth, the pieces may never be put together," argues W-abe getting up and wading along a ramp to his robe with his drink in one hand.

"A hope that brings much comfort with it," says Mim-sey following W-abe out of the pool at an unimposing distance. "But those pre-calc-progeny may hasten us to a cold bath like that of a probability becoming a certainty."

"What are you going to tell me that will close the gap besides generations of mathematicians and multitudes of lifetimes?" W-abe asks tying his belt and looking for a chair.

"Remember your black and white marbles?" Mim-sey asks rhetorically while putting on his robe.

"The white marbles are binary zeros and the black marbles are binary ones...," interrupts W-abe.

"They are the polynomial," resumes Mim-sey settling in, "that describes the verbs and verb modifiers in an English sentence. It is a tabulation of the rows that represent the verb component of the sentence. It tells you how many rows have one black, two black, three black marbles ..."

"And the complementary polynomial for the nouns and noun modifiers in an English sentence are an account of the rows that represent the noun component of the sentence. It tells you how many rows have one black marble, etc.," conveys W-abe moving to a bijou room with mellow light.

"When you add the two polynomials," says Mim-sey following his friend with his drink in hand, "you get a polynomial with the same coefficients as a row in the Pascal/Chu triangle."

"If you add the numbers for any row of the triangle, the sum represents a power of two. The power of two is the same as the row number," imparts W-abe tapping a keypad at the dumbwaiter for a new drink and a snack before settling into a chair. "More proof that you have counted all the black marbles."

"Then, how do you count the white marbles together with the black marbles?" Mim-sey asks settling into a generously upholstered chair for what could be more than just a grunt from his friend.

"Binary numbers consist of various ones and zeros but I do not recall ever having to count both variables," says W-abe taking a chair beside Mim-sey that shares a small table made of hundreds of red, blue, green, gold, opal and clear crystal pieces. "It has been a while since I have been in this neighborhood."

"I will give you a hint from your early years. The row number is the power of two as 'n' and the polynomial can be expressed as an expansion of $(1 + x)^n$. The expansion of $(1 + x)^3$ is: $1 + 3x + 3x^2 + x^3$. These coefficients sum to 8; one subset has no black marbles, three have one black marble, three have two black marbles and one has three black marbles," prompts Mim-sey.

"Oh. You can include the white and black marbles in the tally with the polynomial that comes from the expansion of $(a + b)^n$. Row three of the Pascal/Chu triangle becomes: $(a + b)^3 = a^3b^0 + 3a^2b^1 + 3a^1b^2 + a^0b^3$. Again, the coefficients still sum to eight," formulates W-abe having cleared another of Mim-sey's hurdles.

"But the exponents tell more of the story," counsels Mim-sey while getting up to order himself another drink and hors d'oeuvres from the keypad. "The black marbles are 'b' and the white marbles are 'a'. As the number of black marbles increase the number of white marbles decrease and vice versa."

"But the row number sets a limit on the number of white marbles. In this case you cannot use more than three white marbles even though it is the black marbles that count. What does this mean for math for the masses?" W-abe asks taking the opportunity to pick up his order.

"Remember that number line. The one that you use to count or measure time and space?" Mim-sey asks returning to his chair to wait for his order.

"Just incase that is not a rhetorical question, yes," says W-abe setting his refreshments on the small, multicolored, crystalline table in order to change into a dryer robe.

"Math for the multitudes now has five versions of the number line. There is the natural or decimal: 1, 2, 3.... There is the binary number line: 1, 01, 11.... There is the 2^{n+1} number line that increments in powers of two and...," drones Mim-sey while moving a similar polychromatic table over to his chair.

"There are the two similar expressions of 2^{n+1}," admonishes W-abe while picking up Mim-sey's order to take it to his table. "There is the number line that increments as $(1 + x)^{n+1}$. At the same time there is the number line that increments as $(a + b)^{n+1}$."

"In a word the schemes are 'symmetrical'. You rotate a sphere and it does not change. You draw a line that is eight units long in any of the five dimensions, and it does not matter whether you call it 0001, 2^3, $(1 + x)^3$, $(a + b)^3$ or 8. The five counting schemes are symmetrical," says Mim-sey feeling a little surprised and very flattered that W-abe would pick up his order and bring it to him.

"The ability to work with symmetry is the first hurdle for quantum mechanics, wave mechanics, string theory...," says W-abe settling into his chair with the intention of staying awhile.

"You surprise me because I thought you would say calculus," crows Mim-sey changing his robe.

"I surprise myself," says W-abe. "You remind me of what I had to learn besides higher math."

"And then there is the ultimate surprise," Mim-sey declares. "The easy employment of a few dimensions and some symmetries may be enough to focus on this part of the universe in our lifetime."

"The protection of elite knowledge and sublime coherence will be penetrated by ordinary familiarity," declaims W-abe watching his friend sit while reaching for his drink.

"Very poetic - If everyone had telescopes in Galileo's day he would not have been suppressed by the Inquisition," rants Mim-sey. "Math and science would have advanced all the quicker."

"You restore my faith," imposes W-abe. "A little interference goes a long way."

"Yes, and if Mr. Abbit does not publish, it could be a long time, if ever, before anyone connects the dots that lead to our part of the universe," ordains Mim-sey picking up the remote laser control.

"Obfuscation trumps assassination," belittles W-abe.

"Our hands are clean and no one suspects that there is anyone behind the curtain," says Mim-sey reducing the room light until a familiar star pattern on the ceiling becomes visible from wall to wall.

"Mr. Abbit makes a lifetime of bad choices, with some help from his friends. Whose fault is that?" W-abe asks realizing that the millions and millions of specks are galaxies not just stars.

Adding a straight red line appearing to begin at their galaxy and tracing to a dark patch of space, Mim-sey decrees, "As long as Mr. Abbit does not publish, we do not have to do anything drastic."

"But only as long...," frets W-abe certain that Earth's galaxy is at the end of the line.

Episode 50

How do you carve a community out of a swamp? First, you dig some canals to drain the boggy environs. Then, you subdivide the land and sell it while you wait for the muck to dry. It does not hurt to quarry all the stone you can use or sell. You can call the holes that are left swimming pools or lakes and charge accordingly. It may take a hundred years or so to forget what the pools and lakes really are but forgetting is inevitable and irreversible.

Roy and Ali-ce have a March day that they can spend at one of the famous water attractions in the county. The pool has the advantage of trees, grass and Mediterranean architectural accouterments. You have to adore stucco; the perpetual shriek of children; a paucity or refreshments and seating.

The pool is less crowded than the train and bus that they take to get there and the sparse shade trees provide some protection from the subtropical sun, so you can count on staying for a little more than a half an hour. If that is not enough motivation, you only have to think of the throng at the beaches and hotel pools to be glad you are here. Between swims under the din of the children and during the stay in the shade they take time to talk.

"How are your father's plans for my murder coming along?" Roy asks without considering the effect.

"If I was not completely convinced of his motives that would be a very painful remark," relates Ali-ce a little distracted by the Earthly custom of showing your bellybutton in public.

"Sorry, but the same is true for me," says Roy without including bellybuttons in his habitual surveys.

"His target is more likely to be some form of technology that neither one of us know about but I have next to nothing in the way of clues," says Ali-ce resigned to ignoring the vulgar custom but uncompromising in her commitment to keeping her bellybutton covered in public.

"Has he given up on *Julius Caesar*?" Roy asks with a vague recollection of Ali-ce's sending or giving her father a copy of the book.

"Yes," Ali-ce reports. "He asked what I was reading and I told him *Romeo and Juliet*."

"I do not suppose that could be useful to his purpose?" Roy asks with a certain disdain for the sight of a passing pierced bellybutton.

"No, but it serves ours. If I gave him something useful, it would tell him that I know what he is up to," states Ali-ce wondering if Roy could be persuaded to wear a swimming suit that covers his bellybutton.

"You are so clever," admits Roy who never thinks of bellybuttons as gonads in spite of now knowing better.

"Yes, I am," divulges Ali-ce plucking some grass. "But I did not think of that until later."

"So have you sent it to him?" Roy asks looking out at the pool and its collection of people.

"Along with *Macbeth* and *Coriolanus* - I have asked him to keep them for me," adds Ali-ce pleased to notice the obvious that bellybuttons are less obvious in the pool.

"So he can read those without asking you for them?" Roy asks.

"More and more clever she grows," orates Ali-ce turning to Roy with a smile.

"Is that Shakespeare?" Roy asks as Ali-ce seems to relax for the first time.

"No, that is Ali-ce Mim-sey. Have you heard of her?" Ali-ce asks with a facetious arch to her eyebrows.

"One of my favorites," proclaims Roy returning the smile.

"One of your favorites?" Ali-ce asks with a burlesqued expression of disappointment.

"Why, Ali-ce, are you jealous of someone?" Roy asks.

"Not unless I have a reason to be," discloses Ali-ce with some of the burlesque fading.

"Do not pout," Roy stresses to save the mood. "You have no competition, no equal."

"Then what do you mean, 'one of my favorites'?" Ali-ce asks with genuine curiosity.

"It is a custom," says Roy still smiling. "When someone asks about an author with a reputation, it is a way to imply that he is familiar with him and he is one of many."

"So you have read my works and the works of others?" Ali-ce asks smiling again.

"You have written something?" Roy asks taking his turn at the joke.

"No," prompts Ali-ce wishing she could think of a way to continue the theme.

"But, if you did," urges Roy trumping her hesitation. "You would rank with the greats."

"No wonder you are a cosmic danger," construes Ali-ce pleased with Roy and the easy way he keeps his better moods. "You covered a lot of ground with just one remark."

"Actually, it is a cliche," explains Roy sitting up.

"So have you read Shakespeare?" Ali-ce asks leaning back on one arm.

"I have read about him," Roy clarifies. "He is easier for me to listen to than to read."

"I thought he was dead," scoffs Ali-ce steering Roy away from a sense of seriousness.

"He is?" Asks Roy. "What ever became of living three or four hundred years?"

"Thank goodness that does not happen here," gripes Ali-ce still leading.

"Why?" Roy asks pleased to follow and share in the amusement.

"Can you imagine how long his plays would be?" Ali-ce asks.

"Can you imagine the prestige of saying you read one?" Roy asks in pseudo earnest.

"So you have not read any of his plays?" Ali-ce asks.

"I read *Macbeth* in school. That was forty years ago. Do you suppose your father's target is still a head of state?" Roy asks embarrassed to return to the dreaded topic so easily.

"Duncan was a king," says Ali-ce trying to miss the point without changing the subject.

"Is the choice of weapons significant?" Roy asks self-consciously.

"Not to my father," concedes Ali-ce confidently. "I think it is more typical of the period."

"Your father does not have a penchant for knives?" Roy asks rhetorically.

"I do not know. But I think he is more concerned with the circumstances that attend the murders," connotes Ali-ce more to make the point than to answer Roy's question.

"Is he trying to avoid blame or responsibility?" Roy asks turning to look at the pool again.

"I think not," says Ali-ce to the pool. "The General is the one to blame not the troops."

"So what do you think he is trying to do?" Roy asks standing up to go to the pool.

"I think he is measuring the scale of the blame for murder," intimates Ali-ce joining him.

"He wants to justify murder or he wants to get a feeling for the guilt that goes with it?" Roy asks taking Ali-ce by the hand and walking down the steps into the chilly water.

"If he can justify it, maybe it mitigates his guilt," says Ali-ce putting her back against the side of the pool. "Maybe he wants to know what kind of burden that guilt is and if can bear it."

"Do you think he is looking for a way to avoid murder?" Roy asks.

"He does not need anyone's help to do that. He would avoid murder if he could," defends Ali-ce timing the waves that hit under her chin and canceling (*pii*) them with her throat muscles.

"And If he cannot avoid murder as an option, I could be on the list of targets?" Roy asks.

"Only because we do not know who is on the list," strums Ali-ce exercising her throat and imposing a vibrato on her voice.

"What would it take to find out who his target is?" Roy asks to an ever calming pool surface especially at their end.

"I do not think anyone here would know," modulates Ali-ce unconcerned with her pool leveling habit that goes back to her childhood.

"Why?" Roy asks becoming aware of the anxiety spreading among the confused revelers.

"I guess no one will know," says Ali-ce in an innocent staccato, "until murder is the only option."

"So we will remain clueless," says Roy glad, at least, that the source of the calm water is unknown while the psychological agitation it creates drives people from the pool.

"And we will stay that way as long as I am here," thrums Ali-ce taking some pride in her aquatic skill.

"You need to go home again," says Roy with the suspicion that Ali-ce should not *pii* in the pool.

Episode 51

Dr. W-abe did not know, until now, that they had one, but there is no rule that says the Robes Chamber cannot have a solarium. There is no rule that says a Robe cannot do business in the solarium. So it is that Dr. W-abe is going there to talk to 1101th-Robe.

Another thing Dr. W-abe does not know is where this amenity is. This is the disadvantage or, maybe advantage of the matrix elevator. He could be in any quadrant and on any floor. He does not have to know 'where it is' to press the button marked D-V.

Dr. W-abe finds 1101th-Robe wearing dark glasses and light shorts. The light spectrum in the room varies according to 1101th-Robe's prescription. Dr. W-abe accepts a pair of protective glasses but declines the opportunity for a vitamin boost. The effect of a boost, of course, depends on the duration of the conversation in this case and a few other things in almost every other circumstance.

"I suppose ocean exploration teaches a lot about contained environments," murmurs 1101th-Robe disturbing the lotion on his abdomen enough to chase an itch, "and that translates into some of the practical requirements of space exploration."

"Of course, the details are endless and progress is slow and uneven," discloses Dr. W-abe wondering why the process of boosting is couched in such a cavernous room.

"Any key information?" 1101th-Robe asks plucking an aromatic leaf and sniffing at it.

"Our combination of metal and glass serves us well but the earliest dive and recovery was very telling," says Dr. W-abe trying to estimate the size of the room and the height of the ceiling.

"I recall it had its tragedy," evinces 1101th-Robe trying not to move his limbs or disturb his coating of lotion.

"Yes, we used the same combinations of metal and glass for the hatch as well as the hull of the dive vessel," says Dr. W-abe reaching back across time to reconstruct a more specific memory.

"That would seem reasonable," suggests 1101th-Robe facing a new point on the compass.

"It was - until the hatch would not open," advises Dr. W-abe jumping to the important fact.

"There was not enough life support before you got the hatch off?" Inquires 1101th-Robe.

"Maybe, but they decided to drill through the hull to provide oxygen while they solved the hatch problem," explains Dr. W-abe anticipating the next question.

"I suppose this gave you, uh them, more time to find a solution without stressing the ship, the explorer and the recovery team," inserts 1101th-Robe skipping a myriad of details as well.

"That was the thinking but we did not know how the enormous pressure compressed the ship and that it had not recovered to its normal size," declares Dr. W-abe expecting incredulity.

"That is why the hatch would not open?" 1101th-Robe asks.

"It shrank by the same amount as the rest of the ship," says Dr. W-abe again avoiding specifics.

"And the argonaut?" 1101th-Robe asks shifting to a new point of the compass.

"Presumably, the explorer shrank by the same amount," maintains Dr. W-abe hoping that this much information on a complex point would do for the moment, "so his relative size was the same."

"It could not have been much shrinkage," contends 1101th-Robe anticipating the gap between ordinary and scientific reality. "That would have been visible and something of a shock."

"Not nearly the shock we got when we drilled though the hull," rallies Dr. W-abe.

"Oh?" 1101th-Robe asks supposing he had not anticipated enough.

"The ship imploded," recounts Dr. W-abe with a wince. "I do not know how much of it turned to gas, heat and dust but what was left was shiny sand and sparkly gravel."

"The passenger?" 1101th-Robe asks harvesting a new leaf for its feeble distraction.

"He was pureed," discloses Dr. W-abe trying to be responsive without being graphic.

"What would cause an implosion or an explosion like that?" 1101th-Robe asks moving on verbally.

"It took many years and lots of theory. But the extreme pressure at the depths of the ocean has its parallels with extreme acceleration and gravity," offers Dr. W-abe. "Part of the mass and energy of the ship compresses, in this case, slightly into dimensions beyond the normal three that we know."

"So this was compression on the subatomic scale?" 1101th-Robe asks following the theme tentatively.

"Yes, but an incredibly small amount. In fact, it cannot be measured by any instrument, and it is distributed among all the atoms of the ship and its interior," commits Dr. W-abe intending to reassure 1101th-Robe of the quality of his supposition.

"And the decompression?" asks 1101th-Robe turning to a new point of the compass literally and figuratively.

"Like a bell ringing itself apart. Like drilling a lightbulb, sawing an egg," tenders Dr. W-abe with a sense of finality.

"What was the solution?" 1101th-Robe asks gingerly moving the conversation forward.

"Part of the solution was to use dissimilar metals in the glass hatch and glass hull," proffers Dr. W-abe wishing for something to drink but settling for plucking another leaf. "That way physical and atomic changes in the smaller dimensions would not make the hatch 'grab' in a normal environment."

"I suppose that was the easy solution," opines 1101th-Robe following the emerging ideas.

"Actually, no. The shrinkage problem was easier to solve," says Dr. W-abe relieved by the leaf's fragrance and the turn of conversation. "We did that with microwave noise and laser noise."

"You can guess that I do not know what that is," says 1101th-Robe. now facing Dr. W-abe.

"That just means that there was a lot of variety in the microwave and laser energy directed at the ship," reveals Dr. W-abe wondering how he could find one of these plants for his balcony.

"And that ironed out the wrinkles in the atomic shrinkage?" 1101th-Robe asks.

"Yes," Dr. W-abe offers, "and we could test it without sending anyone down with the ship."

"Now that you knew what to look for you could send probes down, recover and test them," proposes 1101th-Robe wishing to move the conversation on to his next concern.

"And apply the atomic-relaxer from the outside," extends Dr. W-abe patiently.

"So now we were ready to explore space?" 1101th-Robe asks almost impatiently.

"Actually we already did that," discloses Dr. W-abe while offering 1101th-Robe a new leaf. "This just prepared us to travel at higher speeds and closer to black holes and white holes."

"I see. I think," judges 1101th-Robe pondering the nexus of the elements of pressure, acceleration and gravity.

$$1$$
$$1\ 1$$
$$1\ 2\ 1$$
$$1\ 3\ 3\ 1$$
$$1\ 4\ 6\ 4\ 1$$
$$1\ 5\ 10\ 10\ 5\ 1$$
$$1\ 6\ 15\ 20\ 15\ 6\ 1$$

Chapter 18

Chapter 01001000....

Chapter BE

Chapter $(1 + 5x + 8x^2 + 4x^3 + x^4)$

Chapter $(a^4b^0 + 5a^3b^1 + 8a^2b^2 + 4a^1b^3 + a^0b^4)$

Chapter $2^1 + 2^4$)

Episode 52

It is early evening and Dr. W-abe is still at the office. Dr. Mim-sey has just arrived and reminds him of the rare book show. It is in a museum in the building and today is the last day. W-abe knows about the show because the media has mentioned some early originals in math and physics. He would like to see them if it does not take too long.

Mim-sey has seen the show twice already and wants to go one last time. Dr. W-abe is counting on Dr. Mim-sey's saturation to keep the visit short. They can go home from there and it will not turn into an expedition. A quiet dinner and drinks, with or without Dr. Mim-sey is always an option.

If the museum visit is brief, then Dr. W-abe will not mind dinner and drinks at a convenient hotel. Otherwise, he can plan on making excuses and have dinner and drinks alone. Mim-sey keeps W-abe's prerogatives in mind like no one else. They are not alone on the matrix elevator going to the museum. But the museum is virtually empty so conversation, there, is easy.

"Even if Mr. Abbit publishes his collection of number lines or strings," says W-abe, "the pedestrian mathematicians that succeed him will be in the same boat as our erudite theoreticians."

"No doubt that includes you. But where is the similarity? His threat is just the proliferation of key notions. Is it not?" Mim-sey asks at an antique book binding display.

"A collection of similar theories tells us that we do not have 'the' theory. Whatever that is. If we want to knit together time, space, gravity, the particles and the forces that comprise one universe or another we will have to knit together the theories that describe them - T. O. E. - a Theory of Everything. Can Mr. Abbit do that?" W-abe asks while looking up and down for the antique editions of math and physics he came for without success.

"He will never take a cruse on the Great Ship T. O. E. but he **does** have an overarching definition, if you will, that applies to all five versions of the strings or number lines he describes," acknowledges Mim-sey confident that this stimulates W-abe's curiosity enough to sustain the conversation.

"His task is significantly simpler but seems similar - defining an overarching symmetry. This would put the five forms of number lines in a family," says W-abe looking at a book on foamed glass foundries and mills. "The patriarch would depend on which form is most useful to one application or another. It is that family of uncles that may hasten our demise and our cry of 'uncle' that comes with it."

"Let go of your dread for a minute," rants Mim-sey wondering if W-abe likes the esoteric book enough to buy it. "Take a look at his overarching arithmetic."

"Without that dread, I could be amused by the concept. So tell me, "What is his 'overarching arithmetic'?'" W-abe asks trying to remember whether or not he has this book on foamed glass.

"All of his strings are counting models," says Mim-sey as he launches his explanation. "The most basic form is binary."

"Except for my latter day habituation to base ten, I can agree," abides W-abe deciding not to buy the book and replacing it on its shelf. "There are just the two numbers - one and zero."

"But every decimal number can be represented as a binary number. As you count, you get row after row of binary numbers," continues Mim-sey. "Each row is one unit larger than the last."

"There is nothing new in that," prevails W-abe still hoping to see the antique math books.

"What is new is that ones and zeros are categories," argues Mim-sey dragging his feet.

"No," says W-abe trying to move Mim-sey along by stepping away from him. "Categories are not new."

"A thing is a chair or it is not a chair. There is nothing new in that. A thing is 'A' or it is 'not-A' is the logical expression. This is not a new concept," says Mim-sey to a thin red volume.

"So what is new?" W-abe asks giving up on herding his friend along.

"In the history of math and logic, the combination of things is represented as a combined category," delineates Mim-sey while handing W-abe a book on gas impregnated glass. "A noun and its noun modifiers can be represented as one combined thing - the noun."

"A true statement together with a false statement is the equivalent of either a true statement or a false statement depending on the rules for combining the two. This is done by assigning a truth value of 'one' to true statements and a value of 'zero' to false statements. What is new there?" W-abe asks returning the book to Mim-sey. "The sum is a truth value of one or zero no matter how many statements are combined."

"By assigning a category value of one or zero to things, you can use the same logic and arithmetic for the words in a statement," claims Mim-sey returning the book to the shelf.

"That takes guts," affirms W-abe grateful to finish the circuit of this room. "But all you have to do is prove that a thing is 'A' and 'not-A' at the same time and this theory of categories falls apart."

"If you say category-A is equal to one in binary numbers then what is the logical, binary, category value of anything that is not-A?" Mim-sey asks moving to the next room reluctantly.

"It would have to be zero in binary numbers," asserts W-abe impatiently surveying a room full of travel books. "But you could just as well say that category-A has a category value of zero."

"You cannot change the value of category-A...," avers Mim-sey looking around hopefully.

"Yes I can," predicates W-abe, on his way to the next room with or without Mim-sey. "It can be either one or zero."

"Listen without interrupting," implores Mim-sey not more than a couple of steps behind. "You cannot change the value of 'category-A' without changing the category value of 'not-A,' If you make category-A equal to zero then by definition 'not-A' becomes a value of 'one'."

"This is so new it makes me dizzy. A category can be assigned a value of one or zero arbitrarily. If you tried that with true and false statements no one would know what you are talking about," lectures W-abe without slowing down for the gardening gallery. "Sometimes a true statement has a truth value of one and sometimes it has a truth value of zero - impossible."

"So let us keep the distinction between category values and truth values like Roy Abbit does," says Mim-sey.

"There is nothing true or false about rocks. You can make true or false statements about rocks. You may even be able to make some statements about rocks that are difficult say whether they are true or false," indicates W-abe gliding through the anthropology and geography exhibit rooms.

"So it seems a little simpler or easier to say rocks are rocks and things that are not rocks are not rocks," poses Mim-sey at least able to walk beside W-abe as he slows his pace.

"In that case, it is not so difficult to say that if the category value for rocks is one or zero," offers W-abe looking for an exit, "then the category value for everything else is the opposite value."

"This is new but not impossible," Mim-sey advances, feeling grateful that there are no exits handy.

"You do not get very much when you add these ones and zeros together," alleges W-abe.

"If you combine the ones and zeros logically you have something," Mim-sey says hoping W-abe is finally on the right track.

"You have two ways to combine the numbers," presses W-abe slowing down for some visibly older editions but finding only theologies and metaphysics. 'One **AND** zero' is different from 'one **OR** zero'."

"Just like the truth values, there are four combinations of ones and zeros when you **AND** them," says Mim-sey with the temptation to linger, "and there are four combinations when you **OR** them."

"Let me see: (1 and 1) = 1; (1 and 0) = 0;" declares W-abe, "as well as (0 and 1) = 0 ...(0 and 0) = 0."

If we use the plus sign for 'and' and the asterisk for 'or', they look like and work like mathematical formulas. "When you 'or' ones and zero you get: (1*1) = 1; (1*0) = 1; (0*1) = 1 ...(0*0) = 0," illustrates Mim-sey pleased to enter the stately literature room.

"I would be more impressed if we could get them to perform like mathematical formulas," vents W-abe afraid that he is going to be here awhile as Mim-sey focuses on the titles.

"It is as easy as 1, 2, 3. When you count with binary numbers, it is a combination of ones and zeros. If you call the ones 'A' and the zeros 'not-A,' can you think of a formula that is true for all possible combinations of ones and zeros?" Mim-sey asks as an open challenge to W-abe's ego.

"Then, Can you make the formula work when you reverse the ones for the zeros and the zeros for the ones?" W-abe asks responding to the challenge without relinquishing a mote of his original position.

Episode 53

There are still a few hours of sun but nowhere near a few more hours of ambience at the pool. If Roy and Ali-ce return to this historic attraction, it will be a while. Roy and Ali-ce are not hungry yet. That makes it hard to decide what to do for dinner.

One of the last southern cafeterias left in Miami is near the transfer point for the bus and train. It is a vestige of old southern style. The restaurant is like a cafeteria. You get in line. Pick up a tray.

There is no missing the counter with dessert. If you have lived in the south all your life, you have no trouble making up your mind. You choose one of the deserts you have always had fruit, pie, cake. Then you choose an entree that is entirely familiar, beef, chicken, pork, then the vegetables, your lifelong favorite, of course, potatoes, green beans, corn-on-the-cob. The breads and beverages are also eternally typical - dinner rolls, ice tea, coffee.

A cashier tallies your choices and finally the southern touch - a waiter. This is a cafeteria where you do not carry your tray, the waiter does. If you have been there before you know to go to a table of your choosing, if not, you follow the waiter. He sets your table, you tip him.

The charm is in the familiarity, the style of cooking, the size of the servings, the impossibly low price and the touch of service that does not require any sophistication on the part of the patrons. Almost everyone takes too much and it leaves them overfed. The entire experience is almost as if you were at home. Dinner is tranquil and full of conversation.

"You have to find an excuse to go home," says Roy taking a tray and giving one to Ali-ce.

"I know," offers Ali-ce looking for something that is not too sweet. "But I do not want my father to connect my return with my prying into what he considers a complete secret."

"You do not have to know who he must murder," says Roy choosing a slice of cherry pie.

"I thought I did," says Ali-ce picking a green gelatin with whipped cream. "We both do."

"If I am not his target," says Roy taking some brisket, "we do not need to know who is."

"Oh, that could make it easier," prompts Ali-ce eyeing mashed-potatoes, corn-on-the-cob, beans, beats and yellow rice. "But if it is about you, he could be very sensitive to my probing."

"That would tend to support the possibility that I am the target; not some head of state or technocrat," articulates Roy planning on mashed potatoes, succotash, extra bread and extra butter.

"I suppose I could find a way to talk about heads of state and technology," says Ali-ce taking ice water and unsweetened tea. "But I am not good at small talk. And he is impossible."

"Impossible?" Roy asks while choosing unsweetened iced tea and an extra glass of ice.

"He never makes small talk. He speaks to conclusions," says Ali-ce as the waiter presumes to give her some hot sauce. "Every word that comes out of his mouth is intended to support a conclusion."

"So, How do you know what he is talking about half of the time?" Roy asks.

"You learn to look for the conclusion first. If it is not obvious, you can ask him," declares Ali-ce putting some hot sauce on her beans. "I ask, 'Where is this going?' or something like that."

"Does it bug him when you do?" Roy asks putting some hot sauce on his brisket.

"Bug him?" Ali-ce asks taking some of Roy's bread and butter.

"Does it distract or irritate him when you do not know where he is going with his remarks?" Roy asks while buttering his bread and adding some sugar and ice to his iced tea.

"No," says Ali-ce, "because it happens so often that he has to take responsibility for it."

"So you will have to approach the subject from some angle that does not point to the topic of murder," urges Roy as he is ready to begin eating.

"And I have to get him to call me home," reasons Ali-ce putting some lemon in her water.

"That will take a miracle. He wants you on my tail," says Roy before a bite of brisket.

"That is it - Friar Laurence," concludes Ali-ce while buttering her bread.

"What is it?" Roy asks taking a sip of iced tea. "What about Friar Laurence?"

"You said we need a miracle," deliberates Ali-ce as she begins to eat. "It was Friar Laurence that married Romeo and Juliet."

"He thought it would put an end to the conflict between the Capulets and Montegues," recounts Roy with a taste of mashed potatoes.

"It did," says Ali-ce turning to her beets. "But not with the happy ending he hoped for."

"Are you proposing marriage?" Roy asks dabbing his gravy and potatoes with some buttered bread. "That would have your father call you back really quick."

"I think we are ready. But as you say, my father is not ready for us to marry," laments Ali-ce buttering and salting her corn-on-the-cob.

"You are going to become a Friar?" Roy asks cutting a bite of brisket.

"I do not need to change gender," says Ali-ce dabbing her fingers with a napkin. "I can become a nun."

"Sister Ali-ce. It sounds good. But how is that going to help?" Roy asks before sipping his tea.

"I will still have a pretext for associating with you on my father's terms," remarks Ali-ce mixing her rice and beans together before taking a bite. "So that will not raise any suspicions."

"By playing his game, maybe you will see some of his cards?" Roy asks salting his brisket.

"Maybe not, but it will give him and me a chance to argue," relates Ali-ce. "If we are lucky, many chances."

"Each or most of the arguments can be used to probe his motives regarding me and murder," imparts Roy taking a bite of succotash and bread.

"If you are not the target," says Ali-ce, "he and I can still argue about murder and such."

"So your motives with regard to me will remain undisclosed," marks Roy.

"He and I can argue at length; up close and at any distance," says Ali-ce with mashed potatoes on her fork.

"From time to time you can turn up at home and continue the arguments," conveys Roy.

"And whether or not he calls me back," declares Ali-ce, "will not raise suspicions."

"And you can make the excuse to return to me anytime that you want," asserts Roy.

"That is the best part," stipulates Ali-ce turning to her dish of beets.

"He might even be grateful for your return to Earth," says Roy.

"The earth that is nature's mother is her tomb; what is her burying grave, that is her womb; and from her womb children of diverse kind, we sucking on her natural bosom find, many for many virtues excellent, none but for some, and yet all different," recites Ali-ce.

"Friar Laurence?" Roy asks returning to his succotash.

"Act two, scene three," cites Ali-ce with a bite of mashed potatoes.

"Do you care to tell me what he means?" Roy asks.

"I do not know all of what he means," deliberates Ali-ce. "But I know what I want it to mean."

"Okay?" Roy asks putting some brisket and succotash on his fork.

"Earth does not have to be our tomb," says Ali-ce taking a turn at her beans and rice.

"At least not prematurely," urges Roy with a bite of brisket.

"Earth is nature's womb and so will sustain us," says Ali-ce choosing to taste her beets.

"Like it naturally sustains various kinds of plants," weighs Roy sipping some tea.

"The same nature or earth that supports the plants supports their many virtues...," reflects Ali-ce dunking her buttered bread into her mashed potatoes.

"...regardless of differences," spurs Roy reaching for his cherry pie.

"I like that," says Ali-ce ready to give her undivided attention to her corn-on-the-cob.

Episode 54

Before responding to Dr. Mim-sey's challenge, Dr. W-abe takes advantage of the hiatus in the conversation. Dr. Mim-sey is rethinking his advantage specifically whether or not he has one. Dr. W-abe is recalling a meeting with 1001th-Robe.

Just because the circumstances are tenuous, it does not mean the occasion or the conversation is not substantial. Both Dr. W-abe and 1001th-Robe are on time and at the Robe's office. But Dr. W-abe is soon asked to join 1001th-Robe while he makes the connection to his out of town appointment. At least it is not a phone chat. That might make one or the other wonder whether the meeting is significant. That speculation soon disappears.

"So the benefits of ocean exploration include a ship that is worthy of the environment and one that is not limited by gravity and acceleration," prompts 1001th-Robe at the matrix elevator.

"That put us down the road a long way. It is possible that the issue of dimensional changes would never have been solved without ocean exploration," relates Dr. W-abe.

"I suppose ships would have otherwise just disappeared out in space and there would be no clue to the cause," states 1001th-Robe as they enter the elevator and 1001th-Robe presses 'Lev-train'.

"We would have had to settle for a threshold below the speed of light and far from black holes, worm holes and white holes," declares Dr. W-abe hoping for enough time to explain even a small part of what he is leaving out.

"But I am not sure I know enough about these effects to follow any discussion that involves them," submits 1001th-Robe aware of only a few minutes to the magnetic levitated train station.

"I will try to keep it simple, offers Dr. W-abe, "when we get to those topics."

"Thanks," says 1001th-Robe while counting on a few minutes wait before boarding.

"No doubt you could appreciate a great deal of the detail but we do not have enough time to do them justice, so to speak," says Dr. W-abe just as mindful of how little time they have to talk.

"I presume we are still at the stage," theorizes 1001th-Robe looking at the path indicator, "where we send out a ship, leave equipment and return with a small ship that saves the astronaut."

"We travel great distances, out and back. We achieve great speeds and see very many stars but see very few planets because of the enormous scale of the universe," summarizes Dr. W-abe.

"How did this prepare you for a solution to the scale of the universe?" 1001th-Robe asks.

"In order to go farther, faster," says Dr. W-abe wondering if anyone will get on the elevator on the way to the lev-train station, "we had to slingshot past large stars and black holes to accelerate to higher speeds with less onboard energy."

"We had to get the astronauts back," says 1001th-Robe moving the conversation along inconveniently fast. "I suppose they could not use the same stars and black holes both ways because each neighborhood of space is unique."

"It took lots of planning and math to get out and back but it was done for years; for several hundred years but without recognizing Earth's environs," says Dr. W-abe disappointed to see the elevator doors open. "That gave us a huge inventory of black holes and a good sense of the groups of galaxies in the universe."

"I suppose, if you did not know where you were exactly, you could still tell, more or less, how to get back," submits 1001th-Robe allowing Dr. W-abe off of the elevator first.

"Actually you are right," prompts Dr. W-abe. "When we got to a new part of space. We could count on looking around until we found a galaxy grouping that was recognizable, backtrack to it and then backtrack to even more familiar surroundings until we found home."

"Did we ever go fast enough to need the lasers and microwaves to resize the ship?" 1001th-Robe asks on the way to the Lev-train station.

"Yes and it turns out that changes in the size of the ship, especially in the direction of travel, while we were at top speed became a measure of how fast we were going," reveals Dr. W-abe.

"I can guess that the better we could measure the speed then the better we could measure the distance," surmises 1001th-Robe entering the station, "and therefore the better we knew where we were."

"But we never could go fast enough or far enough to visit much of the 12 billion light-years of the universe that we could see," discloses Dr.W-abe as he wonders how many more seconds of conversation he has.

"So, eventually, we knew we would have to test the wormhole theory," weighs 1001th-Robe trying not to walk any faster than necessary in order not to impose on the conversation.

"If the wormhole could be survived," states Dr. W-abe watching the passengers' platform, just ahead, "it still would be difficult to know where we were at distances of thousands or millions of light years."

"That was a problem we could solve, no doubt," says 1001th-Robe looking at the clock.

"We have a twelve billion-light year view with many landmark galaxies and clusters," delivers Dr. W-abe. "The problem became how to find a wormhole that would bring us back."

"We only had the experience of deep oceans at that time," proposes 1001th-Robe.

"So we seized on that experience with a twist. We knew that we would exit the wormhole at a high speed," uttered Dr. W-abe disappointed that there is a train at the platform. "But, at the same time it would take too much energy to slow down and return through the same one."

"So we sent back our only option - a smaller ship - through the same worm hole?" 1001th-Robe asks.

"Actually a series of small ships. But none with astronauts aboard," says Dr. W-abe observing the blue illumination which signals time to board.

"So the one-way trip was born," says 1001th-Robe nearing the middle of the long platform.

"It is humbling to think about; but it was the price to pay. If the astronaut found a habitable planet, he could live out his life there," says Dr. W-abe looking at the hotel size, polychromatic, glass cars and hoping Mim-sey does not see something like this. "The eggs he launched back along the way would delineate his path."

"The successes galvanized our world into one culture," says 1001th-Robe stopping at what must be his car, "and we were able to benefit from a network of ships and habitable planets."

"But the planets we found were like our deep oceans. They included amazing creatures, plants and physical environments but no civilizations. Their environments could support us for the stay there but they could not contribute to our travel or return. Except in the form of replenishment," says Dr. W-abe when the illumination turned to red signaling last call to board.

1001th-Robe steps on board, faces Dr. W-abe and says, "Do not be disappointed. Your contribution is invaluable and essential. Your next appointment is already scheduled. Change it if it is inconvenient," as the door closes.

1

1 1

1 2 1

1 3 3 1

1 4 6 4 1

1 5 10 10 5 1

1 6 15 20 15 6 1

Chapter 19

Chapter 11001000....

Chapter ABE

Chapter $(1 + 5x + 8x^2 + 5x^3 + x^4)$

Chapter $(a^4b^0 + 5a^3b^1 + 8a^2b^2 + 5a^1b^3 + a^0b^4)$

Chapter $2^0 + 2^1 + 2^4$)

Episode 55

Dr. W-abe knows which books on mathematics and physics he can expect to be at the book show. He is sure it will take longer to find the books he wants to see than to look at them. The thrill will be over in a moment, if it happens at all. Confirmation of disappointment will replace his expectations. Disappointment will eventually eclipse hope for delight.

He will remember the show and the museum with the same dull disappointment whenever and every time he thinks of it. Yet another certainty will be that it will be that much harder to get him to a museum or to a book show or to a book-anything the next time, if there is a next time.

Dr. Mim-sey is trying to convince him of the quality of the show but Dr. W-abe is already thinking about home, food and drinks. Dr. W-abe is steadily working his way toward a display area that should be his last disappointment. His duty is done to his friend and his predecessors.

He is not disappointed in his anticipated disappointment. It is a shame to be right about a negative expectation. Dr. W-abe searches the walls for the one display he knows should be there and he cannot find it. This can only mean they do not have it. W-abe is used to finding what he wants. He turns to leave. Dr. Mim-sey can follow or not.

Then, there it is, out on the floor. In a cabinet all its own. It is so well lit the glass envelope does not inhibit viewing. Someone good got the job to show this one. You can see the two pages of formulas that describe and predict the existence of the Q-particle.

There are Bodumnloi notes in the margin in the author's hand. He is still tinkering even after the book is published. Dr. W-abe knows the burden. He reaches 600 years through time and sees the author's struggle.

There is a partial separation at the top of the page. He can feel the author searching for that page and the moment he finds it. The satisfaction of finding it is sullied by the accidental tear and it will nag at him. But he goes on with his intention to firm up the formulas and writes right there.

Now Dr. W-abe is in a rare good mood. The show is finally worth it. Dr. Mim-sey waits until Dr. W-abe loosens his concentration on the book before he resumes the conversation with his challenge of an overarching formula for the five counting schemes.

Dr. Mim-sey asks, "Yes, given logical 'and' and given logical 'or,' can you think of a formula that is true for all binary numbers?"

"Well, let us see 'A' equals 'something, something' which equals one," verbalizes W-abe while walking around the book tower. "'Not-A' equals 'something, something' which equals zero."

"That will take care of the category values of column 'A'," utters Mim-sey matching Dr. W-abe's circuit around the prized book. "Decimal two involves column 'B' but we can put that off until we finish the definition of 'A'."

"We can replace the word 'not', if I write 'not-A' as '|A' then the formulas begin to take the shape of: $A = \{\text{something, something,}\} = N$. And: $|A = \{\text{something, something,}\} = |N$," expresses W-abe. "'N' and '|N' are opposite values."

"You are onto something. I think," says Mim-sey stopping with W-abe at the cabinet front.

"I am not done, yet," conveys W-abe staring at the historic book. "A thing is equal to itself, no matter what, so I can make the formulas: $A = \{ A, \text{something}\} = N$ and $|A = \{|A, \text{something}\} = |N$."

"The logic of 'and' and 'or' are symmetrical and opposite just like 'A' and '|A' so you can use that," formulates Mim-sey waiting for his friend to move away from the cabinet.

"No doubt, I will use 'and, +' and 'or, *' more than once: $A = \{[A*...] \text{something}\} = N$. $|A = \{[|A +...] \text{something}\} = |N$," attests W-abe moving from the cabinet but not leaving the room.

"I wish your friends could hear you," trifles Mim-sey keeping a conversational distance.

"They cannot and they will not," vows W-abe looking at technical books lining the walls. "They will never know we had this conversation."

"Okay, since binary numbers are ones and zeros, part of the definition must include some form of (A + |A) or (A * |A)," admits Mim-sey as W-abe picks up one book, then another.

"The two formulas are logical complements of each other, so let us try A = {[A * (A + |A) something} = N," says W-abe opening an antique volume. "|A = {[|A + (A * |A) something} = |N."

"The two formulas are symmetrical. Which is what they have to be to get symmetrical results. What else can they have in common?" Mim-ey asks looking at W-abe's selections.

"Picture this. If you draw a circle and label it 'A' what have you got?" W-abe asks without opening the book in his hand.

"Everything outside the circle is 'not-A.' What is the distinction?" Mim-sey asks unable to guess the significance of his friend's book find.

"That area outside the circle can be infinite," says W-abe noting that the binding is leather and therefore must be from Earth.

"What does that mean for binary numbers?" Mim-sey asks trying to see the pages now that W-abe has opened the book.

"You can attach as many zeros to end of a binary number as you want without changing its value. Remember?" W-abe asks estimating that the book is about 120 Earth years old.

"Then it does not matter whether you 'and' or 'or' all those zeros they still equal zero," says Mim-sey supposing that the book is an English translation of an early work on mathematics.

"We can represent those combined zeros as '0...' the ellipsis indicates an infinite series of zeros and it does not matter how you sum them. The formulas become A = {[A * (A + |A)] something 0...} = N," says W-abe turning pages. "|A = {[|A + (A * |A)] something 0...} = |N."

"That last something is '+' or '*'. All we have to do is plug in ones and zeros for the 'A' and the '|A' and whatever works; that is it. Go on. Try it," agitates Mim-sey becoming jealous of the book Dr. W-abe is holding.

"The two formulas only work with '*' as the last logical operator," advises W-abe.

"But to work for any value of 'A', one or zero. You just have to remember to make '|A' the opposite value," says Mim-sey. "Now that you have the formula. You have the theory."

"Abbit's theory of the several number lines?" W-abe asks picking up the prospectus for the book he is still holding.

"You are mean, if not jealous," quibbles Mim-sey.

"I will never admit to being jealous," says W-abe as Mim-sey picks up the same prospectus to see if, perhaps, there is another copy of this book. A book that Dr. W-abe is almost certain to buy.

"Well, 'Dr. Qapu,' how about a 'Theory of Categories,'" wrangles Mim-sey finding to his frustration that there is only one copy for sale.

"Let me see," says W-abe. "Category = {[Category and/or (something)] or Universe of zeros}. That 'something' needs a name."

"That 'something' in the middle is the set or subset that is the topic of discussion," says Mim-sey reading that Dr. W-abe's pending purchase is translated from Greek. "In this case 'A' and '|A' are 'the set of discourse'."

"The theory of categories becomes: Category = {[Category and/or (Set of discourse)] or Universe of zeros}," iterates W-abe learning from the prospectus that the original text was lost for more than 1,000 Earth years.

"When translated from math-speak, the theory becomes 'A category is equal to itself, its set of discourse and a universe of zeros'," articulates Mim-sey despairing of learning Greek or finding an original version of that book to trump the one W-abe is holding.

"It would be easier for me to like this theory of categories, if it were just a little more relevant," contends W-abe wondering what those missing years must mean to Earth, if not his world.

"On the one hand, you are right. If we are going to define a category with a decimal value of two or more, we will have to add more rows and columns," says Mim-sey. "We have provided for that with the set of discourse. A larger of set of discourse will include 'A,' 'B,' 'C' and so on."

"That is a help," assents W-abe inserting the prospectus into the sales register and keying in his personal data. "That means that the theory of categories applies to any number line of any length."

"But a number line, by definition, can satisfy only one dimension," Mim-sey advances as a caveat. "It will take a combination of number lines to include two or more dimensions."

"I think Mr. Abbit is asking a lot of his five counting schemes," declares W-abe.

"And his success is our doom," protests Mim-sey while he resolves without much hope for success to send a message to Ali-ce about the next book she must send him.

"Neither his success nor our doom, if we can help it," maintains W-abe who now has his own copy of *Archimedes Method* in English. A language he just happens to understand almost as well as his native Bodumnloi.

Episode 56

It is just sunset and the day's heat is already leaving. Without deciding to, Roy and Ali-ce decide to window shop on the way to the train. Some of the better stores were there before the train came to town. So the way to pedestrian conveyance is studded with haute couture et., haute etc.

They both ate too much but neither of them feels like they spent too much. Neither one likes to throw away food. He is always conscious of how much of his food depends on money. She is always conscious of the seriousness of never running out of food on the ship.

This part of the city has been renovated. There are more trees which are a relief to the eye and spirit. But somehow they managed to add more concrete and asphalt. The heat, nine months out of the year, is too much for most green spaces. That takes too much water.

Water from a rainy season, six months of the year, is not collected in cisterns for the green spaces. Up to 60 inches of rain a year, is efficiently directed into storm sewers and then to the bay. Therefore the trees - they can survive on the little amount of rainwater that penetrates the iron grate surrounding their trunks.

Of course extra expertise and expense are justified for the iron grates. They must be just as aesthetic as the flowers and ornamental plantings they are supplanting. The fascination of the store windows quickly displaces any interest in the landscaping. The will to conversation eclipses the distraction of the windows.

"So what was Friar Laurence doing at the time of his discourse on Earth as both tomb and womb?" Roy asks as they cross the street to the book store that used to be a 10 cent store.

"He was collecting herbs and medicinal plants. It was early in the morning. Romeo had just come from Juliet's garden to tell Friar Laurence about his plan to marry," says Ali-ce.

"An excellent piece of stage craft," says Roy recalling that the 10 cent store, here, was closed during a downsizing. "The audience does not have trouble accepting what seems to be a natural event."

"It was, in reality, neither believable nor natural. A Capulet marrying a Montegue?" Ali-ce asks remembering that the urban legend has the store manager holding a 'going out of business sale' by mistake.

"I guess, you are right," says Roy. "That is a total reversal of the story up to that point."

"The closest of ties for the greatest of enemies," says Ali-ce recalling that the 'sale' was discovered too late by corporate managers. "A feud of generations ended in the few minutes of a marriage ceremony."

"And Shakespeare does it with the convergence of the two circumstances. Romeo did not waste any time getting to Friar Laurence. Friar Laurence was already up and about. He naturally had to be," credits Roy with the little he knows. "Collecting cuttings has to be done in the cool of the morning."

"There was nothing to overcome," says Ali-ce admiring the larger books in the window but doubting the urban legend of the former tenant. "Romeo did not have to work up his nerve. And Friar Laurence did not have to shake off sleep or postpone any of the day's business to focus on the plan."

"He did not have to set up the audience," says Roy lingering at the window with genuine interest.

"They were already there with him," affirms Ali-ce knowing Roy will not go in the store.

"So by becoming a nun you are going to reverse our story?" Roy asks openly admiring the books and resisting the urge to buy any of them.

"Not just me," says Ali-ce feeling Roy's conflict. "You are going to reverse the story by not promoting your manuscript until we find out what is behind my father's mission of murder."

"It is a kind of marriage," says Roy.

"We are uniting to settle our father's strife," says Ali-ce resolving to get Roy more books.

"Our 'father's strife' being our competing worlds?" Roy asks.

"Is this not what this is all about?" Ali-ce asks looking at a bridal store window.

"It cannot be about us?" Roy asks looking at the tuxedos for a few seconds before exhausting his interest.

Ali-ce says, unable to choose which dress she likes the best, "Not if our two worlds are going to annihilate each other."

Roy acknowledges, ready to move on, "It was 'all' the Capulets against 'all' the Montegues."

"That was the story," says Ali-ce, considering the improvement that a few bags of diamonds would make on one of the gowns.

"But how can it not be about us?" Roy asks deciding that the dress with rhinestones is just too expensive.

"You are right," says Ali-ce reading the card in the window that says 'Mrs. Xie by appointment'. "But if we are going to have a happy ending, we have to make sure you are out of danger."

"There is another advantage to this," says Roy glancing at the Cuban flag next to the card.

"What is that?" Ali-ce asks wondering if Mrs. Xie speaks English, Spanish and Chinese.

"Suppose your father and whoever he is working for becomes convinced that I am not a threat. Your career as a nun will be an excuse to stay here," says Roy as they approach the Porsche dealership.

"You are right," confirms Ali-ce who is always amused to see a car inside a building.

"And getting married will not be the excuse," vexes Roy.

"But if I have a good reason to stay," defends Ali-ce as she hears the rain begin, "then getting married along the way can be understood."

"He may never accept it but he will see it as a tributary rather than a consequence," coaxes Roy turning with Ali-ce to watch the rain.

"It will be part of the bigger picture," says Ali-ce delighting in the rain and misty breezes.

Roy says, enjoying the rain that much more by knowing Ali-ce's pleasure in it, "He probably will not find much support for sacrificing the bigger picture because of a marriage he does not like."

"No matter how much he does not like it," collogues Ali-ce as the rain falls harder.

"But you cannot marry me or any mortal if you become a nun," presses Roy wondering how long the rain will last.

"I am not married now," observes Ali-ce hoping it will rain for a while.

"That is true," concedes Roy hoping it does not rain too long.

"My father will be glad to know that as a nun I am not getting married in the conventional sense," says Ali-ce watching the rain.

"Will we ever get married?" Roy asks guessing this is the warm side of the storm front.

"Perhaps," says Ali-ce watching the rain lighten, "I cannot say I will be a nun all my life."

"We can be married in a lot of ways without the ceremony," allows Roy gazing at the rain.

"If we are very lucky," suggests Ali-ce gauging the rainfall, "we will have that ceremony."

"But first you must become Sister Ali-ce," concedes Roy guessing the rain is stopping.

"And Sister Ali-ce," advises Ali-ce, "has to become a thorn in my father's side."

"But one that he needs here," asserts Roy focusing on puddles to measure the rate of rain.

"I do not think we have to worry about that," grants Ali-ce watching the puddles too.

"What I mean is he has to want to keep you here," says Roy walking to the east end of the portico and listening to the train a couple of blocks to the south. "You have to be needed here."

"So for the time being, keep working on your manuscript," says Ali-ce, "rewrite, revise."

"That could force his resolve to kill me," says Roy leaving the shelter of the portico.

"I do not think so," confers Ali-ce joining Roy and turning south with him to get to the train station. "Just do not circulate your manuscript until we can find out if that is the cause of his mission."

"How long will that take?" Roy asks as they pass a loading dock for a sheet metal shop.

"I do not know. Maybe not too long," says Ali-ce feeling a little impressed by the large machinery in the metal shop along the way. "You do not have to plan on living the next thirty years looking over your shoulder."

"If I have to, I will abandon my manuscript," says Roy without feeling the least impressed by the shop.

"Do not do it, unless you have to," confides Ali-ce considering what that could mean. "In any case, we will spend the rest of our lives together."

"That will be wonderful," affirms Roy with affection.

Episode 57

Dr. W-abe's is in good spirits from the book show and does not want to end it by immersing himself in public transportation. Mim-sey, always ready for an opportunity, suggests going to a hotel floor and trying a new restaurant. W-abe likes that idea because he can get drinks and soon. If he does not want to eat, it will be understandable and excusable there. Hotel restaurants are used to that sort of thing.

Mim-sey can take a room, if he gets too drunk which is much easier when Dr. W-abe is in a good mood. It does not matter to either one of them that hotels are the most expensive watering holes in business.

Dr. Mim-sey, for the moment, is unable to promote a conversation while he feels the sting of Dr. W-abe's successes in meeting his challenge for the mathematical theorem. Not to mention his jealousy of W-abe's new ownership of a book that just may be unique.

W-abe, without an impulse to gloat, fills the gap of silence by reflecting on an encounter, this time, with 0101th Robe. His recollection is the impression of a perimeter office. One wall could be an exterior wall.

But it is hard to tell. The room lighting brings the other walls to a similar glow. It takes the sunset to remove the illusion or at least the distraction. Dr. W-abe intends to ask if this office is near the atrium under the triangle of Robe's Chamber or on the perimeter of one of the towers. But despite W-abe's intention the conversation starts somewhere else.

"A one-way trip is a high price to pay for exploration," declares 0101th-Robe leaning back in his chair.

Dr. W-abe says, taking a cue from 0101th-Robe and settling in for the talk, "The eggs that the explorers sent back gave us hope that others could follow and that the first would not be last."

"Along the way, navigating the worm holes and finding a few habitable planets is certainly a benefit to our civilization," maintains 0101th-Robe moving a file to the side of his desk.

"I imagine that you know as well as I that exploration brings benefits and changes to the whole civilization," says Dr. W-abe. "No matter what our differences, we all unite behind the astronauts."

"But there is two-way traffic now. How and when did we, uh, turn the corner on the one-way trips?" 0101th-Robe asks overlapping his hands and resting them on his desk.

"The cost of exploration, as you might guess, put severe limitations on the frequency of explorations," says Dr. W-abe. "But the one-way trips began with returning information via the eggs."

"I suppose after an egg or two we could return to the new planet with more explorers," 0101th-Robe contends while slowly oscillating one hand just above the surface of the desk.

"Yes but only after we learned a lot more. We learned to build ships that we could accelerate to nearly the speed of light. At that speed a precursor to a worm hole forms as a shockwave ahead of the ship. This skews the prominence of the white holes," continues Dr. W-abe cautiously ushering the Robe through the physics of high speed travel.

"What do they do?" 0101th-Robe asks resting his hands to the desktop.

"Did you ever fall in the swimming pool wrong?" Dr. W-abe asks deferentially.

"I wish I could say only once," conveys 0101th-Robe with chagrin for his honesty, "but it was twice that I can remember."

"What do you remember?" Dr. W-abe asks without any indication of condescension.

"It felt like a wall, solid," relates 0101th-Robe reflecting across time to one of the incidents.

"The white hole is a way to avoid that kind of event. Resistance like that would destroy the ship," imparts Dr. W-abe contemplating some of the implications of that idea.

"So you aim for the white hole while the worm hole develops in front of you," renders 0101th-Robe leaning on an arm of the chair.

"Exactly," confirms Dr. W-abe folding his arms to avoid imitating the Robe's gesture. "And we emerge, without harm, in a part of space that permits reducing speed at a manageable rate."

"What happens then?" 0101th-Robe asks clasping his hands on his abdomen.

"Space is a big place. The white hole is a feature of the fabric of space. Its location is recorded by the returning eggs," imparts Dr. W-abe hoping the details are not too sparse for the purpose of the conversation. "Then while still at or near the speed of light there is time to pick a likely galaxy to approach."

"I guess then you put your ears out," supposes 0101th-Robe, "and look for signs of technology."

"Actually, water first," acknowledges Dr. W-abe regretting the redirection.

"You need to land right away?" 0101th-Robe asks grabbing at the thread of the conversation.

"No. We make a list of water-worlds, if any. If the decision is made to land, it is a one time landing," reveals Dr. W-abe deliberately giving 0101th-Robe some credit for his supposition.

"Of course," reasons 0101th-Robe centering on the gist of the conversation again. "You will want to land on the planet that has water and the best possible conditions that complement survival."

"Exactly. That priority steers us away from a technologically advanced planet because they might feel threatened by our arrival," clarifies Dr. W-abe gratefully confirming 0101th-Robe's estimation of the facts without estimating how much of it was the Robe's guess.

"I do not suppose there are a lot of either," features 0101th-Robe moving a piece of paper along with a quick read.

"There maybe millions of habitable planets and more than a few with an advanced civilization," explains Dr. W-abe, "but in a universe that is billions of light years across, they are very far apart."

"Have we found very many?" 0101th-Robe asks.

"Less than a thousand with water. Only one with what you would call a civilization on it, Earth," responds Dr. W-abe afraid that his time is about up for this interview.

"So we must have lost a lot of explorers," says 0101th-Robe with one hand still on the nagging paper.

"Actually, none of them," triumphs Dr. W-abe hoping for the time to explain.

"How is that to be believed?" 0101th-Robe asks pushing the piece of paper aside.

"When the explorers got to the end of their resources, they chose a moon or a planet to set down on. Before they landed, they launched the last of their eggs for home with that information. We know where each and everyone is," apprises Dr. W-abe rigid with tension. "All of their remains have been returned home."

"That is overwhelming," quakes 0101th-Robe without moving.

"That is what we owe Earth. Without Earth, Roy Abbit's planet, there would still be no return trips of any kind," declares Dr. W-abe unable to relax into what moments before had been a generous chair.

1

1 1

1 2 1

1 3 3 1

1 4 6 4 1

1 5 10 10 5 1

1 6 15 20 15 6 1

Chapter 20

Chapter 00101000....

Chapter CE)

Chapter $(1 + 5x + 9x^2 + 5x^3 + x^4)$

Chapter $(a^4b^0 + 5a^3b^1 + 9a^2b^2 + 5a^1b^3 + a^0b^4)$

Chapter $2^2 + 2^4$

Episode 58

    Dr. W-abe is not willing to let his current recollection take over his mood. Dr. Mim-sey is left to his own devices only as long as it takes to go from the book show to the restaurant. This gives Mim-sey time to recover from the chagrin of W-abe's construction of the counting theorem and his purchase of possibly the rarest book on this planet or Earth.

W-abe, for his part is unaware of both the mortification of Mim-sey and his latest agenda. The restaurant they have chosen is a little too dark for Dr. W-abe but he does not hesitate to accept a table. He does not need to see a menu to order drinks. After a few drinks, if he is hungry, he will order dinner and if the wait service has to turn up the light so he can see the menu that is the waiter's problem.

Dr. Mim-sey has already solved the menu problem for himself. He will have what Dr. W-abe is having. That way he does not have to fuss with a menu and the waiter will give Dr. W-abe the check. Of course, Dr. Mim-sey will express indignation and regret when Dr. W-abe gets the check and will offer to pay. But the die is cast. Dr. W-abe will pay without surprise or hesitation.

So the only thing left is Mim-sey's idea of a conversation. While Dr. W-abe orders drinks with a keypad, Mim-sey makes a quick sketch on a napkin and pushes it into a pool of light next to Dr. W-abe.

Dr. Mim-sey says, "Counting gives eight rows of binary numbers:

$$
\begin{array}{ccc}
A & B & C \\
0\ 0\ 0 & = & 0 \\
1\ 0\ 0 & = & 1 \\
0\ 1\ 0 & = & 2 \\
1\ 1\ 0 & = & 3 \\
0\ 0\ 1 & = & 4 \\
1\ 0\ 1 & = & 5 \\
0\ 1\ 1 & = & 6 \\
1\ 1\ 1 & = & 7.
\end{array}
$$
"

Shifting his chair into better light, Dr. W-abe responds, "We have established that the logical definition of the first two rows is: A = {[A * (A + |A)] * 0....}. Its logical complement is |A = {[|A + (|A * A)] * 0....}."

"Extending the logical definition to column B has the effect of defining the decimal numbers of two and three," says Mim-sey. "This paves the way for column C and so on."

"This requires revising the definition of category-A to include column-B," says W-abe.

"Now, this definition of category-A is: $A = \{[\ A * (A + |B)] * 0....\}$. Its logical complement is $|A = \{[|A + (|A * B)] * 0....\}$," concedes Mim-sey as he transcribes the formulas on a napkin of his own.

"This revised version of the formula for category-A looks a lot like the original formula for category-A," acknowledges W-abe. "Category-B is: $B = \{[\ B * (|A + B)] * 0....\}$. Its complement is $|B = \{[|B + (A * |B)] * 0....\}$."

"The formulas begin to differentiate themselves with category-C and afterwards," grumbles Mim-sey reflecting on a fact of mathematical life. "But that is the way it is for mathematics that are very close to zero or very close to infinity."

"Category-C must include columns A, B and C so it is $C = \{[\ C * (|A + |B + C)] * 0....\}$," says W-abe without putting pen to paper. "Its complement is $|C = \{[|C + (A * B * |C)] * 0....\}$."

"The revision of the definition of category-A that includes columns B and C is: $A = \{[\ A * (A + |B + |C)] * 0....\}$," says Mim-sey accepting the drinks and two wireless menu pads from the wait service. "Its complement is $|A = \{[|A + (|A * B * C)] * 0....\}$."

"Negative numbers are the arithmetic opposite of positive numbers with identical absolute values," says W-abe extending the axiom. "It is not a stretch of arithmetic conventions to allocate the not-ed category formulas to the negative numbers on a number line."

"The number three includes both columns A and B. Category-AB is a category name just like any category name. It just has two coefficients in the name. So, the formula for category-AB must be AB = {[AB * (Set of discourse)] * 0....}," prescribes Mim-sey.

"The set of discourse for category-A is (A + |B) and the set of discourse for category-B is (|A + B)," says W-abe surveying the dinner options on his menu pad.

"So the set of discourse for category-AB must be ((A + |B) something (|A + B))," infers Mim-sey linking to W-abe's rummaging on the menu pad to his.

"In that case by substituting ones and zeros the logical operation that completes the formula is logical '+'," maintains W-abe. "AB = {[AB * ((A + |B) + (|A + B))] * 0...}."

"And |AB is -3 on the number line," affirms Mim-sey hoping his friend is hungry.

"The logical operation that completes that formula is logical '*'," says W-abe finally choosing an entree and transmitting his order. "|AB = {[|AB + ((|A * B) * (A * |B))] * 0...}."

"So the numbers that populate the number line are defined by binary, decimal numbers and the category formulas," says Mim-sey copying W-abe's order.

"What does Mr. Abbit say about a category formula for an alternative number line as a second axis?" asks W-abe without missing Mim-sey's maneuver with the menu pad.

"All the work is done," says Mim-sey ordering the same drink and entree as W-abe.

"The logical operators provide the opportunity for two more equivalent equations."

"Then we do not have to validate a completely new set of formulas for a second axis?" W-abe asks.

"You can if you want to but it will be enough to prove that *A = {[A+ (A * |A)] * 0...}* is an equivalent expression of A = {[A * (A + |A)] * 0...}," contends Mim-sey.

"The only difference comes from transposing the first two logical operators," says W-abe.

"This time the fewest possible variables offer the most benefit," opines Mim-sey. "In this case that of an x- and y-axis."

"So |A = {[|A + (A * |A)] * 0....} is equivalent to *|A = {[|A * (A + |A)] * 0....}*," says W-abe allowing that the identical meal orders will make service more efficient and that much faster.

"And," Mim-sey offers, "the door is not closed to imaginary numbers on the y-axis."

"Will we be using imaginary numbers?" W-abe asks trying to muster the interest.

"Mr. Abbot covers a lot of ground without resorting to that," relents Mim-sey.

"That takes some of the urgency out of his threat," judges W-abe.

"If we can limit our intervention to the occult action of dispatching Mr. Abbit," tenders Mim-sey in a ploy to cultivate W-abe's confederacy, "we will not be revealed."

"But if his work gets published and acknowledged, the revolution in the evolution of mathematics could mean the end of our civilization," contends W-abe without considering merits of Mim-sey's hypothesis beyond its naivety.

"That threat justifies the most drastic intervention," says Mim-sey.

"But that intervention could mean a terrible price including revealing the proximity of an alternative and accessible civilization - ours," says W-abe more as a warning to Mim-sey than as a hypothesis.

"Any lesser price must be paid," colors Mim-sey with grave conviction. "Mr. Abbit's life, our lives and the lives of others could be forfeited and this is justified against the cost of one or both civilizations."

Episode 59

The plane from Rome is on time but the connection from New York to Miami is delayed. Ali-ce is traveling with a U.S. passport so there are no delays in Customs. Inconveniently, Roy and Ali-ce have to go to an attorney's office on the way home from the airport. Roy's attorney leaves Miami today and will not be back before Ali-ce returns to Rome.

The office reminds Ali-ce of her father's office. But only marginally. The office is smaller and the clutter is not as deep. Roy found this attorney in a saloon - a friend of a friend. His trust, however easy, is well placed though. His friends do not use any other attorney for personal business and Roy has a lot of respect for their judgement.

The attorney for his part does not represent hopes as promises and always delivers what he promises. Reassuringly, he charges a respectable fee for his work. An attorney who does things for gratis is unsettling. They are a little late but so is the attorney, so Roy and Ali-ce have time to talk.

"Sister Ali-ce," hums Roy looking at Ali-ce and her black and white habit.

Ali-ce surveys the office that is half the size of Roy's apartment. "You like the title."

"Yes. How do you like Rome?" Roy asks wondering how to fit three clients in this room.

"It is one place where I could spend the rest of my life," posits Ali-ce realizing that there are also no windows here.

"I am glad you can spend some of it with me," says Roy with some urge to leave the office.

"If you would move to Rome, you would not have anything to complain about," proposes Ali-ce now supposing he might not want to learn Italian or like living without air conditioning.

"We cannot let your father think your involvement with me is anything but his bidding," says Roy.

"Oh yes," affirms Ali-ce. "I am here to mine the libraries and contain your manuscript."

"So," says Roy handing her some papers, "put down your pick and shovel and read this."

"What is it?" Ali-ce asks reading quickly and easily but with some doubt as to its significance.

"It is not a marriage certificate but it is the next best thing. It is a power of attorney. That is why we are here, to finalize it," advances Roy feeling a little self-satisfied with his initiative.

"You told me you were going to do this. But is it not a little too soon?" Ali-ce asks.

"Maybe not for your father," prompts Roy. "He needs reassurance from time to time."

"So when I tell him there are no more copies of your notes or manuscript he can believe it," proposes Ali-ce reasonably comfortable with Roy's contribution to the solution.

"With this you have full access to my life during and after," says Roy. "Whatever he wants, you can do. Because of that, he does not have to wonder if you are telling him the truth."

"I do not have to lie to him anyway," declares Ali-ce. "If I cannot do something, I will tell him."

"This just makes it easier to do the things you can do," offers Roy hoping for her assent.

"As long as this serves to keep my father in the dark about us, that is good," says Ali-ce.

"You can tell him that this allows you to act as my agent," presses Roy. "Tell your father that the prospective publisher will not take you seriously without this power of attorney."

"You can kite query letters through me and your manuscript never circulates," says Ali-ce adding her hypothesis. "I can negotiate any publishing contract right out of existence."

"You can assure him that you stand more completely than ever between me and any chance of publishing," says Roy wondering if this will change whether and how often they see each other.

"He will consider this an improvement and the justification of our relationship," says Ali-ce warming to the idea for Roy's sake.

"This might relegate his current plan to a backup plan," speculates Roy still selling his idea.

"To put off the option of murder? I suppose he could wait until you published to change the time line at some point before you put pen to paper," reasons Ali-ce considering the specifics.

"Change the time line?" Roy asks trying to guess what this means.

"It is more than a euphemism for murder," says Ali-ce wondering what, if anything, would stop her father.

"He does not want to commit to murder unless he has to and this way he is less likely to have to," says Roy getting back to their single best hope and wishing the attorney would get here.

"He does not have to worry about my motives this way either," reassures Ali-ce.

"Is he afraid that a different kind of relationship will change the course of events?" Roy asks trying to poke holes in Dr. Mim-sey's hypothetical options.

"That has been his main fear," chafes Ali-ce focusing on a more domestic version of the conundrum. "You and I could become his family version of the clash of civilizations."

"So now you are a nun and a literary agent who can work at a safe distance," rebuts Roy.

"Is this a safe distance?" Ali-ce asks taking Roy by the hand.

"Not even for a nun," says Roy and pleased to say so.

Episode 60

The Robes of the Robe's Chamber are not in the habit of explaining things. Mostly they are in the business of collecting and verifying information. So when Dr. W-abe finds that he has been directed to a large room with one impossibly large table and twelve very comfortable looking chairs, he is not surprised when no one mentions what the room is for or anything else for that matter.

Accouterments not withstanding Dr. W-abe is confident that he will not see all twelve Robes this time. His expectations are confirmed when 101th-Robe enters with just an intern. The intern is carrying a tray with two cups and a carafe. Dr. W-abe and 101th-Robe sit in the two chairs that are conveniently close. The tray and cups are forgotten once the conversation begins.

"I need to know how Earth has come to such a prominent place in our lives," says 101th-Robe.

"The first spaceflights that approach Earth, of course, do not land on it. They send back the eggs of data for others to use after them. Then they choose another planet to land on where their remains are ultimately retrieved once two-way travel becomes possible," narrates Dr. W-abe using the conference table as an armrest. "There is a planet that Earth calls 'Mars' that is only ten minutes away, uh, at the speed of light."

"How many of these explorations of Earth are made initially?" 101th-Robe asks leaning back into his chair.

"I think five," states Dr. W-abe looking at the conference table's patina.

"Do any of those explorers land on Earth?" 101th-Robe asks swiveling to face Dr. W-abe more directly.

"Yes, the fifth. He is equipped to launch a few eggs from the surface with data from there," relates Dr. W-abe. "These are picked up by subsequent explorers and forwarded back to us."

"What is the value of those eggs? It must be considerable compared to the lives of the astronauts," 101th-Robe says tracing a small oval on the table with his finger.

"As you might guess they contain information that is not bandied about in our media," responds Dr. W-abe while watching 101th-Robe orbit his ovals.

"I suppose, at least, the first four explorers studied the languages in Earth's electronic media," speculates 101th-Robe as he looks up from the table. That much must have been in some of the eggs returned to our planet.

"Yes," declares Dr. W-abe wondering how specific he needs to be.

"The fifth explorer arrives - fluent in English - to what appears to be an important urban center."

"I suppose that is where the best information and the most danger are at the same time," suggests 101th-Robe moving the focus along faster than Dr. W-abe might have.

"If anything could be more dangerous," says Dr. W-abe, "I do not know what it would be."

"What happens to this intrepid soul?" 101th-Robe asks anticipating Dr. W-abe's contribution.

Dr. W-abe steepls his fingers and rests his elbows on the arms of his chair. "The astronaut looks for and finds a library that is open day and night."

"I suppose the priority must be to collect the most useful information possible," submits 101th-Robe signaling his interest for more details on this point.

"Mathematics is chosen," indicates Dr. W-abe again wondering about the need for details.

"I suppose we can plan on taking as long as necessary to unpack that information after we have it," says 101th-Robe resting an arm on the conference table. "Meanwhile we can continue to siphon off Earth's broadcast media."

"Yes, it is also decided to let the explorer choose what to send back. He does not have to limit his research to the 'modern' mathematics," summarizes Dr. W-abe volunteering that detail.

"Modern work in mathematics can be gathered at any time in the future?" 101th-Robe asks guessing that modern mathematics on both planets might resemble each other more and offer less of what is new.

"Yes," says Dr. W-abe not feeling very far ahead of 101th-Robe's thinking. "What is hoped is to find something or anything we are missing in our own remote or independent development."

"But not in time to save the fifth Earth explorer?" 101th-Robe asks moving on.

"He spends three days compressing information before he feels so conspicuous that he cannot continue," synopsizes Dr. W-abe glossing over deep details and the stuff of novels.

"What becomes of this effort?" 101th-Robe asks moving on anyway.

"He launches the eggs that include his final location," W-abe sums up not expecting to return to specifics about the library or its environment. "It is, I think, in a desert in the west."

"I suppose the climate could preserve his remains," supports 101th-Robe with confidence.

"Yes, and he is eventually recovered," adds Dr. W-abe confirming the Robe's confidence.

"What does he transmit that is valuable?" 101th-Robe asks touching the edge of the tray.

Avoiding a glance at the tray to avoid refusing an offer for a drink, Dr. W-abe volunteers, "The modern mathematics is so idiomatic it is not immediately useful."

"I am not sure I know what modern mathematic is," says 101th-Robe dreading technicalities.

"Definitions vary but that includes mathematical developments from the present to three hundred Earth years before Roy Abbit," apprises Dr. W-abe saving more specific details for the pointed questions, if they cannot be avoided.

"What can we use from Earth's earlier mathematicians?" 101th-Robe asks laying his hand flatly on the conference table.

"Earlier development is uneven. Archimedes has done work on the volume of geometric shapes that is lost for 3,000 Earth years. About 350 Earth years before it is rediscovered, it is reinvented," says Dr. W-abe.

"But his work is superior to the modern incarnation because his involves infinities and geometric volumes of spheres," Discloses W-abe. "Which is what ultimately makes travel at the highest speeds possible for us and will, eventually, for Earthlians for that matter. Earth just does not realize it yet."

"Archimedes, what kind of name is that? Is it early English?" 101th-Robe asks.

"Archimedes is Greek," divulges Dr. W-abe resting his hands in his lap and smiling to welcome further questions. "The Greek language is different even with a different alphabet."

"We are lucky Earth has had those thousands of years to translate Greek math into English," relates 101th-Robe.

"Both languages are still spoken, now, but that is not why we are so lucky," says Dr. W-abe. "His calculus, which is so valuable to us, comes from an English translation of a palimpsest."

"It is rare that I encounter a new word and I know better than to pretend that I know the word when I do not. What is a palimpsest?" 101th-Robe asks moving his hand to the edge of the table focusing on W-abe's forthcoming response.

"It seems, the only surviving account of 'Archimedes Method' in existence is erased because someone has a need for the 'paper' it is written on. The new text on the erased text is a palimpsest," conveys Dr. W-abe while he watches 101th-Robe become introspective for just a second before he speaks.

"Our ability to make round trips from remote extremes hinges, in part, from an English translation from an erased text?" Asks 101th-Robe trying to focus the facts precisely.

"You can add," imparts Dr. W-abe without missing the substance of those narrowest of chances, "if we had been a few Earth years earlier, we would not have had that."

"Is there anything else that we have learned from the Greeks?" 101th-Robe asks.

"Pythagorus has done work with curves that follows consistently from linear dimensions," collogues Dr. W-abe. "Both of them allow us to rethink some things we have underestimated."

"Do you think this math will filter down to the schools, ever?" 101th-Robe asks.

"As far as the public on our planet is concerned," says Dr. W-abe, "Earth is a fantasy. It has not been discovered yet. But we can exploit the math without attributing its origin to Earth."

"These two mathematicians focus our calculus so that two-way trips become possible," treats 101th-Robe as he tries to bridge the connections among the facts.

"The calculus leads to deeper compressions of space beyond the three dimensions of ordinary space. This takes the theoretical limits off of speed. It requires more recent, modern mathematics, to manage time and communications," says Dr. W-abe not prepared fill in the specifics among those facts and bracing himself to disappoint the Robe.

"But the first achievement is the compression of space and attaining higher speeds," conveys 101th-Robe.

"None of it is easy but it can be tested and proven before ever returning to Earth," instills Dr. W-abe acknowledging the importance of that aspect of development.

"So, why are we visiting Earth instead of them visiting us?" 101th-Robe asks.

"One big reason is Archimedes calculus is lost, to Earth, for those twenty-five hundred Earth years," imparts Dr. W-abe delineating the disadvantage.

"What about Pythagorus?" 101th-Robe asks looking for the significance.

"Modern mathematicians on Earth do not mix his work with Archimedes for centuries," Dr. W-abe communicates trying to respond to the question. "They underestimate its value for inventing or even simplifying the calculus."

"It sounds like Pythagoras is too sublime to be believed," proposes 101th-Robe.

Anticipating some surprise, Dr. W-abe injects, "Actually, he is common to all of us."

"How is that?" 101th-Robe asks without a guess as to the answer.

"Our batball fields are applications of his theorem for triangles. If you call the side opposite the 90^0 angle of home base the hypotenuse, his theorem becomes, 'The square on the hypotenuse side is equal to the sum the squares on the other two sides'," says Dr. W-abe.

"The very footprint of the Robe's Chamber," weighs 101th-Robe genuinely surprised.

"That theorem is easy to assimilate so it is introduced long years ago on a time displaced return trip from Earth. Batball is assumed to have evolved from the schools," summarizes Dr. W-abe.

"There is not anyone who will argue that Batball is not our own invention. This is a consequence of the time shift that comes with high speed travel," says Dr. W-abe successfully introducing a point of information that he knows to be essential.

"We can return to the past?" 101th-Robe asks unable to disguise incredulity by now.

"And the future," discloses Dr. W-abe with yet another fact that cannot be left out.

1

1 1

1 2 1

1 3 3 1

1 4 6 4 1

1 5 10 10 5 1

1 6 15 20 15 6 1

Chapter 21

Chapter 10101000....

Chapter ACE

Chapter $(1 + 5x + 9x^2 + 6x^3 + x^4)$

Chapter $(a^4b^0 + 5a^3b^1 + 9a^2b^2 + 5a^1b^3 + a^0b^4)$

Chapter $2^0 + 2^2 + 2^4$

Episode 61

Dr. W-abe can work anytime. This means he can take any day and as many days off from work as he wants. Dr. W-abe simply does not take time off from work. Eight or ten hours is a short day. Five days is a short week. Two days in a row is unusual for time off. The Board of Director's annual meeting is no different in a lot of ways.

Here, someone has his attention night or day and its meetings last for hours. The burlesque of a vacation comes with the location of the director's meetings. Once a year, they meet at an island resort in the gulf. It is always the same island but the resort campus changes from year to year. Dr. W-abe can afford this as a vacation. It is even more affordable with the company's money.

If it were a real vacation he would have to use his own money, but in some sense it is a vacation for Dr. W-abe. He can bring his wife but she quit going a long time ago. Extravagance and indolence she can get anywhere and in any quantity. The island offers, potentially more tropical heat and boredom than she likes to attribute to true luxury. This is fine with Dr. W-abe he is comfortable as long as she is satisfied with her choices.

Most of the time Dr. Mim-sey goes instead. He does not go to any of the meetings. Mim-sey goes to all of the dinners. The rest of the time, He spends going from the suite, to the pool, to the beach. He can sleep and eat when he wants. Like Dr. W-abe, Dr. Mim-sey does not have to spend any money. He can count on Dr. W-abe's company account for that. Dr. W-abe's personal assets go unmolested.

There are no meetings on the first day with the different arrival times for board members. The first day for W-abe always includes a trip to the island equivalent of a grocery and intoxicants stores to stock the suite's kitchen. Along the way, there is time to talk. Dr. Mim-sey can count on Dr. W-abe not discussing board business.

W-abe asks, while strolling down the aisle for shelf stable food, "Can two number lines work for any and all categories?"

"The lines are categorically distinct, so yes," declares Mim-sey pretending to look for a pate that both of them can enjoy. "The example have been using is that of nouns and verbs."

"I can see the arithmetic on one line. It is one dimensional. Take a sentence like, 'The villain must die or we will.' On a single number line you can use vectors for the words. The arrows for the nouns will point one way and the vectors for the verbs will point the other way. All the ground we have covered is intact but if you add a second axis, the categories become two dimensional," grants W-abe picking up three jars of pate while looking at the packaged fungus.

"Nouns and verbs in two dimensions - I can accommodate that. The two categories are distinct like time and space. I never confuse nouns for verbs or verbs for nouns. Those I keep separate," conveys Mim-sey trying to see what is so good about Dr. W-abe's choices. "I will admit that since adverbs can modify a noun or a verb that adverbs can be, uh, confusing."

"Remember that if you square the length of one side of a right triangle and add that to the square of the adjacent side of the triangle that the sum is equal to the long, hypotenuse, side when it is squared, $a^2 + b^2 = c^2$," adds W-abe surveying the packaged beverages. Those little exponents are the price we pay for that next dimension.

"You are unusually kind to make it so simple," affirms Mim-sey. "Of course I remember. And I know you go around squaring E's, M's, C's, D's and lots of other things, as well, all the time."

"It is the jumping off place for circles, ellipses, and hyperbola which are conic sections," professes W-abe choosing his drinks without looking at the price.

"A conic section is just a slice of a cone," imparts Mim-sey hoping this is not the first year he has to pay his share at the store. "You cut through a cone and the resulting shapes depend on whether you are parallel, perpendicular or somewhere in between in terms of the cone's axis."

"Conic sections raise the terms to three dimensions," clues W-abe while grabbing some green oil. "The march on to more dimensions get easier to represent mathematically if not graphically."

"The formulas for higher dimensions derive from $a^2 + b^2 = c^2$? So categories must be describable in even more than two dimensions," says Mim-sey looking at the red oils unable to choose among them.

"But this is where our villain has cornered himself. How does he explain a noun-squared or a verb-squared? What is a $(hospital)^2$ or an $(is)^2$?" W-abe asks.

"Are you looking for a reason not to destroy Roy Abbit?" Mim-sey asks.

"I think we are past that," grants W-abe picking up the red oil he wants. "There is no saving him. It is only a matter of how or when. In the meantime, it would be nice, though, to put holes in his theory."

"Are you being more professional than personal?" Mim-sey asks with a mix of hope and confidence that he has W-abe where he wants him.

"It is part of what I do for a living," says W-abe taking the blue oil, the yellow sauce next to it and turning toward the cashier. "I take apart my theories along with everyone else's."

"Well if you were being more professional than personal you would not have missed the point about categories-squared and two square root dimensional categories. What are square roots for?" Mim-sey asks hoping W-abe got everything they need.

"Most of the time they are used to weed out the arithmetically challenged so that the rest of us can work with more freedom," says W-abe choosing artisan bread and some indigenous vegetables.

"You never disappoint me. You know what the real purpose of square roots is," alleges Mim-sey preparing to make the perfunctory offer to pay at the check out.

"Yes, it makes it easier to do arithmetic without exponents," says W-abe shopping the candy. "Roots get rid of all those two's three's and so on that superscript so many of our numbers."

"So square roots are a mathematical convenience?" Mim-sey asks looking at magazine banners.

"Some would say, a not so convenient convenience, but yes," acknowledges W-abe.

"Then you would not have much trouble with the idea that a category squared is just a mathematical convenience," contends Mim-sey as he slows down at the fresh fungus.

"When you put it that way, I have to," says W-abe. "But I do not like it. I do not like two-dimensional nouns or verbs which are as hard to swallow as nouns-squared and verbs-squared."

"Of course, there are potentially more dimensions," says Mim-sey studying the red fungus.

"That will not make the notion any easier," warrants W-abe bagging some white fungus.

"What will help is that ordinary algebra, statistics and probability can be used to describe ordinary sentences," deliberates Mim-sey selecting some of the larger red fungi.

"Even a sentence like, 'The villain must die or we will.'?" W-abe asks approaching the checkout.

"I am uncomfortable with the idea even though it is inescapable," claims Mim-sey.

"So am I, but let us not dodge the focus it provides. While we explore his category math, we will commit ourselves to the cause," asserts W-abe keying in his local payee information.

"Perhaps we will find a way to avoid our fate," advances Mim-sey picking up the bags.

"Or perhaps, Mr. Abbit will hand us the rope for his," submits W-abe heading for the door.

Episode 62

The smallness of the apartment makes even a light rain intimate. It is no wonder that even if it wakes you it is easy to go back to sleep. This has been Roy's experience more than once. He does not begrudge it when it happens even though the nuance of newness is long gone.

Ali-ce has been in Roy's apartment a few times when it rains and the experience is still novel for her. Sometimes, the rain is quite heavy. There have been tropical storms and Roy has seen his share of hurricanes in the apartment and even the windows have survived the danger.

His penchant for solitude, silence and petulance has insulated him from gratuitous encroachments of the neighbors. The methodical harangue of television programming is usually replaced with recorded music and the remote control keeps that from being too loud, too soft or monotonous. Getting his mail at a post office box means he has no home address and no solicitors at his door. The only exception is the telemarketer. Caller-ID frustrates that option. As a result Roy likes it at home, small as it is.

He sleeps comfortably as far as the demands of work and other responsibilities do not compete. A call in the middle of the night is rare and, therefore, guaranteed to disturb. This time when it happens, he does not look for his glasses before reaching for the phone. This could be a legitimate call.

"Do you know what time it is?" Roy asks trying to remember where he left his glasses.

"Oops," deplores Ali-ce looking down at a busy neighborhood street. "It is about nine A.M. here in Rome. I suppose, now that I think about it, it is the middle of the night there."

"I was sleeping but you can wake me up," chats Roy delaying turning on his lamp.

"If you were here in Rome," coaxes Ali-ce, "this would not have happened."

"By telepathy maybe but not by phone at least," tantalizes Roy finally turning on his lamp.

"You have become quite casual with telepathy when we are together, you know," chimes in Ali-ce turning away from her rustic window.

"I do not think I could manage it with anyone else," claims Roy sitting up in his bed.

"You are not a telepath," explains Ali-ce, "but you could do it with anyone who is."

"That is not what I mean," clarifies Roy. "No one else would tolerate my chaotic mind."

"It is a riot in there but you manage to focus," coaxes Ali-ce.

"What if I offend you?" Roy wheedles as he finally finds and puts on his glasses.

"I know I am not always a priority," prevails Ali-ce pulling her coffee cup closer. "I know what you think about my father and the sacrifice you are making by not publishing."

"I want to explain to you that since I do not know your father you sometimes become him when I resent his agenda," discloses Roy while stacking his pillows.

"I know when you are focusing on him and his agenda," Ali-ce warmly remarks before taking a sip. "As soon as you realize that you are including me, you change your line of thought immediately."

"So you know I am telling you the truth," infers Roy planning on making some coffee.

"Not only that but as soon as the chagrin hits," divulges Ali-ce leaning back into her chair, "you flood your thoughts with the most wonderful thoughts of us."

"That is not you planting seeds?" Roy asks.

"I would not get any joy out of that," says Ali-ce pulling her cup a little closer to the edge of the table.

"It would be like looking at a mirror?" Roy asks leaning back into his pillows.

"Yes and it would become a vain burlesque very soon," admits Ali-ce giving up on what is left of her coffee.

"But you can tell when I am not interested and lots of other things. How do you manage with that?" Roy asks.

"Actually, it is easier than you think. I just wait for those other things to surface as open remarks," acknowledges Ali-ce carrying her cup to the sink. "If it is my business you say so in so many words."

"That is because I do not know what you know and it is the human condition of not being telepathic," advances Roy. "Our understanding may be deep and our speculation precise but if I do not trot the horse around the track you have to pretend you do not know the horse."

"When you get good," says Ali-ce, "you can ignore the elephant in the room as they say."

"So what I do not discuss become the elephants in the room," surmises Roy getting out of bed and carrying the phone foot of the bed.

"Yes," says Ali-ce rinsing her cup and drying her hands, "but I put them in another room."

"Now that you mention it, I have to go to the other room," says Roy leaning toward the phone base in anticipation of ending the call. "Tell me why you called and I will call you back."

"I found your birth certificate - here in Verona," says Ali-ce.

"My parents met there," substantiates Roy with a smile.

"You told me," says Ali-ce returning his smile. "I love you. Ciao."

"I love you more. Ciao bella," claims Roy before hanging up the phone for the moment.

Episode 63

For 111th-Robe it is just the most convenient door on his side of the tower that his office occupies. For officer Gim-bal it is also the sixteen-floor atrium that occupies the space under the triangle of the Robe's Chamber. All that empty space would have become real estate in any other building. But in this case, the decision to add towers includes the decision to raise the three floors of the original triangular building and create an atrium under it.

As the towers grow, so grows the atrium. Along with the mounting space comes the landscaping, lighting, matrix elevators, escalators, R. A. T. systems, and climate control. Without climate control, the atrium is permanently warm and rainy. So now it stays sunny, dry and cool, day and night, century after century.

Officer Gim-bal does not have to worry about being missed or lost in this cavern. The ground that occupies the original triangle has fountains, plants and benches. It is arranged so that anyone waiting in front of one of the towers can be seen, even if they are sitting. He does not have to worry about what to do with his time either. 111th-Robe is right on time. Officer Gim-bal stands.

"Officer Gim-bal. We have had a few conversations with Dr. W-abe," advises 111th-Robe sitting down and gesturing for W-abe to do the same.

"Did he give you any new perspective on Dr. Mim-sey's petition?" Gim-bal asks resuming his seat noting that the bench is textured and heated to feel warm and upholstered which is not always the case for crystal.

"To some extent," confers 111th-Robe, "but we have not discussed Dr. Mim-sey."

"I know you prefer not to divulge conversations, specifically," grants Gim-bal, "but you would not hurt my feelings if you told me anything that amounts to progress on this problem of Roy Abbit and Earth."

"Dr. W-abe tells us that mathematics of considerable value has been harvested from Earth," advises 111th-Robe resting an arm on the bench back. "And that its value, in fact, leaves us in debt to Earth."

"So he is reluctant to see any harm imposed on any of Earth's inhabitants?" Gim-bal asks.

"I do not think he is worried about that. In fact, I think he is like most of us in the thinking that neither Roy Abbit nor his mathematics pose any danger," allows 111th-Robe leaning against the back of the bench. "So the real harm is in punishing anyone on a false premise."

"So nothing has changed?" Gim-bal asks straightening his legs and crossing his ankles.

"More than you might think," weighs 111th-Robe as he lowers his arm from the bench back without turning away. "Dr. W-abe's remarks have made me aware of the debt that we owe to Earth."

"So unwarranted harm to Earth is more harmful than ever," says Gim-bal folding his arms.

"And we are more reluctant than ever to impose such harm," acknowledges 111th-Robe turning a little to look at the mirrored finish of an imposing water feature and its flowering plants.

"This makes me aware, and you too, I suppose, that Dr. Mim-sey's efforts and motives along this line are that much more, uh, misplaced," confides Gim-bal realizing how quiet it is at the moment.

"I guess you are reluctant to describe Dr. Mim-sey's actions as tending toward evil or criminal," says 111th-Robe looking at Gim-bal then turning his gaze back to the landscaped pond.

"Earth, Roy Abbit and his mathematics are not the only hostages of his efforts," predicates Gim-bal. "The Robe's Chamber and the fabric of our culture are being threatened too."

"And not by mere protest or supposition. Dr. Mim-sey has filed a formal petition on the issue," declares 111th-Robe while breathing in the fragrance that spreads from the flowering plants.

"That is like the difference between the threat of murder and attempted murder," affirms Gim-bal wondering how pedestrian traffic seems to have stopped or at least keeps away from here.

"Anyone can threaten murder and mean it most sincerely," consults 111th-Robe. "But no laws have been broken until the attempt is made. Of course there is the issue of proof."

"It has been my experience," says Gim-bal taking the moment to sit up, "that the nature of the attempt has everything to do with the kind of punishment that is imposed for the attempt."

"Exactly, If someone threatens you and then pushes you off a bridge, it is attempted murder when you survive," says 111th-Robe feeling free to study the glassy water.

"It does not matter how badly I am injured?" Gim-bal asks leaning back a little.

"You could live decades. When you finally do die, if your death is not related to those injuries," says 111th-Robe standing slowly, "your perpetrator cannot be charged with more than assault and battery."

"But there are still differences in degree. What if the circumstances were identical except that the victim is a significant office holder?" Gim-bal asks feeling obliged to stand.

"Then a judge or jury could impose a more severe sentence," deposes 111th-Robe stepping closer to the pond.

"Can someone be executed for attempted murder?" Gim-bal asks hoping it is not an ignorant question.

"No one ever has," says 111th-Robe turning away from the tower entrance and walking slowly along the edge of the pond. "But we have not found anything to prevent such a sentence."

"Does that mean you are looking just to make sure?" Gim-bal asks while keeping pace with the Robe.

"Only we can issue a summary death warrant. So we have to look at the precedents," resumes 111th-Robe following the curve of the pond rather than crossing the triangle centering the atrium.

"It is not exactly summary," says Gim-bal now observing that the traffic is unabated but somehow circumvents this Robe and his movements. "You have to have all twelve Robes consent to it."

"Technically it is, in this case. Dr. Mim-sey does not leave us any choice," says 111th-Robe walking evenly on. "If we consider his petition at all, we will have to consider issuing a death warrant."

"Are you saying that he put you in that position?" Gim-bal asks.

"It was not mere modesty on his part to make a capital case and omit the request of a capital sentence," says 111th-Robe moving with a pace so even that it seems more like the pond is turning.

"He has maneuvered to make the Robe's Chamber the judge and executioner with the fabric of our civilization as a ransom," says Gim-bal looking down to make sure the Robe's feet are actually moving.

"Either Roy Abbit or our civilization is the victim," warrants 111th-Robe.

"Regardless of which one, the Robe's Chamber, is also a victim," says Gim-bal.

"If Roy Abbit or his mathematics could be proven to be more than an empty threat," says 111th-Robe nearly completing the circuit around the pond. "We would not count ourselves among the victims."

"What are the chances of that?" Gim-bal asks supposing what the end of the circuit means.

"Under the best of circumstances it would take many lifetimes for Earth to develop the technology that would discover us or threaten us," continues 111th-Robe before stopping at the bench.

"So Dr. Mim-sey's petition is a little premature," infers Gim-bal without sitting.

"Not for him, he only has one lifetime to get what he wants," conveys 111th-Robe walking toward his tower without Gim-bal or the necessity of saying goodbye.

1

1 1

1 2 1

1 3 3 1

1 4 6 4 1

1 5 10 10 5 1

1 6 15 20 15 6 1

Chapter 22

Chapter 01101000....

Chapter BCE

Chapter $(1 + 5x + 9x^2 + 7x^3 + x^4)$

Chapter $(a^4b^0 + 5a^3b^1 + 9a^2b^2 + 7a^1b^3 + a^0b^4)$

Chapter $2^1 + 2^2 + 2^4$

Episode 64

This swimming pool is fabulously expensive for a small pool because every drop of water in it is manufactured. The island is so small that it does not have any fresh water, or at least enough. Water for domestic use is collected from rain runoff. All the resorts and commercial users import or desalinate their water.

This does not make the resort noticeably more expensive, though, since everything but the sand is imported to the island. But the cost of filling the pool just once exceeds the total annual income of any of the resort employees, even its director. Evaporation and cleaning mean the pool is filled an equivalent of ten times a year.

This by itself is more than enough charm for Dr. Mim-sey. He never looks at the pool without wondering at its cost. But for Dr. Mim-sey this is not nearly as inspiring as its location. The pool is outside. This, outside, is bigger than even the largest atrium. It makes the largest tree he can find small. It makes the most hulking hotel small. It makes the adjacent miles and miles of beach small. And he, Dr. Mim-sey can sit in this outside and in this beautiful pool.

Before he gets out of the water Dr. W-abe approaches and takes a chair in the shade by the pool. Dr. W-abe is not dressed for the pool. Dr. Mim-sey is pleased. He knows he can sit next to Dr. W-abe and plan on a conversation that is not interrupted by inventing excuses to follow him in and out of the pool. It only takes a little small talk to start the conversation.

After drying his face and arms Mim-sey asks, "Why would you, whose life and career are mathematics, submit to lessons in grammar, arithmetic and elementary algebra?"

"I am already risking being out of character so I can say, or rather acknowledge, that you are a linguist, who has made that his life and career, so I can expect more than a mere lesson," charges W-abe.

"If that is a compliment," says Mim-sey, "you certainly are risking being out of character."

"Do not make me regret my naked generosity. Let us get past the flattery and get to the point," warns W-abe squinting at the pool's glare. "Even if I have to endure your method of getting there."

"I hope I do not disappoint you. Mathematics that expresses the dynamics of the universe from the largest to the smallest is what you prefer," counsels Mim-sey as he finishes his drink. "So, you even join in on work and discussions of string theory despite your doubts of its methods and conclusions."

"I can put that aside, briefly," prompts W-abe turning his chair away from the pool's brightness.

"You will not have to," discloses Mim-sey while ordering a new drink. "I point you to your passion because I want you to see the same thing in the math I have to show you,"

"'...the dynamics of the universe from the largest to the smallest'. Is this just a metaphor for the soul of literature?" W-abe asks moving his chair again to make his table more accessible.

"You are nothing if not sarcastic. In a sentence, in any language, like 'The villain must die or we will.' you are moving. There is the concrete subject, 'the villain', and you go from there to his ephemeral fate, 'he must die'. This is not just a time line in transition," says Mim-sey reluctantly turning his chair from the pool to maintain his connection with Dr. W-abe. "You are taken from a focus of the finite object, this man, to its categorical opposite his fate."

"Putting it that way, it is like the transformation from a particle to a wave," says W-abe. "When being specific about the location of something in time and space, we refer to it as a particle."

"And that same particle can be described as a wave when you refer to its momentum, velocity or energy. The price that is paid in the transformation is that you cannot be specific about both at the same time," imparts Mim-sey while he watches the wait-service deliver W-abe's drink in a hematite cup on a iolite tray.

I lay the blame on physics not mathematics," vents W-abe testing his drink with a sip.

"For the people that use language there is another kind of transformation," renders Mim-sey. "The field of information is brought to a focus by the statement or sentence fragment."

"You mean there is a range of possible meaning that the statement could include but once read or said that field of possibilities is eliminated or reduced to a more specific meaning," interprets W-abe pleased to see he has an perfect view of the velvety beach and rolling, iolite sea beyond.

Wishing for a view of the pool instead, Mim-sey says, "The speaker and the listener, the author and the reader, have a common experience and the statement is a very specific part in that."

"Like the electron traveling through space that hits the phosphorescent screen. It could be anywhere, and still be an electron," discourses W-abe uninterrupted by the arrival of Mim-sey's drink and the unhesitating exit of the waiter. "Hitting the screen is a specific expression of the electron in a universe of possibilities."

"I guess so. I am not very good with the differences between waves and particles," admits Mim-sey focusing his consolation on his drink with both hands. "But specifically, in this sentence, 'The villain must die or we will.' the seven words are going somewhere."

"How do you plan to put that into numbers?" W-abe asks.

"If you were counting with binary numbers, every possible combination of nouns and verbs for a seven-word sentence, the numbers would go from 0000000 to 1111111 which has a decimal value of zero to 127," says Mim-sey as he watches W-abe adjust his chair again to enhance his view of the sea.

"A graph of the sentence," says W-abe, "would have to include that range: 0 to 127 units."

"A graph of any seven-word sentence would have to include that range," says Mim-sey.

"How would I differentiate one seven-word sentence from another, on a graph?" W-abe asks wondering how long Mim-sey will resist turning his chair completely away from the pool.

"I am going to apply the 'blade of grammar'," bombinates Mim-sey now with his drink on his right, W-abe to his left and a choice to make.

"Leave the sarcasm to me," snorts W-abe watching Mim-sey turn away from his drink.

Turning from W-abe to pick up his drink Mim-sey explains, "Most seven word sentences differentiate among themselves by the order of occurrence of the nouns and verbs."

"So instead of seven ones for a seven-word sentence," formulates W-abe smiling as Mim-sey turns back to his drink table, "let us use a one for the nouns and a zero for the verbs."

"The noun component of the sentence, 'The villain must die or we will.' is represented by 1100110," grants Mim-sey finally setting his drink on W-abe's table and moving his chair again.

"I do not know whether 'must' and 'or' are nouns or verbs," concedes W-abe, still smiling.

"Let words that modify the verbs be treated as verbs," says Mim-sey settling into his chair without distraction for the first time, "and let the ones that modify the nouns be treated as nouns."

"If we change the numbers, we change everything," says W-abe.

"So for the time being let us not change the numbers," craves Mim-sey smiling as well.

"We can still change things without changing the numbers," says W-abe. "We can say, 'Mr. Abbit must die or we will.' and we will come up with the same combination of ones and zeros."

"So for the time being," says Mim-sey in a bid for W-abe's cooperation, "let us not change the words of the sentence."

"The binary number for the nouns is 1100110," says W-abe watching the sea turn grey and the sky turn red with the setting sun, "which is 1 +2 + 16 + 32; which is 51 in decimals."

"So the binary number for the verbs must be equal to 127 minus 51 which is 76," says Mim-sey relieved to see a horizon divide the sea and sky into space he can almost comprehend.

"Why not 128? Since 2^7 is 128?" Asks W-abe keying in a new drink order for them both.

"If you count to 127 in binary numbers you will count 128 rows because you do not count from 1 to 127. You count from 0 to 127. But a sentence with zero words is not a sentence," expounds Mim-sey gazing at the navigation star that was so important in the early days of ocean going ships. "The combination of seven word sentences includes sentences with one word (either a noun or a verb) up to seven words. These seven words can be any combination of nouns and verbs."

"The binary number for the verbs is 0011001 which is 4 + 8 + 64 which is 76," says W-abe recalling that the navigation star is really a distant moon with an equatorial orbit marking 0^0 latitude.

"Now suppose we plan to rendezvous at a particular location. Suppose we want to get there at the same time so neither one of us is seen 'waiting' for someone," says Mim-sey to his now empty glass. "But to disguise the coincidence we travel from different starting locations."

"We must travel at different speeds," depicts W-abe. "The one that has the farthest to go, must travel faster than the one traveling the shorter distance."

"Assume there is a hotel 127 miles north and 127 miles east of our building. Suppose my second home is 76 miles north of our office. Let us say your other house is 51 miles north of our office building," says Mim-sey. "We could leave our individual houses at 9 a.m., travel at different speeds and arrive at the hotel at the same time."

"What does this have to do with our nouns and verbs?" Asks W-abe thinking that this place is perfect for a circumnavigating rail-ship.

"For this seven-word sentence the nouns and verbs have a different velocity. But they get us to the conclusion together," says Mim-sey trying to imagine a way to bring development to this combination of island, beaches and sky. "In this sentence, the nouns have a velocity of 51 and the verbs have a velocity of 76. They meet at 127 which just happens to be the end of the sentence."

"That is not a formal expression of the terms but is a good approximation," says W-abe ordering a new round of drinks for himself and the ever hopeful Dr. Mim-sey.

Episode 65

Ali-ce left a light rain behind in Rome. She learned from a friend of her father to travel only with carry-on luggage to navigate the airports.

Ali-ce changes planes in New York. The delay for the connecting flight to Miami gives her time to watch the ground crews through the oversize airport windows as they cope with the snow and ice. She arrives in Miami less than two hours later than planned and takes the train south to a hotel that is conveniently close to the southern end of the rail line. She tells the desk clerk to provide a key for Roy Abbit when he gets there.

He lives only two stops away. So after she calls him she still has time for a shower and to put on something he likes. While she waits for his key card in the latch, she looks out of the window and the sun is shining.

That too is incidental until she realizes that beginning with her taxi in Rome she has made the entire trip under a roof or ceiling. Fair weather or not so fair, night or day, her way is lit and protected from Rome to Miami. She plans to tell Roy how much this resembles home.

"I have been meaning to do that," says Roy carrying a bottle of wine to the kitchenette.

"Open the wine?" Asks Ali-ce taking a briefcase out of her closet and putting it on the bed.

"That," allows Roy rummaging through a drawer, "and look up my birth certificate."

"How do you manage?" Ali-ce asks looking through the papers in her briefcase.

"They are not competing or concurrent plans. It has been my plan to get a copy of my birth certificate for years," says Roy. "The intention weaves in and out of my plans all the time."

"It cannot be very important then," contends Ali-ce finding what she is looking for.

"It is important," corrects Roy upon finding the opener, "but I keep working around it."

"I brought it with me," reveals Ali-ce turning to sit on the bed while looking at the papers.

"The wine?" Roy asks winding the corkscrew into the wine stopper.

"Now you are being funny," says Ali-ce holding the papers up to Roy and waving them like a flag. "Here are a couple copies of your birth certificate along with the bottles of wine."

"Okay," says Roy handing the bottle to Ali-ce. "You open the wine while I look at my birth certificate."

"There is nothing there that you do not know already," replies Ali-ce.

"True, but this is like being there. I have never known the doctor," says Roy, "and now I wonder about him."

"No doubt he is long gone by now," presumes Ali-ce barely focusing on her task with the wine bottle.

"It would seem so. He would have been older than my parents and they are both gone," says Roy studying everything about the papers, the smell, the typefaces, the signatures...."

"A hospital number and a register number. What do you think they mean?" Ali-ce asks.

"They probably do not mean anything anymore," surmises Roy moving down the page.

"Oh, Look," declares Ali-ce pointing at Roy's papers with the dry end of the cork. 'Sex of the child, Male'. We would never have that on our birth certificates."

"What about your death certificates?" Roy asks looking up from his paper.

"It would be gender," asserts Ali-ce. "We make a distinction between sex and gender."

"That is odd," poses Roy. "You distinguish sex and gender but you are intrinsically both."

"Not culturally," protests Ali-ce putting the cork in the wine bottle.

"Not even in death?" Roy asks watching Ali-ce set two glasses on the table.

"Not even then, especially then," rallies Ali-ce returning to Roy's perch on the bed.

"You do not want an error or stigma to be a part of such a permanent record," reflects Roy moving a little to make a more comfortable place for Ali-ce.

"Look," remarks Ali-ce. "It says what color your parents were."

"Were you expecting grey?" Asks Roy reclining on his back while holding up the papers.

"I am always expecting grey," says Ali-ce as she leans on an elbow.

"I have been spelling my mother's maiden name with only one 't' all these years," discloses Roy while adjusting to share their space.

"Is that bad?" Ali-ce asks without looking away from the papers.

"Not most of the time," says Roy looking at Ali-ce. "I use it for passwords and codes."

"Now you must remember to spell it wrong," notes Ali-ce.

"I was born alive at 9:42 A.M...," relates Roy pointing to that entry. "How convenient."

"To be sure," says Ali-ce sitting up. "Stillborn is more than a little inconvenient."

"That is not what I was thinking - 9:42 A.M. is very convenient," specifies Roy returning to a sitting position to join Ali-ce. "I always imagine babies arriving at some inconvenient time."

"Look at this. Your birthday is May 15?" Ali-ce asks this time pulling the papers closer to herself.

"No. I always celebrate my birthday on August 13," says Roy looking back across the years.

"There must be a mistake," speculates Ali-ce with a tension she hopes to keep to herself.

"I wonder how that happened?" Roy asks laying the paper on the bed.

"There must be other records from the hospital," says Ali-ce falling back on her research reflexes. "They will be dated similarly."

"When you are back in Rome you can get to Verona and look up the old records," yawns Roy dismissing the mystery.

"I have your power of attorney," says Ali-ce. "And I can navigate the bureaucracies."

"You speak Italian now but do not go to any trouble," says Roy expecting Ali-ce to have more significant concerns. "It is probably easier to use this date than to change old records."

"It could be important," says Ali-ce desperately trying to sound merely curious.

Episode 66

For Officer Gim-bal, finding Dr. W-abe on the island is easier than getting extra pickled fungus for his pate. Only slightly more difficult is downplaying his role as Officer Gim-bal which is what he is, here, on Clidim. He and Dr. W-abe are aware of his sham role as a book publisher on Earth, so Officer Gim-bal uses the obvious as subterfuge.

Gim-bal is on the island with a mix of motives and circumstances. He is nearing the end of leave from his assignment on Earth. While he is here, he gives Dr. W-abe a chance to buy a first edition copy of 'I-Xohe ta I-Sliunou Qumnwe K-Here' (*The Physics of the Hyper-space Q-Particle*) at a very good price. If Dr. W-abe does not buy the book, Officer Gim-bal can be trusted to return it safely to its owner. The otherwise invulnerable Dr. W-abe is mellowed to merely cautious which is enough.

"Thank you for giving me some time," says Gim-bal sitting on a luxuriously cushioned chair in Dr. W-abe's suite.

"I am not the one that has to be in two places at once," smiles W-abe while getting refreshments for them both.

"In my case," says Gim-bal returning the smile with the intention of keeping the mood informal, "it is a convenience, not a measure of my capacity for work."

"I am amused by your audacious simplification of such a complex situation," says W-abe.

"It just takes two backup J. R. Gim-bals to make it look like I am always at home," says Gim-bal as Dr. W-abe sets a tray with drinks, salt, sugar and spices on the table between them.

"But, I assume," concedes W-abe putting salt in his empty cup, "you are here to talk about some other significant simplification."

"Yes, if that is what it is," says Gim-bal serving himself a drink and preempting the formality of being served. "One of the Robes was telling me that we found some mathematics on Earth that proved to be very helpful to us."

"And you want something more specific?" W-abe asks setting his spoon on the tray duly warned to avoid introducing any formalities.

"Actually a couple of things. I was wondering what Earth offers that we do not already know and why they do not use it themselves?" Gim-bal asks before taking a light sip.

"I am sure Earth will," says W-abe setting his cup on the tray, "once they have the capacity."

"All right how did they help us?" Gim-bal asks adding some sugar to his drink.

"It takes a lot of power to launch and maneuver a spaceship," advises W-abe while watching Gim-bal take another sip. "Once it leaves the planet, the sources of power are extremely limited."

"Whatever power there is, has to be used for life support as well," Gim-bal adds.

"Life support does not divert a lot of power," acknowledges W-abe. "Even unpiloted ships are hard to get back because of the problem of supplying the power for the return trip."

"And Earth has such an ability?" Gim-bal asks setting his drink down to cool.

"More precisely, Earth made it possible for us to return to our planet," apprises W-abe.

"This is not about interrupting gratuitously," discloses Gim-bal while standing and walking to the patio doors. "I really do not know what powering a ship involves."

"I understand. Most of the onboard power for a ship is used for maneuvering - launching, landing, pointing them at heavy objects," divulges W-abe joining Gim-bal at the patio doors.

"I am listening," affirms Gim-bal opening the doors and stepping outside.

"The heavy objects are stars, planets and black holes, even galaxies. They are free rides, so to speak. We use them to speed up and slow down," says W-abe surveying the tropical vegetation that borders the sitting area. "But they are far apart and not conveniently placed for round trip purposes."

"The best hope of using the heavy objects depends on resupplying onboard power along the way," supposes Gim-bal stepping off of the patio and ambling toward the pool.

"When we can count on that," says W-abe following Gim-bal's lead and walking along with him, "we can use maneuvering power and the heavy objects to get up to light speed and beyond."

"You do not have to conserve power for the return trip and you can make longer trips out and back and, I suppose, in less time," argues Gim-bal focusing on a storm gnarled tree with waxy leaves

"Yes, there are a lot of advantages in being able to use onboard power and resupplying it along the way," articulates W-abe wondering what the interest Gim-bal has in the weatherbeaten tree.

"By the time we discovered Earth, I understand that we were very good at traveling very far and very fast. But Earth was so far away that we could not carry enough power to return," cites Gim-bal.

"There were no inhabited planets found up to then," adds W-abe.

"Or since?" Gim-bal asks abandoning the tree and walking toward the swimming pool.

"No," vouches W-abe. "So we were hoping Earth might have some useful information."

"If not for that problem then some others?" Gim-bal asks peering into the pool.

"Who would not appreciate finding a tree bark that makes a good medicine?" W-abe asks.

"So what was the solution to the power problem that Earth supplied?" Asks Gim-bal still peering into the pool.

"Do you know what a tank circuit is?" W-abe asks glancing into the pool as well.

"No," says Gim-bal looking up from the pool. "Just incase that is not a rhetorical question."

"Since I already know the answer, I guess it is," says W-abe wondering if there might be some defect in the pool bottom or side to justify Gim-bal's apparent study. "No one here had considered tank circuits significant until they saw Earth's use of large coils and large capacitors to manage high voltage and current."

"What do the coils and capacitors have to do with the voltage and current?" Gim-bal asks.

"When the components are matched," declares W-abe, "it is possible to circulate electrons back and forth between the coils on one end of the ship to the capacitors on the other end of the ship."

"This must do something besides generate heat," says Gim-bal easing away from the pool.

"Your suspicions are well placed. Most of the heat is inefficiently diverted for cooling, heating and a minor source of electricity as circumstances require," says W-abe with a glance at the horizon, "but the movement of electrons means a change in inertia. The change in inertia is what we use to move the ship."

"But electrons do not have much mass, do they?" Gim-bal asks.

"No, they do not. But the current flow causes molecules to shift in the direction of the flow," defends W-abe stepping onto the sand along with Gim-bal, "and they do have more mass."

"So the mass of the ship becomes an advantage for the first time," concedes Gim-bal.

"Redundant tank circuits allow some circuits to charge while others are discharging," continues W-abe. "That way the ship does not have to be rotated to take advantage of the bump in inertia."

"And the refueling issue?" Gim-bal asks looking at the horizon while he crosses the sand.

"I am sure that the coil circuits can generate electricity when they are moving through a magnetic field," says W-abe wondering what Gim-bal's presumed interest could be in the ocean. "So even though I do not know the specifics, I do not doubt the capacity to generate and store electricity on the way."

"Is this scheme efficient?" Gim-bal asks stopping well away from the water's edge.

"Again without knowing the specifics," contends W-abe while considering the possibility that Gim-bal has no interest in the ocean, "history has shown that it is 'sufficient' to our needs."

"Even if it is not efficient, it is no more than a cost issue," confirms Gim-bal as he turns back to the hotel. "The possibility of round trip travel now only depends on the costs."

"Exactly," echos W-abe feeling vaguely confident that Gim-bal found what he was looking for.

$$1$$
$$1\ 1$$
$$1\ 2\ 1$$
$$1\ 3\ 3\ 1$$
$$1\ 4\ 6\ 4\ 1$$
$$1\ 5\ 10\ 10\ 5\ 1$$
$$1\ 6\ 15\ 20\ 15\ 6\ 1$$

Chapter 23

Chapter 11101000....

Chapter ABCE)

Chapter $(1 + 5x + 9x^2 + 7x^3 + 2x^4)$

Chapter $(a^4b^0 + 5a^3b^1 + 9a^2b^2 + 7a^1b^3 + 2a^0b^4)$

Chapter $2^0 + 2^1 + 2^2 + 2^4$)

Episode 67

  For a resort restaurant on a resort island where the homes can cost more than a building-burb home, it is strangely pedestrian. No one has a uniform. The wait-service seems to wear what they have on that day. These are clothes that people wear at the coin laundry or the grocery store. They are clothes you wear on your day off.

You wonder if the convention of wearing shoes is a grudging concession to the work environment. For Dr. Mim-sey, it is something he notices every year. Dr. W-abe's concerns are, as always, elsewhere. But at least the drinks are prompt and he knows most of the other directors and their wives. So he is not easily distracted by them. He does not ask questions about anyone he does not recognize. They could be a friend filling in for the spouse or even a new spouse.

Whatever the real story is he will let them tell him what it is. Dr. W-abe never talks about his colleague's relationships. Certainly he knows their lives in detail. W-abe just does not have a habit of talking about them and their lives. It is only after a few years that Dr. Mim-sey connects faces to the names he knows.

This does not trouble him. Mim-sey is not networking. He is here to eat and if the rest of them take that for granted he does not. Dr. Mim-sey will study the simplest menu to be sure he gets the best the gulf has to offer. If he can derive that it was caught that morning so much the better.

Mim-sey could ask, everyone else does, but he is not sure he will get the truth. He is the last one to order first. Dr. Mim-sey tells the waiter he cannot decide. Mim-sey has to hear what everyone else is ordering so he can divine the secrets of the best choices. The advantages of a table of ten are not to be dismissed lightly.

Dr. Mim-sey does not talk much during these dinners. He is enjoying his dinner. The rest of the table more often than not talks family or shop which he knows the least and cares less about. Besides, there is time after dinner, when everyone goes their separate ways, that he can engage Dr. W-abe in the conversation that is important to himself.

"'The villain must die or we will.' can be represented as a combination of two polynomials or a system of two equations. What else?" W-abe asks idly stirring his fork in small circles on an empty plate.

"Are you in the mood for statistics?" Mim-sey asks pulling a small plate of sweets closer.

"As long as you do the work, I can count on it not being too complicated," says W-abe.

"Simple?" Mim-sey asks spooning his choice of dessert to his dinner plate rather than the unused and waiting desert plate.

"I did not want to say simple. Not just because I know that this is not your field. But because statistics and probability lose their value when they are too far removed from real life," says W-abe keying in an order for drinks and the hors d'oeuvres that he likes after a good dinner.

"I know that you like math even more than your close friends, so I can infer that you are telling the truth. While math is not a priority with me, reality is," says Mim-sey hoping W-abe ordered drinks for two.

"I suppose this operation of translating nouns and verbs to ones and zeros serves to transform qualitative values into quantitative values," drones W-abe stacking dishes to make room for the hors d'oeuvres.

"This makes it possible to graph a sentence in terms of a normal curve. Now we can see what part of the sentence is doing most of the work," says Mim-sey hurrying his dessert to make room in his own way.

"We cannot just assume that in a sentence with mostly nouns, that the nouns do most of the work?" W-abe asks looking across the room to see if the bartender is working on his drink order.

"That would assume that in a two-word sentence like 'Abbit lives.' each word carries 50% of the sentence. But even you can begin to doubt this simplification when you recognize the conspicuous change in the sense of the sentence by transposing the two words to 'Live Abbit'.

"The verb is changed from third person to second person and the statement becomes an imperative. One has a truth value the other can never have a truth value. This contrast could only be more striking if you said these over the prostrate body of Roy Abbit," explains Mim-sey hoping he has Dr. W-abe's attention beyond mere courtesy.

"This two word sentence is composed of a noun with the binary value of 10 which is decimal one. The verb has a binary value of 01 which is decimal two. They sum as 3 with a mean value of 1.5," details W-abe smiling at the arrival of the assortment of sugar cured, salted and acidulated fungi.

"Now, Mr. Abbit has to calculate the standard deviation for the noun but he has to do that with little or no experience or skill in square roots and the sigma that comprises the conventional formula for the standard deviation. The sigma indicates the sum of elements of the formula. Three is the absolute value of the sentence and is the sum, sigma, *romn* of the words," grants Mim-sey eyeing the sugar cured fungus.

"I can help you with this. I will call a sentence of two words: N = 2 where N is the number of words. Squaring the absolute value of each word and then *romn*, uh, summing them is another way of summing the elements of the sentence. For the words of this sentence: $1^2 + 2^2 = 5$," says W-abe as he takes a serving of his favorite acidulated white fungus.

"What do you do with these numbers?" Mim-sey asks finishing his dessert and reaching for the hors d'oeuvres.

"These values are used to calculate some intermediate values. When, the sum of the square of the absolute value of the words is multiplied by N, the number of words, you get the intermediate value of 10 (2 x 5 = 10). Another value comes from the square of the absolute value of the sentence. Remember, the absolute value of the sentence is three. Three squared is nine," says W-abe adding some salt to his snow white fungus.

"What do we do with these?" Mim-sey asks amending his desert with the sugared red fungus.

"The square of the absolute value of the sentence is subtracted from the sum of the square of the absolute value of the words or 10 - 9 = 1. This number is divided by a value that involves the absolute value of the sentence again. In this case the divisor is: [number of words times (number of words - 1)] which is [2(2-1) = 2].

"The result is 1/2. The square root of this number (10-9) / 2(2-1) = 1/2 gives us the standard deviation. In this case the standard deviation is 0.707106781," says W-abe pleased with the fresh crunch of his fungus.

"Now that we have the standard deviation we can calculate the area under the normal curve that the noun represents," inserts Mim-sey finishing his initial sample of sugar cured red fungus.

"This is done when the absolute value of the noun (1) minus the mean value of the sentence (1.5) is divided by the standard deviation of the sentence (0.707106781). This gives us the z-value of the noun component of the sentence," prescribes W-abe moving on to the pickled blue fungus and blue onion.

"So. What is does the z-value indicate?" Mim-sey asks trying to decide which hors d'oeuvres to sample next.

"There are three standard deviations on each side of the mean. The mean is the center of the normal curve. In this case, the z-value is not a whole number so it is a fraction of a standard deviation. The area under the curve that this number represents can be calculated," says W-abe with blueish lips.

"How do you calculate that?" Mim-sey asks taking some more sugar cured red fungus.

"I do not. It is just as easy to lookup in a table of z-values. In this case it is: (1 - 1.5)/0.707106781 which is a z-value of minus 0.7071," says W-abe smiling with blue teeth.

"What does a z-value with a minus sign mean?" Mim-sey asks.

"That minus sign on the z-value locates the area under the normal curve to the left of the mean (center) of the curve. This z-value means that 23.89% of the area under the curve is represented by the noun," says W-abe adding salt and blue oil to his blue fungus and blue onions.

"The noun represents a little under one fourth of the sentence," claims Mim-sey.

"So what is the verb's z-value?" Asks W-abe.

"Whatever it is, it represents 76.11% of the area under the curve," affirms Mim-sey.

"You might be amused to learn that the z-value for that number is the same but without the minus sign, 0.7071," says W-abe smiling with a satisfaction that seems to exceed the merits of the hors d'oeuvres.

"Another symmetry," says Mim-sey dismissing the substance of anything so obviously simple.

"Symmetries, infinities, and zeros are landmarks in mathematics," imposes W-abe.

Episode 68

The morning routine in the hotel is comfortable. It does not seem to matter how long they are separated.

Ali-ce and Roy are in and out of the bathroom for this and that without imposing on each other. What is most distracting, though, is the lights, mirrors and towels.

When you turn on all the bath lights, it is too bright. If you turn on just some of the lights there is not enough light. How do they do that? Not having a wall full of mirrors at home leaves Roy uncomfortable in the view of himself that he gets in the hotel bathroom. Neither Roy nor Ali-ce needs a mirror to see each other from head to toe with or without clothes. At least they can choose to see and be seen. If a fire chased you naked into the street, you would feel less naked.

As for the towels - there are always plenty, even for two people but there is no place to put them - not even the floor has enough room. How can a hotel have a big bathroom with no floor or counter space?

It works as a plan to keep you from stealing the hotel soap, if you do not want to be reminded of a place as alien as the hotel bathroom. Fortunately, for Roy and Ali-ce, like almost everyone else, they have other reasons to rent a room. At the same time almost anything else is more interesting to talk about.

"*Ali-ce?*" Roy reflects while studying his teeth in the bathroom mirror.

"*Mmmmm,*" projects Ali-ce swishing a mouthful of mouthwash.

"*Are you listening?*" Roy reflects pushing his tooth brush behind his upper left molars.

"*A little,*" projects Ali-ce now at the limit she can bear the mouthwash.

"*What is It?*" Roy reflects working his brush toward his front teeth.

"*Your mother and your molar,*" projects Ali-ce anticipating expectorating.

"*I am brushing my teeth,*" reflects Roy scrubbing the upper right molars.

"*You are thinking about your mother and that you might need a root canal,*" projects Ali-ce finally expectorating.

"*I am almost done,*" reflects Roy brushing his lower gums.

"*Almost done thinking about your mother?*" Ali-ce projects inspecting her gums.

"*You love doing that,*" reflects Roy scrubbing his upper gums.

"*Your language does that,*" projects Ali-ce massaging her gums with her finger.

"*Where is the floss?*" Roy reflects wiping his lips while Ali-ce hands him the floss.

"*That is the way YOU always do it. But you have not thought so much about your mother and father that I know of,*" projects Ali-ce washing her face with soap.

"*It is that birth certificate. It makes me think of her before I was born,*" reflects Roy.

"*She was mostly happy. Finding work was a problem,*" projects Ali-ce rinsing away the soap.

"*How do you know?*" Roy reflects flossing forward from his molars.

"*That is how you remember her from around that time,*" projects Ali-ce with her face buried in a towel.

"*Where is the mouthwash?*" Roy reflects flossing his upper teeth.

"Is it in the cabinet?" Ali-ce asks sliding the cabinet mirror open.

"No. That is what I remember her telling me," says Roy dropping the floss into the trash.

"*Look under the sink*," projects Ali-ce closing the cabinet.

"*We can talk after I am done,*" reflects Roy pouring a measure of mouthwash.

"*There is no hurry. Telepathy is fine,*" projects Ali-ce as she leaves the bathroom.

"*Why do you put the mouth wash down here?*" Roy reflects while Ali-ce begins dressing.

"*It is a habit I guess. In your last apartment the medicine cabinet was too small for the mouth wash,*" projects Ali-ce looking through the clothes in the bedroom closet.

"*Oh, yes, You will have to learn some new habits*," reflects Roy rinsing his teeth.

"*And you have to let me talk you into moving to Rome,*" projects Ali-ce surveying her choices of what to wear.

"*Your father would be alarmed if you were keeping me and I insist on earning a living*," reflects Roy unwilling to tolerate the mouthwash any longer and expectorating.

"You changed the subject and I want to know more about your mother," projects Ali-ce.

"Do you know that I can tell when you are listening, telepathically?" Roy reflects.

"How do you do that?" Ali-ce asks as Roy enters the bedroom.

"We talk a lot that way, anyway," maintains Roy, "and I know you are telepathic."

"But you are not supposed to hear or feel anything when I am just listening," says Ali-ce.

"Yes," says Roy. "But if I ask a question without verbalizing it and you always answer."

"Then I answer telepathically before I remember to say it in so many words," says Ali-ce.

"Most of the time," affirms Roy rubbing his cheek hoping forlornly not to shave today.

"It is just as hard for me to deceive you as it is for you to deceive me," protests Ali-ce.

"You do not know how much that makes me love you," protests Roy returning to the bathroom to shave. "But a lot of the emotion is not expressed or thought in words."

"I know but it puts you in a mood that dominates your words and thoughts," says Ali-ce.

"Then I guess you know what I am thinking," reflects Roy.

"*We have been doing this since you were seven,*" projects Ali-ce standing at the door.

"*Fifty years is a long time but we have not spent 20,000 days together,*" reflects Roy.

"*It only makes the few days we share more important,*" projects Ali-ce approaching Roy.

"*Did I fall in love with you when I was seven?*" Roy reflects covering his neck with soap.

"*No. But that was not our first meeting,*" projects Ali-ce watching Roy lather his cheeks.

"*I forgot about that,*" reflects Roy methodically scraping off the shaving cream with his disposable razor.

"*It is not important after a while,*" projects Ali-ce moving closer to Roy.

"*And it is all important after a while,*" reflects Roy nearing the end of his shaving task while looking at Ali-ce in the mirror.

"*I love how you can change the subject without changing the subject,*" projects Ali-ce standing close to Roy.

"*So you will not have much luck talking about my childhood now,*" reflects Roy.

"*I can tell but I am not disappointed,*" projects Ali-ce with a hand on Roy's lower back.

Episode 69

Officer Gim-bal does not use subterfuge to broach business with Dr. Mim-sey. No doubt Mim-sey already knows about the rare book pretext with Dr. W-abe. But the offer of a free lunch and drinks takes the edge off of formality and that is all Officer Gim-bal needs to situate Dr. Mim-sey for the duration of the conversation.

"Fairness and the force of law requires that any discussion of your Robe's Chamber petition cannot be withheld from them," concedes Gim-bal as their vehicle winds up a mountain road.

"Does that mean you are acting in their behalf?" Mim-sey asks studying passing foliage.

"No, it is part of the nature of that petition," allows Gim-bal accommodating the leisurely ascent by pouring a cold drink for them both. "It is as much out of my hands as yours, now."

"Am I to be intimidated or limited in whom I talk to or what I may say?" Mim-sey asks.

"Not in any manner," grants Gim-bal tasting his drink with satisfaction.

"Prudence is recommended?" Mim-sey asks tasting his drink with customary caution.

"Only because not lying is prudent," declares Gim-bal stirring his drink in the drink holder. "And that is always advised with or without a petition to the Robe's Chamber."

"So this is just a causal conversation," says Mim-sey satisfied with the quality of his drink.

"That depends on the topic," says Gim-bal as Mim-sey relaxes into the deep seat and cushioned armrest.

"All I just need to remember," says Mim-sey, "is that, casual or not, whatever I, or anybody, says about topics relating to my petition could turn up as 'hearsay or anecdote' for the Robes."

"That is a good idea," says Gim-bal with a glimpse of the gulf over Mim-sey's shoulder.

"Dr. W-abe told me that he has spoken to the Robes about the petition," says Mim-sey.

"The contents of the petition will eventually become public, so I will not be betraying any confidences by saying yes, he has," says Gim-bal as the waxy canopy envelops the vehicle.

"We both know that but I just want to gauge how close they are to making a decision," admits Mim-sey trying to tolerate the foreignness of the foliage and the dust of the winding, unpaved road.

"You think that if I were being evasive that I would be protecting their ability to gather more information by concealing what they may or may not know?" Gim-bal asks.

"That and I can compare your guile or lack of it to Dr. W-abe's," enlarges Mim-sey.

"I do not think that any of the Robes indulge in misdirection," advises Gim-bal.

"Neither do I," assents Mim-sey wondering if any resort is worth the price of this dizzying trip. "But I am concerned about how I am going to appear when the petition becomes public."

"Dr. W-abe does not seem concerned," deems Gim-bal. "About his image, I mean."

"W-abe did not file the petition," says Mim-sey, "and he would not think of his image anyway."

"You want to be on the winning side even if you do not know what it is?" Gim-bal asks.

"I have to give you more credit than I could ever give Dr. W-abe," recognizes Mim-sey.

"He would not anticipate the necessity or the advantage of being on the winning side of a public issue, uh, once it becomes public?" Gim-bal asks retreating into his drink for distraction.

"I think I know him well enough to say yes," says Mim-sey with his impatience surfacing.

"I thought he is a friend of yours. That is not very complimentary," cautions Gim-bal.

"Dr. W-abe is not one who can be insulted by the truth," apprises Mim-sey rethinking the merits of his friend and the sparse comforts of this casual assent. "He just does not think in those terms."

"He is not a politician?" Gim-bal asks while he pierces a piece of fungus with a small fork.

"No," says Mim-sey watching Gim-bal consider the taste and the quality of his choice.

"And you are?" Gim-bal asks while targeting his next choice of the salty fungus.

"Should I be insulted by that question?" Mim-sey asks reducing the tint on his window.

"I am not a politician," says Gim-bal redirecting his focus from the fungi to the broader view of the tropical foliage. "Political nuances defy my perceptions and as a result its practitioners do too."

"You should be a politician. I think merit, or maybe the lack of it, digs a hole for you and politics is the ladder out of the hole," says Mim-sey while feeling free to help himself to a serving of white fungus.

"History is full of meretricious politicians. How does true merit confine you?" Gim-bal asks without missing the flash of sky and sea breaking through the passing landscape.

"An expert cannot be or do anything else but serve his native expertise," says Mim-sey with a growing appreciation for the almost unbroken cover of tropical flora. "I could not become an opera star now if I wanted to."

"We both know you do not have a talent for opera. But your point is that if you did, your career would not only not support a change in role but it would argue against it," predicates Gim-bal having exhausted his interest in the flora.

"The exception, of course, is politics," imparts Mim-sey with no plan to revisit the white fungus. "No matter what your talent is, it can be parleyed into a successful political career."

"A successful politician is rich and powerful," says Gim-bal stirring the red fungus.

"And retirement only comes with failure or the grave," says Mim-sey stirring in his seat.

"Do you suppose there is any political potential to your petition?" Gim-bal asks.

"It would be impolitic to answer that question," says Mim-sey glancing at the landscape.

"I can see how a 'yes' makes your motives for the petition suspect." Taking a serving of red fungus, Gim-bal asks, "But, motives aside, why deny the political potential of the petition?"

"I could not deny its political value now and trade on it in the future," says Mim-sey taking some more white fungus while he postpones and disguises his preference for anything else.

"Your political opponents would remind you of your inconsistency constantly," proclaims Gim-bal looking at Mim-sey's serving of white fungus as it goes uneaten.

"But I hasten to add," says Mim-sey, "that I do not have political aspirations at this time."

"What would you say to the argument that your petition is just a way to launch your political career?" Gim-bal asks chasing his last bite of red fungus with a sip of his drink.

"A healthy political environment," says Mim-sey, "has every kind of argument, pros and cons."

"Is that another way not to say yes or no?" Gim-bal asks refreshing Mim-sey's drink.

"No. I could say 'no' without foreclosing on a political career," offers Mim-sey. "I could take up a political career and argue that ad hoc circumstances compelled me to change my mind."

"You cannot deny that there is any political potential to your petition but could you maintain that you are unaware of any political significance to your petition?" Gim-bal asks.

"It is a battle," declares Mim-sey, "to start a political career from a position of ignorance."

"The essence of a worthy politician is that he is right and knowledgeable, more informed than his opponents, even his constituency," acknowledges Gim-bal refreshing his own drink.

Swirling his drink in preparation for a sip, Mim-sey says, "No one would argue with that."

"You cannot deny that you are ignorant of at least some political substance to your petition," advances Gim-bal. "You cannot maintain that you will never capitalize on that political fodder and you will not say or do anything to foreclose on that eventuality."

"Who would believe me?" Mim-sey asks.

"The Robes would have an obligation to believe," says Gim-bal, "what you swear to be true."

"And after their decision to either ignore or act on my petition," responds Mim-sey, "I would spend the rest of my life disadvantaged by my own expediency in the public record."

"Where is the disadvantage in your circumstances?" Gim-bal asks.

"I have no second career. If I leave my current career, my resources would evaporate," Mim-sey protests almost mechanically. "I could not reenter my current career with any hope or expectation of resuming where I left off."

"What if this conversation became public?" Gim-bal asks in the glare of a patch of light created by the paved quadrangle in front of the mountain hotel.

"We are both obliged to report these remarks to the Robes to keep the petition process transparent," Mim-sey remarks in a lecturing tone. "This conversation will become part of the petition record and eventually available to the public. So there is no - if."

"All right then, what happens *when* this conversation becomes public?" Gim-bal asks.

"I will be persuaded, despite my reluctance, to take up the gauntlet in order to preserve or restore the public tranquility," says Mim-sey as the vehicle stops at the stony steps of the hotel.

"Are you expecting some unrest?" Gim-bal asks in the blaze of the treeless courtyard.

"Publication of this petition will transform the legend of that alien Earth into fact," rebuts Mim-sey with a squint resembling a wink. "I would be surprised at any peace ever again."

1

1 1

1 2 1

1 3 3 1

1 4 6 4 1

1 5 10 10 5 1

1 6 15 20 15 6 1

Chapter 24

Chapter 00011000....

Chapter DE

Chapter $(1 + 5x + 10x^2 + 7x^3 + 2x^4)$

Chapter $(a^4b^0 + 5a^3b^1 + 10a^2b^2 + 7a^1b^3 + 2a^0b^4)$

Chapter $2^3 + 2^4$

Episode 70

    Since Dr. Mim-sey and Dr. W-abe cannot be sure how much they will be drinking while they are at the board of director's retreat, they never provision the suite's kitchen with things for breakfast. When they are up early enough for breakfast, there are plenty of hotel restaurants that serve it.

Breakfast at the best hotels also includes the best service. Dr. Mim-sey, who likes breakfast anyway, likes it all the more with service. Dr. W-abe does not seem to notice. He focuses on his food but he does not luxuriate. Breakfast is breakfast. Dr. Mim-sey is ready to concede without prompting that luxury is luxury.

However, extending it to breakfast is rare for him and he is always surprised and delighted when the wait-service puts the napkin in his lap for him. They do it so fast and naturally he wonders how they learn to do that. And just as wonderfully, they disappear without imposing on the tempo of the breakfast or Mim-sey's favorite topic of conversation which begins at the buffet.

"I suppose we are going to have to look at Mr. Abbit's spin on probability, now," protests W-abe studying his choices of chilled pate and weighing them against his appetite.

"No doubt, I am going to have to make you want to look," retorts Mim-sey trying to decide which pate will be hardest to find once he returns home.

"I am already convinced that his mathematics will, if published, promote ideas that will turn Earth's eyes our way," responds W-abe shifting his focus to the buffet and selecting steamed vegetables for part of his breakfast.

"By eliminating dead end choices and supporting good ones?" Mim-sey asks.

"That and the career mathematicians who hold sway now with their dead end choices and bad ones will be replaced," presages W-abe positioning portions of cool pate and warm vegetables on his plate. "They will be eclipsed by mathematicians who have been using category mathematics for a lifetime."

"It seems ironic that the task falls on me to drag you through the math," observes Mim-sey comparing his choice of pate and vegetables to W-abe's. "You have much more experience."

"Well, Mr. Abbit uses language as his jumping off point. That is your field," says W-abe turning toward his table.

"Okay, to put things in your field, have you noticed that Mr. Abbit assigns category values to every category he uses?" Mim-sey asks following W-abe and giving up on cobbling together a better breakfast.

"And elaborately, each word in his iterations has a value of 2^n," says W-abe. "Upon dividing them into subgroups, they retain their assigned values and sum into a net value of 2^n again."

"Each word is a packet and part of a larger packet," says Mim-sey.

"Like electrons or photons," says W-abe stirring a sweetener into his drink. "I suppose you are going to tell me categories have velocity too"

"I will break it to you easily. Do you still like your blue beans?" Mim-sey asks.

"With blue onions. And when they cook down enough I can have blue beans on bread with a big slice of blue onion," confirms W-abe raising his cup to his lips.

"The store where you buy your beans probably orders them once a month," holds Mim-sey replacing his breakfast plate with a smaller one, "even though they may sell some of them every day or so."

"That is the nature of the business," acknowledges W-abe lingering with his main course. "No doubt Mr. Abbit is familiar with this, age old, practice."

"Some things sell 'faster' some things do not. Right?" Mim-sey asks.

"Each and every item in the store sells at some rate," admits W-abe, "and that changes across the year according to cycles like school, holidays and vacations - lots of things."

"Would you call it a corruption of the term to say that each item has a 'velocity'?" Mim-sey asks surveying the selection of sweet and sour sauces among the condiments.

"It is analogous but not similar mathematically," protests W-abe.

"Would you attribute 'velocity' to Mr. Abbit's treatment of words in a string or iteration?" Mim-sey asks trying to decide if the red fungus will be as good as he hopes.

"Because they drive you to a conclusion?" W-abe asks.

"Even though the nouns and verbs have different velocities, they convey you to the sense of the iteration," presses Mim-sey while wondering which fungus W-abe would choose.

"I cannot say that one word or another is faster but they all sum to the net category value of the string. The combination of words is equal to 2^n where 'n' is the number of words in the string," conveys W-abe pushing the red fungus toward Mim-sey with sympathy for his current quandary.

"The velocity never can exceed 2^n?" Mim-sey asks pulling the red fungus closer to himself.

"No," utters W-abe without looking at the red fungus or Dr. Mim-sey. "You only succeed in adding more words and doubling the subsets with each word you add."

"So, you have a string that is composed of substrings," formulates Mim-sey serving himself generously, "that never exceed the value or velocity of the string itself."

"Are you trying to mimic the notion of 'the speed of light'?" W-abe asks.

"Among other things," says Mim-sey turning his attention to the greying hors d'oeuvres and deciding against any of them.

"Are you saying that this analogy will lead Earth's scientists to the idea that the speed of light is a quantum function?" W-abe asks spooning some acidulated white fungus over his vegetables.

"What do you say to the idea that Mr. Abbit points to that door?" Mim-sey asks.

"The mathematicians with real skills," elucidates W-abe, "would be able to define circumstances for speeds that are faster than that of light."

"And that will eliminate our advantages of distance and time" says Mim-sey stirring his red fungus without the temptation to eat any of it.

"It will not take millions of years," agitates W-abe, "to travel millions of light years."

"The oceans of time and distance will not hide us or protect us," clamors Mim-sey.

"Our super-luminous communications will be tapped," blusters W-abe, "and our location fixed."

"Not just our location but our colonies, ships and weapons," charges Mim-sey.

"And if nothing is hidden then nothing is out of reach," says W-abe savoring his breakfast apart from the possible fate of his universe.

"What is not out of reach is not safe," grieves Mim-sey hopefully testing some smokey sauce on his fading fungus.

"Mr. Abbit must never publish," conveys W-abe completely satisfied, at least, with his breakfast.

"Mr. Abbit must die," charges Mim-sey marrying his disappointment with his breakfast to his cosmic contempt.

Episode 71

Ali-ce always gives Roy a key to her apartment or suite so he can come and go at his convenience. Roy reciprocates with his keys. So Roy is not disappointed when she is not in her suite when he comes in. He is not surprised when she comes in soon after he arrives and she all but expects to find him there.

They find each other when they want to. Otherwise, Roy spends five days a week at the cash register deflecting the perries of customers with an agenda beyond pay-and-go. Ali-ce spends her time compressing information to be transmitted to her father by one of the Gim-bals.

She does not have deadlines, so she does not have to work while Roy is off. But sometimes she does, partly because that is where her interest is at the time and partly because Roy does not need her attention all the time. Sometimes just their mutual company is enough. Sometimes the need for conversation takes the lead.

"I must go back to Rome a few days early," says Ali-ce propping herself up with pillows.

"You must?" Roy asks turning to Ali-ce. "We do not get enough time together as it is."

"That is true for now," says Ali-ce. "But If we are lucky it will not always be that way."

"That is not much consolation now. Why do you have to go back early?" Roy asks.

"I have to arrange some time off to go see my father," says Ali-ce as Roy sits up for the conversation. "And I want to go to Verona to see if there are more records on your birth."

"Were you called back home?" Roy asks leaning into his stack of pillows.

"No," says Ali-ce as she parks her papers on the night stand. "If I were called back I would refuse to return and stay here with you."

"Are you trying to provoke him?" Roy asks covering his feet with part of the bedspread.

"Not that so much," speculates Ali-ce while covering Roy's legs with the bedspread. "I actually think he will not want me here when his victim gets murdered."

"Do you think it will be me?" Roy asks with some minor adjustments of the bedspread.

"I have not been able to eliminate you from the possibilities," divulges Ali-ce.

"If he calls you back, maybe it is to keep you from being connected to the crime," allows Roy as Ali-ce turns on her side and faces Roy.

"Perhaps. At least we have almost two more weeks together," says Ali-ce helping herself to her share of the bedspread. "I want to spend some of that time working on our telepathy."

"Are you going to teach me to telep' some more?" Asks Roy.

"No offence but no," admits Ali-ce. "I have to do this while you are sleeping."

"Okay, but I do not think you are going to learn anything that is new while I am asleep," confesses Roy wondering if he should feel sleepy now even against his will.

"I know a lot about the physiology of sleep. What I need to know is how you wake up," says Ali-ce ignoring Roy's increasing tension.

"I can do that with my eyes closed," grants Roy.

"That is another reason for this telepathy while you are sleeping," confides Ali-ce pretending to be impatient. "You always deflect the subject before I am done with it."

"I was just being funny," declares Roy just incase Ali-ce was not pretending to be frustrated.

"And I love you for it," says Ali-ce soothing Roy's ambivalence. "But we do not have much time."

"What will you be looking for while you are rummaging around in my sleep?" Roy asks.

"Your breathing and heart rate will be at their lowest then," says Ali-ce gathering her papers from the night stand. "I will need to get you out of that state as quickly as possible."

"That sounds harsh," protests Roy a little unconvinced of Ali-ce's attention to her papers.

"It will be," says Ali-ce. "You will have to collect your senses quickly and fight the sense of panic."

"What are you going to do?" Roy asks unconcerned with Ali-ce's interest in her work.

"Do not worry," says Ali-ce pushing his shoulder and tipping him toward the edge of the bed with real effort. "I am not going to push you out of the bed or throw cold water on you."

"But will I wake up as if I were?" Roy asks laughing and resisting Ali-ce's attempt to tip him out of the bed.

"You sleep very well. You will not want to leave it," says Ali-ce giving up on her ingenuine effort to push him over. "But I will want your heart and lungs to work hard and fast."

"Is this something we must practice?" Roy asks hoping for something more to his liking.

"You can do it, no doubt," says Ali-ce. "But you need to be able to do it against your will."

"Does this have to do with your father's plans for murder?" Roy asks more to the point.

"Perhaps," says Ali-ce looking grave. "I might have accidentally played into his plan."

"How is that?" Roy asks leaning down on his elbow to look into her eyes.

"...remember? I sent him *Romeo and Juliet* as a distraction," says Ali-ce disappointedly.

"Your father is going to punish me by committing suicide?" Roy asks too quickly.

"There you go again," says Ali-ce forcing a smile and trying not to discount his charm.

"I see what you mean. So tell me what this has to do with your father," urges Roy.

"Romeo is banished," imparts Ali-ce as Roy returns to the comfort of his pillows, "and Juliet goes to Friar Laurence to prevent her marriage to Paris."

"I think Juliet says that she would rather die than marry Paris," chimes in Roy.

"And Friar Laurence says that if she means it then she will be willing copy or fake her death as a remedy," depicts Ali-ce grateful for Roy's attention to the point.

"Juliet comes up with some extreme options of her own. Do you think your father finds any of them appealing?" Roy asks a little myopically.

"That is my father you are talking about," says Ali-ce matching Roy's near sightedness.

"I am trying to avoid making your father a common killer," says Roy in his own defense.

"So am I," laments Ali-ce defensively.

"What are you getting at then?" Roy asks acknowledging the score to be even.

"Friar Laurence gives her a drug and explains that she will sleep," recounts Ali-ce gripping the beadspread tensely. "That sleep will be so deep that everyone will think she is dead."

"She misses her own funeral, if I remember," relates Roy resting his hand on her fist.

"The first one," relates Ali-ce relaxing her hand. "I want you to be so lucky."

"You think your father plans to poison me?" Roy asks taking Ali-ce's hand in his.

"It is a possibility that I have not thought of before now," says Ali-ce tightening her grip on his.

Episode 72

Officer Gim-bal is beginning to find new respect for Dr. W-abe's capacity for toleration if not indifference. Or perhaps, Dr. Mim-sey curries Dr. W-abe with more caution. At any rate, island culture does not seem to cultivate late evenings when it comes to restaurants. At least until cooks and waiters get paid like bartenders. So dinner and drinks see an inconveniently early end in the restaurant for Dr. Mim-sey. But Dr. Mim-sey is not against propping himself up at the bar.

Officer Gim-bal is secretly relieved to have this option but he makes a feeble protestation about having to catch a morning flight. He wants Dr. Mim-sey to feel like he has the upper hand. Rus-ty Gim-bal plays the role of the captive audience for the moment at least.

"You knew before you filed your petition with the Robe's Chamber that it would cause social and political unrest and more than a little," imposes Gim-bal gauging the distance of the skinny bartender from their seating and his conversation.

"Is there anything wrong with that?" Mim-sey asks with a focus on his half empty glass which in all his years has never lingered at half full.

"Yes," prescribes Gim-bal shifting his attention to Mim-sey. "If that was your motive."

"Are you trying to get me to admit to insurrection or treason?" Mim-sey asks.

"You could advocate such a thing publicly," relates Gim-bal, "but that by itself would not convict you of it."

"Advocating insurrection is not a chargeable offense?" Mim-sey asks picking up his glass.

"In the absence of insurrection, yes," says Gim-bal

"If that does happen, how are you going to connect me to it?" Mim-sey asks wondering why the bartender has not noticed that he urgently needs another drink.

"It is not my job to jury or prosecute a case," says Gim-bal, "but that petition is a conspicuous candidate for insurrection."

"Under the most benign circumstances they would still have to make a very firm connection between my petition and any coincidental unrest," says Mim-sey calculating the seconds the bartender is wasting with other customers. "The unrest or controversy that attends a petition is not necessarily the motivation of the petitioner."

"You are only the messenger," says Gim-bal.

"The official existence of Earth has been in the hands of many of us for many years," says Mim-sey pushing his glass out for a refill as the bartender approaches. "If anyone has betrayed our civilization they, those many, have."

"You are suggesting that the cognoscenti have been standing between us and our ability to prepare for our own defense from the beginning. And now, thanks to you, everyone can openly begin to prepare Clidim's defenses and remedy the error of the past generations?" Gim-bal asks deliberately cultivating Mim-sey's prejudice.

"In a word...yes. Whoever tries to make a case against me would just provoke division of the public," allows Mim-sey retrieving his new drink. "My accusers are guilty of what they are accusing me of. Whether they know it or not."

"His, her or their credibility would be destroyed from then on, even if appointed to office for life. I suppose," says Gim-bal confident that Mim-sey has returned his undivided attention to the conversation and with any luck his motives and prejudices.

"You are referring to the Robes, of course," says Mim-sey turning his glass in increments.

"A lesser authority does not have jurisdiction," says Gim-bal returning part of his attention to the constantly moving bartender, "because the petition is filed with the Robe's Chamber."

"It can only be 'returned' to the Robe's Chamber," says Mim-sey reaching for a napkin.

Gim-bal lectures, "You knew that before you filed your petition."

"Yes," says Mim-sey putting a dry napkin under his glass.

"That makes you look like you are threatening the Robe's Chamber. Does it not?" Gim-bal asks.

"It does not take much thought to realize," disputes Mim-sey, "that such a threat, if it is a threat, does not threaten our entire civilization like the existence of a planet like Earth."

"If the resulting social and political unrest can be managed and our civilization survives, I suppose, the stigmatized Robes we have now will be replaced, through due process, with Robes who are committed to the defense of our civilization," argues Gim-bal as he takes the smallest possible sip from his drink.

"And those 'stigmatized' Robes are the ones who are supposed to convict me of a crime? Mim-sey asks while considering the negative effects of melting ice in his all too expensive drink.

"Then it does not matter whether Roy Abbit is a threat or not?" Gim-bal asks.

"It does not," says Mim-sey. "When the Robe's Chamber issues a death warrant, the public, that they serve, will learn of the threat and the execution of its enemy at the same time."

"And if they do not elect to execute Abbit?" Gim-bal asks getting up to leave and beckoning the bartender in order to pay the tab.

"Someone has to be made the enemy. They can only make things worse if they kill the messenger. They will only have themselves to blame," belittles Mim-sey occupied with the ambivalence that comes with Gim-bal picking up the tab and useing his own money for his next drink.

1

1 1

1 2 1

1 3 3 1

1 4 6 4 1

1 5 10 10 5 1

1 6 15 20 15 6 1

Chapter 25

Chapter 10011000....

Chapter ADE

Chapter $(1 + 5x + 10x^2 + 8x^3 + 2x^4)$

Chapter $(a^4b^0 + 5a^3b^1 + 10a^2b^2 + 8a^1b^3 + 2a^0b^4)$

Chapter $2^0 + 2^3 + 2^4$

Episode 73

Dr. W-abe is consistently casual about lunch in the suite when he is between meetings. But unlike home and every place else where cost is irrelevant, here, he is uncharacteristically pedestrian. Bread and drink are local and without pedigree. Pate, and sauces get the anthropological, "If they eat it, I will." Local fruits, vegetables, and oils complete the menu.

Dr. Mim-sey is afraid that if there were no dishes or utensils, W-abe would get along without them and Mim-sey did not come all this way for soup and sandwiches. Here, like everywhere else, W-abe does not cook or buy things that require cooking.

Dr. Mim-sey for his part does not eat lunch unless Dr. W-abe does. Mim-sey supposes he will not have as good a meal on his own. So he takes his culinary cues from Dr. W-abe. Besides it is an opportunity to engage Dr. W-abe with the looming danger of Roy Abbit and Earth.

If Dr. W-abe suspects this, he never says so. He just assembles a serving or a meal according his mood. The drink, the pate, the complement of vegetables, fruits and seasoning that he wants occur in the order he wants them. Dr. Mim-sey supposes the choices are the best ones and he will have the most satisfying possible meal. The other advantage to this sycophantism is that he has Dr. W-abe where he wants him and the mood he wants him in for the continuing conversation.

"I am convinced," says W-abe pulling food from the pantry and chillers, "that Mr. Abbit must die because that is the surest way to prevent the publication of his Category Mathematics."

"And that is the surest way to make sure that Earth's mathematical establishment misses its chance at simplifying the mathematics that will reveal our existence," says Mim-sey.

"In that case, it could take them centuries to simplify the mathematics for a Theory of Everything to a cerebral or machine scale," drones W-abe collecting utensils and dishes from cupboards, hooks, shelves and drawers.

"Until then," says Mim-sey looking for a seat at the counter that does not impose on W-abe's culinary mission, "Earth will only be able to approximate the physics and mathematics that points to us."

"Earth will not know the quantum effects that provide for light speeds that exceed normal space," explains W-abe positioning his utensils, food, and plates for maximum convenience.

"Earth will not know how matter and the space it occupies compresses into dimensions beyond normal space at those higher speeds," says Mim-sey feeling free to move closer to W-abe and the food.

"You mean a phase transition?" W-abe asks while opening containers and unwrapping packages.

"I do not know the technical term for it," underlines Mim-sey unstacking and positioning empty dishes, "but I know that that is where matter goes when it approaches normal light speed."

"It is also where matter and space came from when light assumed its, now, 'normal speed'," highlights W-abe while he fills the dishes with his choices of pate and fungus.

"Was that a phase transition?" Mim-sey asks while pulling bottles of drink from the coolers.

"Matter, as we know it," asserts W-abe, "is condensed out of a very hot universe."

"Like water from steam?" Mim-sey asks while wondering which bottles W-abe will open.

"That is a phase transition," concurs W-abe placing heated and chilled vegetables on the counter. "The result is the molecules of water take up less space and the mass per unit area is higher."

"I suppose," submits Mim-sey, "water turning to ice is also a phase transition."

"Yes," says W-abe smoothly dishing up heated and chilled sauces. "But ice has a slightly lower mass per unit area than water so it floats in water."

"So the only thing keeping their scientists out of the next quantum neighborhood and ours is knowing where to look," says Mim-sey while wondering how long it will be before it is time to eat.

"You promised to tell me what probability has to do with that," prods W-abe.

"More specifically," prompts Mim-sey while accepting a dining plate from W-abe but waiting first to see what he chooses, "what Mr. Abbit's perspective on probability has to do with that."

"So its going to be simple then," suggests W-abe serving himself some pate and vegetables.

"Just remember," says Mim-sey serving himself similarly, "it is simplicity that seals our doom."

"I do not like to give him credit but I think that it is, uh, **his** version of the simple that seals our doom," imputes W-abe retrieving some glasses and pouring drinks for them both.

"I will give you an example. Suppose you are playing a card game where it would be to your advantage to pull a Jack out of a 52 card deck. If the deck has four suits and each suit has thirteen cards, What do you think the probability is of drawing that Jack?" Mim-sey illuminates without looking up from his plate or putting down his fork.

"Four times thirteen is fifty-two. Four suits mean four Jacks," prescribes W-abe while Mim-sey takes a few eager bites of his lunch. "So my chances are four out of fifty-two."

"Now suppose Mr. Abbit knows that the face cards are marked and you do not," issues Mim-sey while W-abe ladles a warm blue sauce on his vegetables.

"With three face cards and four suits, the Jack is among those twelve cards," enumerates W-abe alternating between bites, drinks and comments. "His chances are four out of twelve."

"He could draw a Queen or a King because he does not know which face cards are Jacks," specifies Mim-sey reaching for a spoonful of sauce, "but he still has a significant advantage."

"Changing what you know changes the probabilities," W-abe says.

"Even though you have not changed the number of cards in the game," says Mim-sey.

"What about our mantra, 'Mr. Abbit must die, or we will.'?" W-abe asks.

"That is more to the point," cites Mim-sey. "The chance element like drawing cards is not eliminated from the probability calculation," while he pours more cold drink into both glasses.

"This is his threat to us," judges W-abe adding some ice to both glasses. "Mere knowledge that does not predict our existence can still point directly to us."

"And that probability is surprisingly high," Mim-sey advocates his point as W-abe takes a drink.

"According to Mr. Abbit a seven word sentence," imposes W-abe, "our seven word sentence, is comprised of seven categories. So its absolute value is equal to 2^7 which is 128."

"For Mr. Abbit, row seven of the Pascal/Chu triangle represent the distribution of the subsets for the seven words or categories," prescribes Mim-sey before taking a drink of his own.

"The entries for that row are 1, 7, 21, 35, 35, 21, 7, 1," defines W-abe with a helping of pate. "Both entry number zero and entry number seven is 1; entry number one and entry six is 7; entry number two and entry five is 21; entry number three and entry four is 35."

"Mr. Abbit begins with the assumption that for any iteration either you understand it or you do not," depicts Mim-sey as he is ready for another round of vegetables, pate and sauces.

"If p = 'understood' then p = 1/2. If q = 'not understood' then q = 1/2," delineates W-abe pushing Mim-sey's favorites closer to his friend. "So p = q = ½."

"You now have all the information you need," represents Mim-sey helping himself unselfconsciously to second servings, "to calculate the probability of understanding a seven-word sentence."

"Let me see. Entry 7 of row 7 is 1," renders W-abe stirring a sweetener into his drink.

"And you have to multiply that by p and q," says Mim-sey rethinking the taste of his drink.

"There are seven words that represent the attempts to make a statement so $p = (1/2)^7$," maintains W-abe following a satisfying sip with his choices of pate and vegetables. "The same is true for 'q' and there is a need to factor in the seven words that the seven attempts represent."

"So one times 'p' times 'q' becomes what?" Mim-sey asks choosing not to sweeten his drink.

"It becomes $1 \times (1/2)^7 \times (1/2)^{7-7}$," says W-abe cutting some island fruit into serving pieces.

"The 'p' and 'q' are equal except for the minus seven in the exponent," affirms Mim-sey.

"There is a difference between seven attempts and seven words. They both have to be counted and this is how you do it," says W-abe. "That number $(1/2)^{7-7}$ which is $(1/2)^0$ which is 1.

"So the probability of understanding a seven-word sentence becomes $1 \times (1/2)^7 \times 1$," claims Mim-sey as W-abe sets out the plates.

"One times one is one. And if you multiply $1 \times 0.5 \times 0.5 \times 0.5 \times 0.5 \times 0.5 \times 0.5 \times 0.5$ you get 0.0078125 which is $1/128$," says W-abe helping himself to some fruit and sweet sauce.

"That is one out of 128 chances that a seven-word statement can be understood," argues Mim-sey anticipating his choices of fruit pending the finish of his pate and vegetables.

"Well that shows you what Mr. Abbit knows," W-abe stresses. "I know just about 200,000 words, phrases and translations in English and I do not have any trouble understanding a seven-word sentence."

"You have anticipated Mr. Abbit's caveat," says Dr. Mim-say. "The small chance that a seven-word sentence can be understood does not take into account the benefit of common experience and context."

"Common experience and context 'mark the cards' so to speak," says W-abe.

"You know what people are saying or trying to say when they say it," says Mim-sey.

"When someone informs you of something, your own lifetime of experience supplies a lot of the information," says W-abe. "You cannot be informed of anything from a perspective of total ignorance."

"What if someone uses a word that you do not know?" Mim-sey asks as W-abe takes a tentative taste of his choice of fruits and chilled sauces.

"Then my chance of understanding the sentence goes down," says W-abe preparing a serving of fruit and sweet sauce for Mim-sey. "But I will have a feeling for what that new word is."

"So while the probability of understanding a word is 50%, the probability of understanding a seven-word sentence is 98.17%," says Mim-sey watching W-abe's choices closely in order to weigh his judgements.

"Where does that number come from?" W-abe asks.

"It is Abbit's," says Mim-sey. "It represents the area under a normal curve where you are likely to understand one to seven words of a seven-word statement as opposed to zero words."

"A normal curve assumes ideal circumstances," says W-abe

"In this case both sides of the curve represent a common experience and a common context," says Mim-say

"Most of the time," says W-abe setting Mim-sey's favorite fruits conveniently close to him.

"So you do not need to know 250,000 words right away," says Mim-sey. "After a certain point you can add words but there is not much gain in the probability of understanding."

"Or there is not much of a reduction of misunderstanding," says W-abe.

"So it is not how much you know about math and physics, it is what you know," says Mim-sey seemingly as much to his plate of fruit as to Dr. W-abe.

"And Mr. Abbit is shining the light on what you know not how much you know," says W-abe seemingly finishing his lunch along with a significant conclusion to the conversation.

Episode 74

Roy never thinks of this anywhere else or at any other time. Perhaps it is just because he lives in the same place for as long as possible. When he thinks about it, it happens even without Ali-ce around. So it is not because she is here or because of her. Perhaps it is because hotels are especially motivated and methodical about making the rooms dark.

This is a challenge during the day, but at night, hotels excel at making a room dark. Light does not leak in from the doors and windows. Usually there is no night-light. Before going to sleep there is only the lamp and the television. When you turn those off, the darkness is complete.

This is when Roy realizes that even though it is so dark, he is still acutely aware of the size of the room. The room does not get larger or smaller. There is no way to confirm the perception. Accuracy is not an issue. But it is curious to him that somehow when the lights were on, he measured the room in his mind without noticing. Then in the immediate first moment of total darkness the dimensions and orientation of the room take first place. If there is any light in the room, no matter how little, the perception is gone. Otherwise, it will be waiting for him when he wakes up. When waking up is an option.

"*Roy,*" projects Ali-ce in the darkness saturated with the motoring air-conditioner.

"*Ummmmm,*" reflects Roy with his back to Ali-ce and a pillow pressed to his chest.

"*Wake up,*" projects Ali-ce facing the darkness where Roy's back would be.

"*Ummmmm,*" reflects Roy pulling his pillow tighter to his chest.

"*If you want to live, wake up,*" projects Ali-ce as Roy pulls his knees up to the pillow.

"*Um,*" reflects Roy with his shoulders and hips pulling away from Ali-ce .

"*You can hear them talking,*" projects Ali-ce as Roy tucks his chin down into his pillow.

"*Yes,*" reflects Roy with his pillow pressing against his chin, chest, stomach and thighs.

"*They are talking about you,*" projects Ali-ce blinking in the darkness without any effect.

"*Me?*" reflects Roy with an interruption to his breathing.

"*You are killing your mother,*" projects Ali-ce.

"*I cannot kill my mother,*" reflects Roy still holding his breath.

"*They are going to stop you,*" projects Ali-ce as Roy begins to exhale.

"*I want to sleep*," reflects Roy relaxing his legs.

"*You cannot*," projects Ali-ce as Roy turns on his back in the darkness.

"*I cannot wake up*," reflects Roy with his right arm across his sternum.

"*The light is getting very bright*," projects Ali-ce belying the darkness.

"*It hurts*," reflects Roy without flinching.

"*Scream*," projects Ali-ce without compromising the silence.

"*The light is getting brighter*," reflects Roy without squinting.

"*Scream*," projects Ali-ce into the silent darkness.

"*There is something in my mouth*," reflects Roy lapsing into apnea.

"*It is nothing. Scream*," projects Ali-ce as Roy's heart begins to pound harder.

"*I cannot*," reflects Roy breathing again with a heave.

"*I will be here when you wake up. Scream*," projects Ali-ce as Roy exhales deeply.

"*She cannot die*," reflects Roy as he begins cycling air deeper and faster.

"*Your mother will live. Scream*," projects Ali-ce into darkness punctuated by breathing.

"*What is that?*" reflects Roy holding his breath.

"*Cold. Scream,*" projects Ali-ce as Roy's hand moves up to his face.

"*Make it stop,*" reflects Roy inhaling as if it hurts.

"*You make it stop. Scream,*" projects Ali-ce as Roy hesitates to exhale.

"*I cannot,*" reflects Roy as heat rises in his arms and legs.

"*Scream,*" projects Ali-ce as Roy's neck begins to sweat.

"What was that?" Roy asks without checking to see if Ali-ce is awake.

"You were dreaming," says Ali-ce in the darkness and without delay.

"I could hear someone but I could not wake up," says Roy into the darkness.

"It was me. I had to wake you up," says Ali-ce placing a hand on Roy's chest.

"Turn on the light and we can talk," says Roy placing his hand on Ali-ce's.

"*Okay but the talk has to be telepathic. You need the habit,*" projects Ali-ce.

"*Is someone listening?*" reflects Roy squinting in the painfully bright light.

"*No and I would know it if there were,*" projects Ali-ce with her eyes still closed.

"*Was that you telling me to scream?*" reflects Roy opening his eyes slowly.

"*Sometimes. It was also you telling yourself to scream,*" projects Ali-ce.

"*But I could not scream,*" reflects Roy sitting up and stacking his pillows.

"*It is not a dream if you can inject your will,*" projects Ali-ce sitting up too.

"*Why scream?*" reflects Roy leaning back on his pillows.

"*In order to wake up you must increase your heart rate and breathing,*" projects Ali-ce.

"*The fear is not enough?*" Roy reflects as he covers his legs with his blanket.

"*Not to wake up. You need more oxygen in your blood. So you have to change your breathing,*" projects Ali-ce propping herself up against her pillows and adjusting her blanket.

"*So did I scream?*" Roy reflects rubbing his eyes.

"*You did not this time,*" projects Ali-ce tenting her covers with her knees.

"*What do you mean this time?*" Roy reflects placing his hand next to Ali-ce's.

"*You will need to practice,*" projects Ali-ce placing her hand on his.

"*Now you are scaring me,*" reflects Roy with Ali-ce squeezing his hand.

Episode 75

A mail-depot lobby may seem to be an unlikely place for a serious conversation. And in most cases it is. But this is a resort island. The visitors want the postmark and they do not want to get home before the letters and postcards are delivered.

The visitors arrive by the hundreds with each flight. They go through customs, gift shop and the nearby and only mail-depot. Like everyone else, you begin your vacation counting your bags, your postage and your change more than once by the time you get to your room.

It is Dr. W-abe's suggestion that Officer Gim-bal should meet him here, in the middle of the post office and in the middle of the week. And to his surprise it is the largest, emptiest lobby on the island. They choose a counter as far from the front door as possible and by happenstance it is in front of a large window with an excellent view of the street. It is easy to admire Dr. W-abe's distinction between a private conversation and a secret one.

"I suppose you told Dr. Mim-sey you were going to the mail-depot," ventures Gim-bal.

"Of course," affirms Dr. W-abe noting that the stores do not seem to be open yet.

"Why did he not offer to come along?" Gim-bal asks.

"It is early for him," says W-abe supposing that the sparse pedestrian traffic is the native population preparing to open offices and stores. "Besides I said I want to go to brunch as soon as I get back."

"I am leaving before the weekend exodus. It is easier to get a flight," lists Gim-bal wondering when things get busy here. "I want to talk to Ali-ce Mim-sey before she leaves here for Earth."

"That circus is not to be missed. Everything about this resort screams resort except the flight-port. I bet 80% of the people here leave here to start the work week on time," says W-abe guessing that it will be at least an hour before sun reaches this side of the street and this window. "This is a curmudgeon resort."

"Do you suppose most of the people here saved half of their working lives in order to say they have been here?" Gim-bal asks privately guessing that most outings happen in the afternoon and evening.

"That and I think they could have done better things with their money," asserts W-abe.

"I have another supposition for you," musters Gim-bal with the intention of changing the subject.

"Go ahead," rallies W-abe hoping he is getting to the real subject.

"Do you suppose Mim-sey is serious about becoming a career politician?" Gim-bal asks.

"Has he been talking to you about that?" W-abe asks feeling a little unprepared for this approach.

"It is my fault," relates Gim-bal a little surprised at W-abe's surprise. "I brought it up."

"Then you know he is serious," notes W-abe resting both elbows on the lobby table.

"Everything about my conversations with him says so," discloses Gim-bal turning away from the window for no particular reason except to and rest his back against the table. "But I have trouble believing that he is serious."

"You do not have trouble believing he is serious. You have trouble believing he is a politician," says W-abe as someone is walking toward the mail-depot with a stack of envelopes.

"That would explain the incongruity," observes Gim-bal glancing up at the chandelier.

"He does not want to spend the years it takes to learn the tried and true ropes," offers W-abe.

"He is a dabbler?" Gim-bal asks watching someone cross the lobby, deposit some letters and leave.

"It is worse than that. He is a follower," opines W-abe while listening to the mail-depot patron come and go. "The most ignominious leader has a better chance at success in politics."

"So, what makes a politician?" Gim-bal asks hoping for W-abe to enlarge on his perspective.

"Politicians are not simple creatures. Each one is unique. But there is a common thread," advises W-abe deciding the patron is not a tourist but someone who knows the mail-depot.

"They have a universal reflex to lead?" proposes Gim-bal with a sense of the conversation.

"The trait is so common it tends to disappear. Their individual circumstances are unique and that prejudices the argument that the politicians are unique too," collogues W-abe. "It is the difference between the forest and the trees."

"What about ruthlessness?" Gim-bal asks as W-abe turns to face the lobby too.

"That is another 'bass-ackwards' characteristic. A politician must be ruthless when it is required of him. Anyone who thinks he, himself, is or can be ruthless is not a politician because he can embrace a ready reflex for ruthlessness without flinching," says W-abe resting an elbow on the lobby table.

"What kind of ruthless is Dr. Mim-sey," asks Gim-bal expecting to confirm a suspicion.

"If ruthless were a religion," persists W-abe, "Mim-sey would be a tireless preceptor."

"He admires ruthlessness that much?" Gim-bal asks to make sure of what he was hearing.

"It is something that he does not admit to even to himself," says W-abe aware of the test.

"What makes you so sure about Dr. Mim-sey's affinity for ruthlessness?" Gim-bal asks.

"It is basic to your career and you will recognize it when I tell you," persuades W-abe.

"Go on," says Gim-bal taking W-abe's familiarity as frankness. "You will not hurt my feelings."

"Friends are like those game boards with squares on them. They have two colors. Your friends let you see themselves on the squares of one color or the other but not both," reasons W-abe watching Gim-bal listening. "It is your job, as a friend, not to force them onto the other colored squares."

"I guess that is the distinction of friendships from every other relationship," says Gim-bal.

"I suppose you are right. You end-up knowing more about your spouse and your children than you want to," philosophizes W-abe polishing the edge of the table with his thumb reflecting on some personal experience.

"But not your friends or your parents, if you do it right, they retain a kind of mystery," depicts Gim-bal wishing for a footrest on the otherwise accommodating lobby table.

"Dr. Mim-sey could spend the rest of his life in political circles and never fail or succeed significantly," predicates W-abe turning again to the window-view of the street.

"So there is a lot of room for mediocrity in politics?" Gim-bal asks leading again.

"It is a harshness that is appropriate," posits W-abe again with a certain personal prejudice , "but it is usually dressed up as the great need for politicians to do a lot of the less than glamourous but necessary social work."

"Why would Dr. Mim-sey want to trade his current position for such a political grind?" Gim-bal asks as he turns to take in Dr. W-abe's view and becomes aware of the slight increase in pedestrian traffic passing by the mail-depot.

"I think, he confuses political success with all the other forms of success. In most of the other realms, success inspires a permanent following of parasites, usually called employees, who support and maintain the corps of success. In most cases the success is built on and lasts more than a generation. Some businesses, as we both know, last hundreds of years before spinning off into a galaxy of corporations," homilizes W-abe as he strikes a slightly less cynical cord.

"So it is how many times he can kill Roy Abbit?" Gim-bal asks with more feeling for politics.

"If he does not kill Roy Abbit regularly and often his political career is that of the one trick dog," prescribes W-abe polishing the point to something equal to the edge of the table in front of them.

"But I am not convinced that he plans to kill Roy Abbit," judges Gim-bal.

"I think he needs the demise of Roy Abbit and would prefer that someone else arranges it," says W-abe responding to Gim-bal's lead without any patronizing acknowledgment of the test.

"But the assassin, or the assassination, cannot be connected to Mim-sey without making him a criminal," proffers Gim-bal poking holes in his own potential theory.

"He seems to think that he can represent Roy Abbit as a danger," advances W-abe hoping to prop up the theory again. "Once the danger is removed - regardless of a how or when, he can take credit for being the good sentinel who sounded the alarm, not only in time but ahead of everyone else."

"So he is trying to make himself a leader from the safety of the crowd," tenders Gim-bal.

"And when he is sure there is a corps of sentiment," presses W-abe as he half burlesques the reality, "he jumps out of the crowd and says, 'I am not only just like you, I am ahead of you.'"

"The crowd is left to supply the words, 'So you cannot be anything but our leader.'," proclaims Gim-bal while grimly ignoring the flimsy burlesque.

"At this point in politics and business everybody agrees. But the crowd of followers in a political setting have to keep killing Roy Abbit or they do not have a reason to follow the leader," vents W-abe noting again that the traffic is mostly in one direction - still typical of workers going to work.

"The three-legged chair company is only a company as long as it makes chairs with three legs," ostends Gim-bal supposing such a company, or any company, likely uses a mail-depot later in the afternoon.

"So it does not matter who kills Roy Abbit," claims W-abe, "or whether he deserves it or not."

"And not very long after Abbit is gone so is his political significance," declares Gim-bal.

"And Dr. Mim-sey is a political 'has been'," predicates W-abe confident that the emptiness of the mail-depot and the emptiness of Dr. Mim-sey's aspirations are both as substantial.

1

1 1

1 2 1

1 3 3 1

1 4 6 4 1

1 5 10 10 5 1

1 6 15 20 15 6 1

Chapter 26

Chapter 01011000....

Chapter BDE

Chapter $(1 + 5x + 10x^2 + 9x^3 + 2x^4)$

Chapter $(a^4b^0 + 5a^3b^1 + 10a^2b^2 + 9a^1b^3 + 2a^0b^4)$

Chapter $2^1 + 2^3 + 2^4$

Episode 76

    For exotic, this surpasses even seeing live fish. Dr. Mim-sey marvels at the concept of a barbeque. It is one of his secret fascinations. And, here, this is real. It is big enough to cook for everyone. You cannot have a barbecue in the building-burbs. The smell of smoke will pollute an entire atrium.

You could not get away with even a small one. Punishment would begin with permanent eviction and get worse from there. Besides the danger of fire, it is a huge source of carbon monoxide. The death of choice for suicides everywhere.

He studies the grill-chefs in awe. There they are, working so matter-of-factly, immersed in heat and smoke every time the breeze changes. Their concern for the food, his food, is uninterrupted. What money they must make. How do they learn to do this? Where?

This is one time there is a line that adds to the occasion. The guests get to choose what they want from the grill and flirt with the primeval. Mim-sey studies the scene and immerses himself in wonder. Perhaps a guest will fall ill right in line. Will a smoke impaired chef have to leave on a pretext? No doubt, we will hear the ambulance.

Dr. Mim-sey shares Dr. W-abe's table along with eight others and has maneuvered for a view of the grill. If something happens, he does not want to miss it. He does not want to miss what he knows to be certain.

Fortunately, his food is not a disappointment even though the potential for drama at the grill disappears with the glow of the coals. The chefs leave the grill and gather at a table of their own.

Dr. Mim-sey can count on the rest of his table to move on with Dr. W-abe being eternally unhurried about his dinner. Dr. Mim-sey orders more drinks to keep it that way. It is another chance for conversation with Dr. W-abe.

"Without the end run of Mr. Abbit spawning endless generations of mathematicians with a different perspective on math and physics," advances W-abe with his mood buoyed by food he rarely experiences and never troubles himself to cook, "it could be hundreds of years before Earth gets it right."

"Those years will be spent in developing intellectual skills and a machine capacity to test their theories with incredibly large volumes of calculations," says Mim-sey as he luxuriates on the same lifting tide of cuisine and drinks.

"In the meantime, key concepts will compete with badly formulated ones. Theories will point to useful and useless avenues of thought without distinction," acknowledges W-abe. "Even the good ones will be different descriptions of the same thing with no way to choose best one."

"Mr. Abbit's contention, that the dimensions comprising a Theory of Everything must equal the subset of a power of two, will be lost in the wars of the theories," prompts Mim-sey following W-abe's lead and cutting his food one bite at a time and savoring each one.

"The short route to the notion of light-speeds that are hyper-normal and hypo-normal, will never lead to a theory that is prominent enough to compete with the space/time/matter and energy of a 'normal' universe," says W-abe barely aware of what he is saying.

"And Earth, without a capacity to model a 'normal' universe with light speeds greater than normal, will treat the physics of light, traveling below normal speeds, as just an interesting laboratory trick," goads Mim-sey hoping he is enjoying his food as much as W-abe is enjoying his.

"As Earth-time goes on," predicts W-abe, "hyper-space/time and its significance for matter/energy will become less probable, even as a theory."

"The woods are lovely/dark and deep...," proposes Mim-sey hugging himself with a cool sip from his bottomless glass.

"What is that?" Asks W-abe still concentrating on his plate.

"It is part of a poem my daughter sent me," tenders Mim-sey while trying to decide if it is not too soon to order more drinks. "Which reminds me, if super-luminous speeds really do represent a phase transition how does someone survive such a transition."

"It is what has been done for generations," says W-abe as he keys in an order for drinks and hors d'oeuvres without looking at Mim-sey's glass or plate.

"I know. But I just never thought about what that means. Does one become an ice cube of absolute zero or plasma with no elementary particles and a heat that is impossible to comprehend?" Mim-sey asks feeling glad to know drinks are coming and that he does not have to pay for them.

"There is no advantage of either of those two extremes. There are other phase transitions besides ice to water to steam to plasma," submits W-abe tweaking his plate with a smokey sauce.

"In this case, which is our case, the phase transition is the transition of time and space."

"I do not suppose you can explain that in terms that I and Mr. Abbit could understand?" Mouses Mim-sey almost without expecting an answer.

"For Mr. Abbit everything is strings. Words in a sentence are a string of words," fleers W-abe dismissively. "All of his math depends on the string of numbers for that string."

"Does the string have to be a grammatical sentence? Would the math be consistent with a row of apples and oranges?" Mim-sey asks as he emerges from his focus on his food and a hint in his mind of what he wants next.

"The strings have subsets," rallies W-abe feeling a little more committed to the conversation. "Dimensions have sub-dimensions, particles have sub-particles and sub-particles can be divided, and subdivided further."

"The conventional wisdom is that time slows and space reduces with speeds approaching normal light speed," reflects Mim-sey planting the topic where his concern is.

"And mass becomes infinite. There is the clue. The mass does not become infinite. It is crowded into the ultrasmall spaces of those dimensions beyond the normal four," deliberates W-abe almost matter-of-factly. "It seems almost infinite given the incredibly small space matter crowds into."

"Time appears to stop, but it does not, because of the smallness of the sub-dimensions is so difficult to comprehend. Space seems to disappear for the same reason?" supposes Mim-sey.

"Now location, point-a or point-b, seem so 'infinitely' definite that velocity loses meaning in normal terms," cerebrates W-abe making a new connection to the timeless reciprocity.

"Movement from point-a to point-b, in space, becomes virtually independent of time," speculates Mim-sey with a vague feeling of a concession to the connection.

"Then once at point-b and with the return to normal light speed or below, matter migrates out of the submicroscopic dimensions and assumes the normal dimensions," recites W-abe.

"Then mass becomes less than a supposed infinity and becomes the macroscopic scale of electrons, protons, neutrons, space and time that are measured at normal or sub-light speeds," relates Mim-sey.

"Atoms and molecules seem to appear from nowhere to just about everyone," says W-abe, "because almost no one knows about the incredibly small spaces that ordinary atoms and molecules are packed into."

"It is like putting the genie in the bottle when it serves us and letting her out at will," says Mim-sey.

"Last-in, first-out means whatever is compressed into the dimensions beyond the first four," says W-abe in an attempt to end the conversation. "What comes out is in an order that reassembles the original molecules, atoms and so on."

"And communications?" Mim-sey asks looking at his empty plate for more food; wondering how he ate it all considering what he will have next.

"Words or word parts and pictures or picture parts are at least in serial order," responds W-abe resting his fork for the first time with an emerging interest in the implications of the conversation. "They are defined by time and space again."

"Without knowing the nuts and bolts, I guess that the strings representing the words and pictures (as a stream of electrons or photons) get compressed into dimensions beyond the normal four, moved from point-a to point-b, like any matter and energy. Then they are unpacked as normal information," infers Mim-sey focusing on the arrival of fresh drinks.

"It is not easy but it is not impossible," says W-abe finishing his drink.

Episode 77

Ali-ce and Roy want to spend an important part of their last night together on the beach. The ambiance of an imposing ocean and an equally opposing beach is guaranteed to please them both. But it rains. The hoped for light and passing rain is a thundering and constant deluge with no end in sight.

They take a taxi and plan to wait out the rain with a couple of drinks. They drink slowly and each hopes the best for the other. Both are afraid they will never see each other again. He is afraid she will not return from this trip to see her father but cannot stop her from going. She is afraid that her chance to save Roy's life has passed and may never come again. She can stay and watch it happen, or try to prevent it, without a clue as to how to do either.

It is important to deny their fears without speciously defending them. The promise of distraction that South Beach always offers is broken. The rain drives everyone inside. Everyone is inside the restaurants and bars, most of which do not have the room for their tables, chairs and staff when the crowd is outside.

The will to focus on each other is trumped. If you want to manage in this crowd, you have to be more self-absorbed than everyone else. It is a crowd of egos milling around to make sure no one is any better than you and you are there to show yourself as better than everyone else. Everyone is his own elephant in the room.

A burlesque is easy to enjoy at the right distance or from the right vantage point. You cannot see backstage. You should not see the person behind the persona. No one is forced to judge its substance. You do not look in a mirror to examine the glass. In a rainstorm, South Beach is not a carnival, it is an endless carousel of careless strangers pretending you are not there.

Both Roy and Ali-ce are glad to return to South Miami. They luxuriate in not being imposed upon and enjoy each other without distraction. A pleasant snack and a comfortable view of the storm turns out to be the real joy. Even if the conversation does not always go where they want it to, they do not pay a price for it in the pleasure of the evening and each other.

"Are we going to practice tonight?" Roy asks changing into a robe.

"No," answers Ali-ce filling the teapot with water. "We have done all we can for now."

"How am I doing?" Roy asks in a chair with a better view of the weather than the kitchen.

"You are doing very well for almost two weeks of work," discloses Ali-ce while pulling some pillows off the bed. "The light and my voice help to bring your heart rate up."

"It is just conditioning but you said willpower cannot be used here," returns Roy.

"We will practice from time to time," remarks Ali-ce, "after I get back from Rome."

"What good do you think this will do," protests Roy tipping his slippers off the ottoman.

"You have to wake up enough to call for help," relates Ali-ce. "If my father has you poisoned with one like Friar Laurence's, you have a chance of saving yourself."

"But I cannot spend the rest of my life waking up in fear and suspecting poisons," says Roy while Ali-ce stacks pillows on the couch.

"You are right. This is just a Band-Aid," says Ali-ce leaning into her pillows with enough room to stretch out. "I want to find out more about your parents and pay a visit to my father."

"Will you confront him?" Roy asks blinking at the lightning and waiting for the thunder.

"Not yet. But I am getting better at arguing with him," says Ali-ce adding a throw to her nest. "Maybe I can learn more of what his target is and what your manuscript has to do with it."

"My murderer does not have to use a poison like friar Laurence's. He can kill me any way he wants," complains Roy craning his neck to see the teapot.

"I do not think he can kill you any way he wants," replies Ali-ce pulling the coffee table conveniently close to the couch. "Death does not go unnoticed and murder is not forgotten."

"So he does not want to become the object of an investigation?" Roy asks.

"There are lots of methods for murder that will not point to aliens but his problem must also include your manuscript," states Ali-ce without any suspicion of who will make the tea.

"I will be gone but the circumstances might create an interest in my manuscript?" Roy asks getting up to make the tea and collect some kind of snack.

"I am beginning to think that is why you are still alive," renders Ali-ce in the direction of the kitchen.

"If I am the target," equivocates Roy into the refrigerator.

"Maybe we will know more about that when I get back," says Ali-ce climbing off the ample couch.

"What can I do?" Roy asks as Ali-ce pushes his chair closer to the coffee table.

"There is not much either one of us can do," supposes Ali-ce joining Roy in the kitchen to collect cups and the required Reed and Barton. "Maybe I can look up your parents."

"Both of them are dead," declares Roy. "I do not think they were murdered."

"It does not seem that they were," relates Ali-ce on her way back to the coffee table and her nest. "They lived in Verona at the time you were born."

"I know very little from that time," says Roy now returning with napkins, tea, buttered toast and sugar flavored with clove, nutmeg and cinnamon powder. "We were back here by the time I was four."

"Your birth certificate has their address at the time," responds Ali-ce piling into her nest after doing her part to set up the coffee table. "He was a curator. She was a housewife."

"They both loved paintings," acknowledges Roy. "Maybe he worked in an art museum."

"How did your parents die," probes Ali-ce as Roy returns to the livingroom.

"They were both heavy smokers," reveals Roy pushing the coffee table closer to Ali-ce and her nest. "She died of a heart attack and he had a stroke and died of pneumonia."

"How long ago did they die?" Asks Ali-ce testing her reach to her teacup.

"He was fifty-nine and she was sixty-three," says Roy spooning the spiced sugar on the toast.

"Your father died twenty-four years ago and your mother died twenty years ago," says Ali-ce as Roy dumps the excess sugar from the toast into a serving dish.

"I was in my thirties," divulges Roy setting a serving of toast as close as possible to Ali-ce.

"Did you do any work on your manuscript by then?" Ali-ce asks.

"I wish," announces Roy watching for Ali-ce's approval of the tea and toast.

"An extra fifteen to eighteen years never hurt anybody," says Ali-ce and following with a careful dunk of toast in her tea. "At least you cannot blame yourself for your parents' deaths."

"No doubt any opinion about a writing avocation would be that it is something to fill the vacuum," speculates Roy pleased with Ali-ce's nonverbal approval of the tea and toast.

"No inheritance?" Ali-ce asks as soon as she was free to speak.

"There was life insurance from both of them. I got my share," says Roy before taking his turn at a dunk and bite of toast. "I made it go as far as I could."

"So what was the vacuum, if not financial?" Ali-ce asks while being careful not to hurry her toast.

"Part of it was that I did not have to support their lives. You know, help them with things, spend vacations and holidays with them," says Roy. "A lot of your life goes into that."

"So you had a little spare time on your hands?" Ali-ce asks before taking a sip of tea.

"I could work on a manuscript without short changing it or my family," admits Roy.

"It was too late to go back to school?" Ali-ce asks as Roy has some more tea.

"I tried that," vents Roy. "I cannot assimilate or process the information fast enough."

"What does that mean?" Ali-ce asks adding hot tea to both cups.

"If you toss me a ball I can catch it, maybe," scoffs Roy preparing another couple of pieces of toast, "but the anxiety is inescapable and extremely distracting."

"And you attribute that to the mental effort of following the arc of the ball?" Ali-ce asks.

"That is just an example," says Roy pleased to see Ali-ce's delight with the toast is undiminished. "For me, anxiety is part of anything that involves learning or processing information."

"But that involves just about everything," comments Ali-ce while Roy dunks his toast.

"Everything except habituation," says Roy as Ali-ce chases her bite of toast with some tea.

"How do your habits help?" Ali-ce asks setting her cup on the coffee table.

"I do not have to audit habits," says Roy wondering if she is finished with her tea and toast. "Multiplication tables, symmetries, patterns, routine actions can be exploited out of habit."

"How do you learn anything?" Ali-ce asks while fixing her pillows.

"That is the irony. I can learn anything. It just takes longer," scorns Roy relieved to see Ali-ce resume her dialogue with her tea and toast. "No one but me is willing to take the time."

"You could learn a second language but perhaps only one in a lifetime," supposes Ali-ce.

"But not fast or well enough to use for years," says Roy.

"You could learn to play baseball," says Ali-ce.

"But I would be past my prime and I could not do much else," says Roy.

"So you were over fifty," says Ali-ce, "before you managed your only manuscript, so far."

"I spent a lot of time on other things," says Roy defensively wondering if Ali-ce heard him.

"Subsisting?" Ali-ce asks trying to disguise a compromised attention to the conversation.

"Mostly," says Roy trying not to interrupt Ali-ce's now obvious cogitative preoccupation.

"This is consistent with your earlier birthday," says Ali-ce raising her thoughts to a focus.

"How is that?" Roy asks without a clue to the meaning of her remark.

"The last three months of a full term pregnancy are important to brain development," says Ali-ce explaining things as much for herself as for him. "So are the earliest years of childhood."

"But fifty years ago no one knew this," says Roy dismissing the present day connection.

"No one here knew this," says Ali-ce almost as if neither she nor Roy is in the room.

Episode 78

The only oceangoing rail-ship 'afloat' plies the gulf coast, continental shelf between the river and the space-port. The space-port is one of two that service the kind of craft that Ali-ice Mim-sey and Rus-ty Gim-bal use. It is the pinnacle of irony that this is so and at the same time it could not be any other way. This thousand mile stretch of coast has been the most lawless and dangerous place ever for centuries. The river creates its northwest border a swamp protects its lawlessness. Archaeologists and law enforcement are unanimous in the grief over what has been lost there.

It could have continued until the present day, perhaps, if the space port had not come along. The land is free to the strongest arm. And the space port has the biceps for that. This dries up the lawlessness. Developers dry up the swamp. The coast becomes a natural wonder. And nature has nothing to do with it. Resorts and exclusive homes have everything to do with it.

This rail-ship, with its 27,500 passengers, is the sound of two hands clapping. Everyone on the ship admires the glamour of the coast. Everyone on the coast adores the glitter of all those tons of sculpted glass gliding back and forth, night and day. The dividends and real estate values flatter each other.

Ali-ce is meeting Officer Gim-bal near the river before she leaves for her shuttle and Earth.

"I have to say I would be happier to see you," says Gim-bal, "if you looked less miserable."

"I have not been my own priority lately," says Ali-ce in the shade of the slowly passing rail-ship.

"That has to change, promise me," says Gim-bal as the rail-ship eases silently by.

"I have to do it for Roy's sake anyway, so it is a promise I can make and keep," says Ali-ce noticing for the first time in her life that this is a ship that blocks the sun and wind but leaves almost no wake.

"Is that why you have to see me?" Gim-bal asks mentally reviewing his expectations.

"Yes, my father has Roy's manuscript, here, somewhere," says Ali-ce. "I need to see it."

"Only your father knows where the original is," says Gim-bal with something new for his list of suppositions, "but I can get a copy from his petition at Robe's Chamber."

"I need to see the original. What is his petition?" Ali-ce asks revising her own suppositions.

"He wants a Summary Death Warrant for Roy Abbit," says Gim-bal wondering what else her father has not told her. "The Robe's Chamber is the only authority to do that."

"What are the chances of that?" Ali-ce asks with heightened concerns for Roy's safety.

"It takes a unanimous vote," says Gim-bal frankly, "Even though that is rare, I think he can force it."

"There is no one more independent than the Robes," says Ali-ce.

"The alternative to a unanimous vote is either a vote 'not to consider or a split vote'," says Gim-bal unreassuringly. "But no matter what, the vote and the petition become a public document."

"And the votes against Roy Abbit become my father's allies for some domestic agenda?" Ali-ce asks considering the now larger, danger and the now smaller profile of the ocean bound ship.

"I am afraid it is not quite that innocent," says Gim-bal in an effort to introduce her to the looming danger. "Once the petition becomes public, the existence of Earth becomes official."

"That is not a significant change from the urban legends of Earth," says Ali-ce.

"In terms of speculation you are right. Earthlings, Greylings, will still have three heads and eight legs. But the terror and the threat of a real Earth will be immediately upon us," says Gim-bal more to the point, "no matter how remote Earth is."

"And father can call for the only responsible action - Earth's destruction," says Ali-ce.

"From the comfort and safety of his armchair," says Gim-bal avoiding the emphasis of a derisive tone.

"Self-defense is the only option for Earth at this point," says Ali-ce in the glare of both a real sunset here on Clidim and a figurative one facing Earth. "They cannot get to us, even if they could find us."

"There will be losses on both sides," says Gim-bal turning away from the glint of the sunset if not the glare of the truth. "Those losses will be touted as proof of the vicious animalism of Earth."

"Father will be vindicated. He will tear his tunic, fall to the ground and beg not to be made king," says Ali-ce facing a mounting breeze and a rising contempt.

"Something like that," says Gim-bal with both sympathy and sarcasm. "His efforts to stay out of the controversy will be proof of his modesty and his modesty will be proof of his prescience."

"And his wisdom will be proof of his righteousness to lead," says Ali-ce with disbelief.

"Our global culture will be polarized, possibly fragmented, for the first time in centuries," says Gim-bal with conviction. "This will make him the leader of the largest faction to survive."

"That will be the faction that represents his sole concern which is, 'to save our world'," says Ali-ce.

"A unanimous vote is the only option that has a chance," says Gim-bal expecting Ali-ce to react to this veiled expression of a ransom. That ransom can only be Roy Abbit's life. "A united Robe's Chamber can discount the hypothetical threat of Earth and perhaps mitigate fragmentation."

"As long as that unanimity can be maintained the Robes can stand in the way of adverse actions involving Earth," says Ali-ce seeming to react to a danger that is just as personal but larger.

"I am afraid that making the Robes the executioners of Roy Abbit is the best and only hope," says Gim-bal noting her priority for her home planet.

"I am afraid, even that hope has been removed," says Ali-ce with conflicting priorities.

"How so?" Gim-bal asks suddenly confused by her stony equivocating statement.

"I think my father has already killed Roy," says Ali-ce taking her turn at turning the table.

"He was alive when I left Earth," says Gim-bal tasting the panic of impotence himself.

"And he still is, but in a matter of some weeks Roy has begun to deteriorate. He has aged what seems like fifteen or twenty years. Roy has lost almost half his weight," says Ali-ce. "He does not realize the extent of it or, perhaps, he is in denial. In either case, he cannot survive much longer."

"And denial, or at least the nature of it is the key," says Gim-bal with a glimpse of where he has failed.

"I do not understand," says Ali-ce recognizing a kind of personal fear in Gim-bal.

"Roy's death at the time of the vote will introduce more controversy," says Gim-bal.

"If news of his murder reaches the Robe's before the vote, they cannot vote for the death warrant. They will look impotent; saved only by chance circumstances," says Ali-ce.

"No one will rely on them to save us again," says Gim-bal looking like he missed the boat.

"If news of Roy's murder surfaces after the vote, the Robes roll may well be questioned," says Ali-ce looking at the horizon like they both missed the boat. "It will not matter whether they try to curry credit for the execution or deny any connection to it."

"The attempt to vindicate Roy Abbit's execution will look like justice after the fact," says Gim-bal as neighborhood environmental lights begin to supplement the rapidly disappearing sunlight.

"The Robe's Chamber will be accused of looking for a scapegoat in the minds of the public," says Ali-ce turning her back to the ocean.

"And if that scapegoat is your father, it will not matter what kind of case they have," says Gim-bal. "Justice will not be served by making the putative 'hero of the day' into a seminal criminal."

"What if the Robes do the right thing and punish my father for Roy's murder?" Ali-ce asks as the dusk marries the ocean and the sky into identical shades of dark grey.

"In that case the Rob's Chamber will come out on the side of Earth in the minds of the public. That probably will save your father. The Robe's Chamber will be asked to set aside a death warrant for your father," acknowledges Gim-bal as he watches the silhouette of the rail-ship turn from black to opal as the sunlight is replaced with the ship's powerful sailing lights.

"And if they do not set aside the death warrant?" Ali-ce asks facing the loss of her father and Roy as certainly as the now overwhelming darkness.

"Appeals probably will keep your father alive until a majority vote frees him," says Gim-bal proposing a hope as remote as that horizon bound rail-ship.

"The battle becomes the attrition of Robes who vote against your father and the appointment of new Robes who do not want to shorten their careers by voting against my father's exoneration," says Ali-ce.

"And what saves him serves him," says Gim-bal painfully aware of the metaphor of the rail-ship disappearing over the horizon.

1

1 1

1 2 1

1 3 3 1

1 4 6 4 1

1 5 10 10 5 1

1 6 15 20 15 6 1

Chapter 27

Chapter 11011000....

Chapter ABDE

Chapter $(1 + 5x + 10x^2 + 9x^3 + 3x^4)$

Chapter $(a^4b^0 + 5a^3b^1 + 10a^2b^2 + 9a^1b^3 + 3a^0b^4)$

Chapter $2^0 + 2^1 + 2^3 + 2^4$)

Episode 79

It is less than a mile down the beach from the barbeque to the suite. Dr. W-abe is open to the suggestion from Dr. Mim-sey to walk back. The breeze is inland so they have to walk close to the waterline to avoid blowing sand. This sand has the advantage of being flatter and firmer than it's driest or wettest counterpart.

The neighborhood is quiet and there are enough patches of light from residence and resort environmental lighting to make the walk comfortable and ordinary, ordinarily. The walk is not significant for Dr. W-abe because he is never overwhelmed by anything.

But Dr. Mim-sey, for his part, is transported by the majesty of the moment. This time, big is not sufficient to the perception. Under a black, black sky, salted by sharp points of blue, red and white flecks beyond counting, Mim-sey is walking on one of the few beaches left on this bijou planet.

The Glass Age is the beginning of the end of this, Clidim's beaches. It takes the thousands of centuries that follow the Stone Age to mine them away. The discovery and alchemy of making glass means there never is a Bronze Age or Iron Age. Metals become the additives to smelted glass for the structures and components of its civilizations.

An unlimited demand for and an eventual shortage of sand is assuaged by the replacement of alchemy with chemistry. Now sand is milled to order. The beaches that remain are safe from destruction and immeasurably valuable. Consequently, no one wants to pay the price of going home with sand in their shoes.

Mim-sey, finding himself dwarfed by a vaulting ebony sky and miles of velvet beach inflates his mission and his ego into proportion with his megalomania. Both of which he considers appropriately without limits. This is his overarching sky and nobody is going to take it away from him. With some effort, he is able to contain his inspiration with a charade of indifference. In a gesture of dissimulation he manages the distraction of a dialogue.

"If Mr. Abbit's world is so far from the facts, how is Mr. Abbit going to point them in our direction?" Asks W-abe as he heads straight for the ocean from the barbeque patio.

"He almost certainly will cause the event that has happened rarely in our history and his 'a paradigm shift'," says Mim-sey following W-abe with some difficulty in keeping up on the soft sand.

"You have to convince me," says W-abe waiting for Mim-sey to close the gap between them.

"Suppose Mr. Abbit sometimes flies a small plane from Washington, D.C. to St. Louis, Missouri. He regularly travels the 850 miles in 8.5 hours. Say that the weather does not affect the flight," shares Mim-sey.

"That is 100 miles an hour or 1.66 miles a minute," says W-abe walking along the water at an unhurried pace.

"Now suppose, on one trip, it takes him 8 hours and 31.66 minutes to get to the same longitude. What does that tell you?" Mim-sey asks feeling grateful for the sparse light spilling from his right.

"Mr. Abbit is north or south of his regular flight path," utters W-abe walking at a leisurely pace.

"When Mr. Abbit is traveling to St. Louis without going north or south there is only one spacial dimension," asserts Mim-sey gauging the space between the wettest and driest sand when the intermittent light allows.

"The north-south dimension represents a second dimension," W-abe details as his contribution.

"When Mr. Abbit appears to be going slower, it is because his speed is shared by two dimensions not just one," says Mim-sey satisfied that his path is almost halfway between the softer, wetter sand and the softer, drier sand.

"The two dimensions mean a longer distance and since speed is a function of time and distance, it must be shared by both dimensions," specifies W-abe.

"Time as a dimension is also shared with the two dimensions," stipulates Mim-sey.

"We could include three dimensions of space for Mr. Abbit's trip and still share the dimension of time with it as well," says W-abe occasionally catching the sharp hiss of blowing sand on the waxy foliage in the dimness to his right.

"Now suppose Mr. Abbit is not moving through space and we are sitting in Washington, D.C. with him," says Mim-sey wondering why they are not walking where the light is better.

"There is no relative motion among us so we are only moving through time," says W-abe.

"So we are all aging at the same rate since our clocks are the same," weighs Mim-sey.

"While Mr. Abbit does move, he ages more slowly because his clock slows," acknowledges W-abe with his own emerging appreciation of the vaulting black sky and multitude of red, blue and white stars revealed by the clear evening sky.

"Part of his motion through time is shared by his motion through space," says Mim-sey.

"At slow speeds, his share of time that is diverted to the other three dimensions is very small," indicates W-abe recalling that in the early days of flight, tourist were landed directly on this beach by the hundreds each day.

"The faster he goes the more his motion through time is shared with spacial dimensions," says Mim-sey wishing he had some Beach Stock. "Movement through space subtracts from movement through time."

"When all of his motion through time is allocated to the movement through space, he cannot go faster," reasons W-abe recalling that at one time a single family owned this beach.

"To return the dimension of motion through time to something above zero," deliberates Mim-sey, "means borrowing from the motion through the dimensions of space."

"That means a reduction of motion through the dimensions of space which is called 'slowing down'," says W-abe considering that this beach belongs to almost two billion shareholders now.

"This much is well known to Mr. Abbit's world as well as ours," sputters Mim-sey.

"So what does this have to do with Mr. Abbit's paradigm shift?" Asks W-abe.

"To begin with, he uses his fractions with factorial numerators and denominators and his Pascal/Chu triangle to show how each row begins and with a proper fraction which is one," reflects Mim-sey recalling that Beach Stock is not sold anymore and that it just sits in trust funds paying dividends for all the generations of its heirs.

"The number one in the first row and the at the beginning of each row represents zero subsets and therefore zero dimensions," says W-abe.

"Suppose that the initial digit represents zero movement through space?" Asks Mim-sey supposing that had he been born an heir to a beach trust that he could count the stars like this sand as his.

"So before the 'Big Bang' it is space that is zero not time. There would be no relative movement through any dimensions but time," prompts W-abe trying to remember if the same Beach Trust owns the coastal ocean going rail-ship that is home ported on the river.

"Row zero of the triangle only has the one digit, so there is no other dimension but time," declares Mim-sey deciding if he ever remarries it would only be fair to be married to an heir of a Beach Trust.

"Row one consists of two digits. The first one must be as before, zero dimensions, zero movement through any spacial dimensions." says W-abe estimating that there are a little over 100,000 Beach Trusts with holdings across Clidim. "The second one is a little tricky. It could mean anything."

"Let us say that the terminating digit for any row represents the symmetrical opposite of the initializing digit for the row," says Mim-sey. His new wife would give him the uncounted grains of sand and he would give her the uncounted stars and planets. "The final digit in any row must represent movement through space without any share in the movement in time. The universe is expanding at its fastest with no dimension of time."

"Row two begins and ends with these digits and includes the number two as well," says W-abe deciding that perhaps his building and the oceangoing rail ship are owned by different Beach Trusts.

"The number two in this case means two subsets or, say, two dimensions," says Mim-sey. "If it is two dimensions, then it is something like the two dimensions of latitude and longitude."

"Or it could mean that movement through time is shared with movement through space," says W-abe considering that the Beach Trust that owns his building also owns one of the Trans-Clidim Flight-Transports.

"Let us not complicate things until we have to," delivers Mim-sey. "Let us say the number two in row two represents two dimensions that are common to our experience."

"Well, things are getting complicated anyway because row three has four numbers," says W-abe. "We can keep the first number as movement through time with no movement through space and the last number as movement through space with no movement through time."

"What about the two threes between the ones? Three dimensions would not be inconsistent with our analogy so far," says Mim-sey.

"Okay, that takes care of, say, the first three in the series 1, 3, 3, 1. How are you going to explain the other 'three'?" W-abe ask.

"I doubt if Mr. Abbit has thought in these terms, if you have not. However, we are in an expanding universe. A universe whose spatial dimensions are increasing at least in size," says Mim-sey. "The universe must at some point begin to slow down expanding and allocate more to the dimension of time again. Perhaps by doubling in sizes or dimensions (in a big snap or big bang) thus making room or dimensions for time. What counts as the normal speed of light must change accordingly."

"Are you saying that a light-year is different in an expanding as opposed to a contracting universe?" W-abe asks as he realizes how much Beach Trust property is, now, riverfront property. "Three dimensions in an expanding universe is distinct from three dimensions in a contracting universe. The two threes are not similar enough to sum into six."

"It would be a change in the paradigm but consistent with Abbit's category mathematics. In an expanding universe the number of dimensions intersecting a light year tends to increase. For a contracting universe the number of dimensions intersecting a light year tends to decrease. Combining the two threes would be a category mistake," says Mim-sey.

"Each additional row means 2^n more dimensions intersect each light year. There is no limit on the number of dimensions but with more dimensions there is less time allocated to each dimension," says W-abe realizing that the Beach Trusts have, in fact, traded this beach for the significantly larger river. After all, a river has two sides.

"Time slows and approaches zero without changing speed just by sharing speed with, more dimensions" says Mim-sey.

"The speed of light must change. For an expanding universe the speed of light is reduced each time the number of dimensions doubles," says W-abe. "In a contracting universe the speed of light increases each time the number of dimensions is reduced by half and the naysayers have to admit it. If not now, then in the era that includes enough dimensions to snap the normal speed of light to double or half its current speed."

"In the meantime, adopting this paradigm is enough," vents Mim-sey resolving to become a Beach Trust heir. "When they do, we are exposed by our super-luminous transactions, like communications and travel that employ the physics below and above the normal light speed."

"Mr. Abbit is knocking on the door of a paradigm shift and it opens on a phase transition for the speed of light," says W-abe finding a seminal respect for Abbit's unassuming good sense.

"And finding us is just a matter of seeing us in the right light," submits Mim-sey.

Episode 80

Finding Roy Abbit's father in postwar Italy is easier than Ali-ce expects in some ways and not so in others. Few people know him by name and almost no one knows he speaks English.

Those who know him, do not know him as Edgar Abbit, most of the time, but as the 'il curatore' or 'the man who works in the museum'. His life is segregated between work and home. He speaks Italian at work and, presumably, English at home. Ironically, if Ali-ce's Italian were not so good it would have been a clue. She would have put people in mind of someone whose Italian is good but not native to Verona, if her Italian were the same.

Museums are at their fewest, so soon after the war, since most of Europe's fine arts are in 'storage' in the United States and the Soviet Union; having been recently removed from 'safekeeping' in Germany. Also good for Ali-ce, there is little room in Europe and Italy for museums.

Space, as in interior space, is sparse owing to the dearth floors, roofs and walls that is everywhere. Finding Edgar Abbit is not Ali-ce's primary purpose anyway. She only needs him so she can find his wife, Jenny. It is important to find her but timing is critical.

If Ali-ce is too early, her father will find out and change any adverse plans he might have for Edgar or Jenny. If she is too late, the chance to change things may be too fine a point in time to insinuate herself into. For the infinity of options her father has, Ali-ce has the fewest. She cannot protect Roy from an infinity of possibilities, she must protect him from the one that prevails against him.

An unlimited number of choices makes the right one extremely difficult to choose. She cannot be just a witness to the murder of Roy Abbit or his ancestor. She must prevent it. But the mercenary who works for her father has his job made easier by knowing where Ali-ce is and what she intends. So it is with no small agony or irony that she has to limit her encounters with Edgar and Jenny to being as casual as possible.

"Do you speak Italian?" Ali-ce asks approaching the sole person in the main gallery room.

"Thank you for asking in English," says Jenny. "You must have guessed that I do not and I thank you for sparing me the task of saying so in Italian. It is so embarrassing."

"Are you learning Italian?" Ali-ce asks taking a seat close to Jenny on her ottoman.

"A little," offers Jenny. "If I thought we were going to stay here, I would learn more."

"We?" Ali-ce asks adjusting to a better speaking distance. "So you are married?"

"...more than a year," says Jenny at ease with the chat. "My husband is the curator here."

"So, you must be Mrs. Abbit," says Ali-ce. "I am Sister Ali-ce."

"Nice to meet you. I am Jenny. But I do not mind being called Mrs. Abbit," says Jenny.

"I am the same with Sister Ali-ce," responds Ali-ce. "It reminds of how fortunate I am."

"But you must have been a nun more than a couple of years," suggests Jenny.

"Do I look that old?" Ali-ce asks burlesquing Jenney's sentiment.

"Oh. No. You look very young," says Jenny smiling. "You just do not seem new at it."

"We are always new at something," reveals Ali-ce. "Your husband tells me you are going to be parents for the first time soon."

"You know him?" Jenny asks with a reflex sense of the ground not yet covered.

"He is the one who knows the collection," relates Ali-ce without returning to square one.

"Well we are not Catholic but he, we both, can use good friends," conveys Jenny. "Especially when things turn difficult."

"The job is not going well?" Ali-ce asks choosing to follow Jenny by leading her.

"That part of our life is a consolation," advises Jenny. "It is my pregnancy."

"He never mentioned a problem," says Ali-ce for part of the truth.

"He does not think there is a problem," says Jenny unaware of Ali-ce's anxiety. "But the doctors say the baby has a weak heartbeat."

"So, you are afraid?" Ali-ce asks including herself and sounding all the more sincere.

"Yes," says Jenny unaware of Ali-ce's personal fears. "But I do not want him to worry."

"Would you be disappointed if I told you not to worry?" Ali-ce asks.

"No, of course not," weighs Jenny. "But part of me will worry just the same."

"Do the doctors think that birth will be too stressful for you or the baby?" Ali-ce asks.

"They were a little too quick tell me about the safety of a cesarian section," vents Jenny.

"What do you mean?" Ali-ce asks as if she does not suspect the alternatives.

"My husband did not notice," says Jenny. "But I think they propose that as an abortion."

"They were thinking about both your safety and the baby's," urges Ali-ce.

"He never considers the possibility of danger for either of us," upbraids Jenny.

"Perhaps he does," prompts Ali-ce mixing sympathy and doubt into Jenny's certainty.

"Maybe," declares Jenny including herself as well. "It is just too soon for him to face."

"I do not think you should try to prepare him for something that may never happen," councils Ali-ce while telegraphing her own best hopes.

"If the fetus lives a couple more months, it may survive the birth," ventures Jenny trying to believe what she has been told so far. "The doctors cannot be sure about how long after that."

"Most babies do very well and you know it," says Ali-ce hoping to change the topic.

"I know," concedes Jenny unsuspectingly. "But I do not want to ignore real alternatives."

"Would it help to know some of the new mothers here?" Ali-ce asks as an option.

"I guess so," admits Jenny. "They can at least try to keep me from being too sure of things I really cannot be sure of."

"Can I see you again, soon?" Ali-ce asks.

"Yes, of course. I am grateful that you already know my husband," accentuates Jenny sincerely. "I think you will be a great help."

They stand up and Ali-ce asks, "Is there anything else we can do?"

"Yes. I am going to see the doctor while my husband is at work. If the news is bad, I will need you and the doctor to convince him. Then I do not know what we will do," acknowledges Jenny.

Episode 81

It is difficult, at first, for Officer Gim-bal to tell whether this visit to 010nd-Robe's personal office is indicative of the importance of Dr. Mim-sey's petition or the press of all the concerns of the Robe's Court. Unlike the greeting rooms, sitting rooms and conference rooms this office is little more than utilitarian.

If his desk were seen in a used-furniture store no one would believe its provenance. The books on his shelves have no obvious organization by the usual clues of colors, bindings and titles. Organization of his books is derived from the content of the cases and a lifetime of experience.

Books, folders and papers on the tables make one wonder if 010nd-Robe even has a staff regardless of the nearly timeless tradition of a permanent and never shrinking bureaucracy.

But with predictable irony 010nd-Robe can move through it all comfortably and effectively. Less predictably the scene of disarray disarms what would have been his surprise at seeing 010nd-Robe. There is one chair behind the desk and two in front. 010nd-Robe is sitting in one of the chairs in front of the desk when Officer Gim-bal comes into the office. He takes his seat in the one next to 010nd-Robe.

"Nobody, can get any closer to Dr. Mim-sey or Roy Abbit than Ali-ce Mim-sey," admits Gim-bal privately wondering why they both are sitting on the same side of the desk.

"Why is she not here?" 010nd-Robe asks.

"She has a deep affection for Roy Abbit and would not be kept from returning to Earth," allows Gim-bal noticing that the intern stayed only long enough to remove a stack of folders.

"You will have to be more specific," says 010nd-Robe picking up a folder and holding it.

"She thinks her father has killed Mr. Abbit," says Gim-bal. "She wants to be there with him."

"How soon can she be back after the funeral?" 010nd-Robe asks.

"Roy Abbit is still alive but she cannot say for how long," deposes Gim-bal.

"Then why does she say her father killed him?" 010nd-Robe asks.

"The way she explains it to me," says Gim-bal as 010nd-Robe initials the first page of the folder and closes it without any more reviews, "her father took advantage of the time shift that is a problem with high speed travel and had him killed at some point in his past."

"But Mr. Abbit is still here, uh, back there, on Earth?" 010nd-Robe asks placing the folder on his desk.

"As I understand it, changing the historic time line does not effect an instantaneous change," informs Gim-bal as 010nd-Robe picks up another folder. "The life of Roy Abbit and his artifacts fade or erode in the stream of time. He becomes a memory to his survivors as his existence becomes fixed in the past. Even his past fades as it becomes more remote. Just like all the other facts of history."

"So he is fading?" 010nd-Robe asks while initialing the first page without reading further.

"His atoms and molecules are taking a different path," attests Gim-bal as 010nd-Robe places the initialed folder on top of the others on his desk. "To us he is just aging but at a rapid rate."

"If he was 'killed' before he wrote his manuscript then that artifact must be 'aging', uh, rapidly, too," concedes 010nd-Robe as he initials another folder from a previous reading and places it on the stack.

"His originals and notes are aging at an accelerated rate," offers Gim-bal as 010nd-Robe picks up a folder initials it and places it on its stack. "But copies have their own time line or history."

"So our copies will remain unaffected?" 010nd-Robe asks picking up another folder.

"They will age in the normal sense," grants Gim-bal as 010nd-Robe initials another folder.

"More significantly," declares 010nd-Robe placing the initialed folder on the stack and picking up another one, "if Mr. Abbit dies just before or after the vote, we get the credit, deserved or not."

"Specifically," reflects Gim-bal, "you get credit for the resulting chaos, rightly or wrongly."

"And the chaos will serve to disguise Dr. Mim-sey's role in all of this," offers 010nd-Robe, "except that to the public perception, he was ahead of all of us in the warning of danger."

"The Robe's Chamber begins its descent into history and Dr. Mim-sey begins his ascent," imparts Gim-bal while 010nd-Robe places the initialed folder on its stack.

"It would appear," says 010nd-Robe picking up a folder, "that there is nothing we can do."

"I think anything the Robes Chamber does will make Mim-sey's job more difficult," says Gim-bal. "That will take away from his success."

"If we vote not to consider the petition, he wins without a fight and his victory is not compromised," says 010nd-Robe initialing the folder.

"If you have a mixed vote, the public will at least take sides on the issue along the lines of the vote," acknowledges Gim-bal as 010nd-Robe picks another folder to initial.

"The Robes that vote for the death of Roy Abbit will appear to be on Mim-sey's side," says 010nd-Robe initialing the first page of the folder. "All the other votes discount Mim-sey's constituency."

"That counter-constituency together with Earth will resist Mim-sey's machinations and mitigate any success he wants to claim," says Gim-bal with 010nd-Robe placing the folder on his desk.

"Historic records will, at least, have some version of the truth," says 010nd-Robe.

"If there is anyone to read it," says Gim-bal.

"It could be the end of both civilizations," blazons 010nd-Robe.

"A unanimous vote for the death of Roy Abbit will put the Robes on the winning side," says Gim-bal as 010nd-Robe initials the first page in a folder without re-reading and then closes the folder.

"The Robes Chamber will be intact and can mitigate Dr. Mim-sey's version of the history that follows. But the 'Dr. Mim-seys' will multiply beyond number," says 010nd-Robe picking up a folder. "Our die will be cast for us, if not by us. The fight of the century will become one that persists for centuries."

"The Robe's Chamber maybe the only body that will live long enough for that kind of fight," says Gim-bal. while 010nd-Robe initials the folder's first page before closing it.

"A unanimous vote is not certain," says 010nd-Robe stacking the initialed folder on top of the others. "There is the possibility that Dr. Mim-sey could have corrupted at least one of the Robes for future needs."

"There will be time, after all of us start down that slippery slope, to clean house," says Gim-bal as an intern approaches the desk; stopping only long enough to remove the stack of folders.

"I must ask you one final kindness. Not for me. So you will not be refusing me. Will you execute the death warrant, if there is one?" 010nd-Robe entreats.

"I do not know who would be harder to refuse. So I will not try. The Robes Chamber's will is mine," assents Gim-bal a little self-conscious at possibly interrupting 010nd-Robe's work any further.

1

1 1

1 2 1

1 3 3 1

1 4 6 4 1

1 5 10 10 5 1

1 6 15 20 15 6 1

Chapter 28

Chapter 00111000....

Chapter CDE

Chapter $(1 + 5x + 10x^2 + 10x^3 + 3x^4)$

Chapter $(a^4b^0 + 5a^3b^1 + 10a^2b^2 + 10a^1b^3 + 3a^0b^4)$

Chapter $2^2 + 2^3 + 2^4$

Episode 82

The egress from the Director's Retreat is through the commercial flight-port on the other side of what are called the mountains of the island. If you have to travel over them or around them, they can be perceived as mountains. But to look at them or fly over them, the perception is strained. There are no stony limits beyond the trees.

The mountains are low and rolling with trees and foliage on the side where the prevailing storms do not reach. Anyone who wants to spend the money can put a home or a resort there. In stark contrast, the storm scoured land around the flight-port is flat and inches above sea level which is made all too obvious by the lack of trees and foliage.

Predictably, everyone must leave at the same time or shorten their vacation. A transport is launched once every half hour with a capacity of six hundred passengers and baggage. The last ones out that day can spend eighteen hours waiting. Directors, their spouses and their baggage must compete for seats with the tourists.

While the tourists and director's entourages mix in the waiting/departure area, they are easy to distinguish. Tourists are in beach clothes. They are usually younger, travel with children and have diving equipment tied to their bags. The director's group is dressed for the office or cocktails. Their baggage is ample enough to put everything inside. These things become significant when you are competing for seats.

These features are more obvious when viewed from the elevation of the Banker's Lounge. The Banker's Lounge is for the few passengers taking the banker's transports. Their fifty-four flights a day move money, papers and clients to the larger ports and usually have less than four passengers on a trip. This suits Dr. W-abe so well that he is the banking officer for the board directors. Nothing is free and this is the price he pays for travel at his convenience. The price, in terms of his labor, is cheap as long as everyone is happy with his numbers and everyone has been happy with his numbers for many years.

As long as there is room, Dr. Mim-sey does not have to fly commercially. This year, like most years, he returns with Dr. W-abe. Flights are quick but not always direct. A touch and go landing to drop-off or pick up something is not unusual. At least it will not be more passengers for established privacy reasons.

The two things they can count on are not knowing the flight plan and flight time. They leave their timepieces upon entering the Banker's Lounge and pick them up on arrival at their building-port. Getting used to this idea and protocol means getting used to never having to answer any questions about the trip, cargo or passengers. There is not a significant question that would get a significant answer. This pleases Dr. W-abe. What please Dr. Mim-sey is the occasion for uninterrupted conversation.

"This change, in the paradigm, goes a long way to simplify a lot of math," says W-abe.

"This is my turn to be defensive if not sarcastic. Please do not start dropping equations," says Mim-sey sitting well below a hedge of narrow windows. "It is too much information at a time."

"All right, but while you are in a defensive mood I think I can tell you for the first time, without a lot of math, about the phase transition for speeds beyond the normal speed of light," says W-abe with the random punctuation of incoherent sound penetrating the hedge of windows above their heads.

"Then we will both be on the same page when it comes to the threat that Mr. Abbit and Earth represent," declares Mim-sey, "since it is he who will make comprehending such things simple."

"If not the math that goes with it," says W-abe cringing with the thought of the frankness of the other lobby.

"Do not try it with me," protests Mim-sey comfortably above the roiling lobby.

"Okay then," acknowledges W-abe pressing the button for refreshments in the absence of a menu. "A so-called black hole is characterized by a lot of matter compressed into a very small space."

"That is a gentle start," maintains Mim-sey. "Even Mr. Abbit's Earth knows that much."

"Part of traveling at speeds greater than the normal speed of light is like that, uh, a black hole," predicates W-abe. "At those high speeds there is significant compression of space in the direction of travel."

"What does that mean in conventional terms?" Mim-sey asks. .

"What that means for atoms is that the space normally occupied by electrons shrinks significantly," broaches W-abe similarly unconcerned by there being no menu but for different reasons.

"So the atoms do not decompose?" Mim-sey asks as he anticipates the rarest of drinks.

"It is more like freezing water. The protons and neutrons do not shuffle easily and the nucleus of the atom expands," proposes W-abe. "The protons are causing that."

"What happens to the electrons?" Asks Mim-sey distracted by the arrival of two bowls of ice.

"They sit and wiggle as close to the atom as they can," says W-abe.

"Sit and wiggle?" Mim-sey asks eyeing the empty cone shaped glass supported in the ice.

"If you smirk," says W-abe removing the glass cones to pour a small measure of coarse salt on the ice, "I will give you the math."

"I suppose the 'sit and wiggle' of the electrons keeps the atoms differentiated," suggests Mim-sey suppressing a wiggle of his own while Dr. W-abe places the cones in the rapidly melting ice.

"Atoms stay what they are at normal speeds and temperatures. And because of that molecules maintain their structure," conveys W-abe pouring an emerald green liquid into the frosty cones of glass.

"Do the molecules sit and wiggle?" Mim-sey asks.

"I will take that as an honest question. The molecules keep their kinetic energy and do not freeze the way that water does," chats W-abe with a sip of his drink. "Electrons just reduce the space they occupy."

"How is this possible?" Mim-sey asks wondering how the drink is made as well.

"Because acceleration is the same as gravity. The electrons increase in mass more than two hundred times. This allows them to orbit their nucleus closer. Even below the horizon of the nucleus," utters W-abe.

"Traveling at speeds greater than the normal speed of light is the same as being subjected to a strong force of gravity?" Mim-sey asks concerned more with the cost of the drink than with its taste.

"Acceleration for particles is the same as gravity for objects," says W-abe feeling the anesthetic charm of his first sip of the rare toxin. "In this case it is atoms and molecules."

"So something like a person or a spaceship will become very small just like its constituent atoms and molecules?" Mim-sey asks.

"Microscopic," says W-abe unwilling to apprise Mim-sey of how really small 'very small' is.

"Hold on," clamors Mim-sey now halfway through his icy, dark emerald drink. "I have traveled at greater than normal light speeds and I have not been microscopic."

"For you, time and space are normal. But for the rest of the universe you are microscopic," clarifies W-abe taking possibly his third sip and reflecting on the rising numbness of his face.

"But I do not just sit and wiggle while I am on my ship," protests Mim-sey.

"If an outside observer were possible," explains W-abe, "you would appear to just sit and vibrate."

"How does anyone live to tell about it?" Mim-sey asks.

"Deceleration, when it is done right is the same as relaxing the gravitational field," states W-abe as he pushes the button for more refreshments which means new bowls of ice and glasses.

"My electrons begin to move away from their atomic nuclei. Space and time return to normal?" Mim-sey asks with the confidence that he does not have to wait for his second drink.

"Yes but the trick is to do it at the right rate. A fish or a frog can freeze for the winter," coaxes W-abe risking another emerald sip. "If the thaw is at the proper rate he gets his wiggle back."

"If I decelerate too fast, I explode?" Mim-sey asks.

"Not chemically," says W-abe as if to answer an unasked question.

"What?" Mim-sey asks struck by W-abe's possible intrusion into his mind.

"Normal explosions are the result of a rapid chemical reaction," informs W-abe choosing to ignore Mim-sey's rising intoxication. "It can be oxidation or decomposition."

"So with deceleration the wormhole begins to pinch off, the leading shockwave retreats and the white hole begins to shrink," Mim-sey supposes.

"And below the speed of light you find yourself on he other side of the white hole," says W-abe.

"So the wormhole is just curve of space that you follow to the white hole?" Mim-sey asks forcing his focus.

"More than that," says W-abe glad to get out of the quicksand of subjective experience and onto the terra firma of cosmology. "The combination of a black hole/white hole pair is a worm hole. The black hole in this case is the ship accelerating to light speed and beyond. The worm hole and white hole comprise the path of least resistance."

"And with a worm hole you can go anywhere?" Asks Mim-sey emptying his tiny glass.

"We can and do," confers W-abe making room for the new bowls of ice and rock salt.

"And if Mr. Abbit's Earth can, they will," claims Mim-sey pulling his new bowl closer.

"Earth can bring competition and conflict to us," observes W-abe while pushing the salt cellar into easy reach for Dr. Mim-sey.

"And Earth will," says Mim-sey pouring a generous amount of salt on his ice.

"The paradigm shift that simplifies the math for an accelerating or a decelerating object," says W-abe, "also makes the math simpler for an expanding universe or a shrinking universe."

"You look worried," says Mim-sey waiting a few seconds to add the emerald toxin to his icy cone of glass.

"I just realized what I am saying. Most physicists concern themselves with the math for accelerating and decelerating objects. Fewer physicists labor with the math for an expanding universe. Fewer still ever think about the math for a shrinking universe," reveals W-abe now occupied with a train of thought that is both familiar and new.

"Why is that?" Mim-sey asks having waited sufficiently long to pour his new drink.

"Our universe, which includes Earth, is expanding," communicates W-abe working his way to the end of his first drink. "When the math for a shrinking universe can be avoided, it is."

"That does not sound like a disadvantage," imparts Mim-sey.

"If there were no such thing as symmetry you would be right. The math for a shrinking universe tells us a lot about the math for an expanding universe," discloses W-abe before taking a slightly larger sip of his ice cold emerald green drink. "Some of the possibilities for an expanding universe can be eliminated simply because of a lack of symmetry with a shrinking universe."

"Eliminating possibilities, that way, means chucking dead end theories, dead end research and all that goes with it," concedes Mim-sey before taking the first sip of his new drink.

"The advantage goes to those who do just that," grants W-abe finishing his first drink while frost still lingers on his glass.

"The disadvantage accrues to us if we do not," says Mim-sey.

Episode 83

Hospitals at night can be very quiet. Small hospitals at night can be even quieter. A small hospital near Verona, Italy, sixty years ago can be palpably quiet. It is the night before the 'delivery' of Jenny Abbit's first baby. Quiet is prescriptive, in this case, but not the ultimate prescription.

Her doctors and nurses know what they have to do. While it is not routine, they are confident that, at least, the mother can be saved and she can have other children. It is the hope that Edgar and Jenny hold onto. It is not only the hope, it is the justification.

The heartbeat that has never been strong has become imperceptible. This baby could still be alive. The question is how much longer - hours, days? The baby that is born alive is not an abortion. Legally, this is an effort to save the baby. Ethically, this is an effort to save the mother. Even though little can be done for the survival of such a baby, waiting means the difference between a birth and an abortion. Waiting means the difference between a legal and illegal procedure. Soon the difference will disappear.

It is two hours before sunrise. With the door closed, it is as dark and as quiet as can be. Jenny is asleep. Ali-ce is sitting in the only chair. She is not moving or doing anything to disturb the quiet. She needs it and Roy's attention. He needs to practice.

"*Roy*," Ali-ce projects without betraying the urgency of the moment.

"*Ummmmm*," Roy reflects selfishly.

"*Roy*," Ali-ce projects with a little more force.

"*Um*," Roy reflects a little less grudgingly.

"*Roy. Do you remember me?*" Ali-ce projects with guarded confidence.

"*Um,*" Roy reflects with an emerging cognition.

"*It is Ali-ce,*" Ali-ce projects fearing too much haste.

"*Um,*" Roy reflects without commitment.

"*Remember?*" Ali-ce projects trusting her persistence.

"*Um,*" Roy reflects with recognition.

"*It is dark, now,*" Ali-ce projects looking up at a small dew covered window high on the wall.

"*Um,*" Roy reflects acknowledging Ali-ce.

"*You can hear a heart beat,*" Ali-ce projects to orient Roy.

"*Ub...um...ub...um...ub..,*" Roy reflects in a dialogue with the pulse.

"*You have a heart beat,*" Ali-ce projects to focus his comprehension.

"*Um,*" Roy reflects with a sense of the obvious.

"*Dub...um...dub...um...dub...,*" Ali-ce projects to reinforce his recognition.

"*Um,*" Roy reflects with a sense of ownership.

"*You have a heart beat. You hear a heart beat,*" Ali-ce projects exclusively.

"*Ub, dub... um... ub, dub... um... ub, dub...,*" Roy reflects without distraction.

"*More light and light it grows,*" Ali-ce projects while looking at the glow of the small, fogged, window above.

"*Um,*" Roy reflects as he hugs himself into a ball to escape the faint, luminous intrusion.

"*Listen. These hearts beat, faster,*" Ali-ce projects observing the thinning dew.

"*Ub, dub; ub, dub... um... ub, dub; ub, dub... um...,*" Roy reflects unable to hide from the beat and emerging light.

"*Roy. Do you remember me? It is Ali-ce,*" Ali-ce projects forcing his attention again.

"*Um,*" Roy reflects forgetting himself a little and coming to terms with the rising light.

"*Remember me,*" Ali-ce projects prying Roy out of himself and his efforts to retreat.

"*Um,*" Roy reflects sharing himself with the light and the familiarity of Ali-ce's voice.

"*More light and light it grows,*" Ali-ce projects as the window's bottom edge glints.

"*Um,*" Roy reflects coalescing into a literal and cognitive dawn.

"*You have a heart beat.... Hear your heart beat,*" Ali-ce projects in her race to dawn.

"*ubdub, ubdub, ubdub, um... ubdub, ubdub, ubdub, um...,*" Roy reflects confirming his awareness.

"*More light and light it grows,*" Ali-ce projects as the dew erodes into islands of droplets.

"*Ali-ce?*" Roy reflects as the fog of a past without history erodes into an insular present.

"*Remember?*" Ali-ce projects hoping for a present with a presence.

"*The light?*" Roy reflects with the emerging suspicion that he is not the light.

"*Soon, you will hear your mother,*" Ali-ce projects hinting at another difference.

"*The light,*" Roy reflects with a clue that Ali-ce is not the light either.

"*Soon, you will hear others,*" Ali-ce projects hinting at still more differences.

"*The light,*" Roy puzzles at what the light is.

"*Soon, more light,*" Ali-ce projects without a hint of its ultimate fierceness.

"*Sleep,*" Roy reflects wishing with more urge to sleep than hope to sleep.

"*No sleep. More light,*" Ali-ce projects without consoling his hope.

"*Sleep,*" Roy reflects more demanding than hoping.

"*No sleep. More light and light,*" Ali-ce projects without a hint of sympathy.

"*Ummmmmm,*" Roy reflects with a dawning anger.

Episode 84

If a cluttered room is indicative of a cluttered mind then 1st-Robe's mind is decidedly uncluttered. If his office is larger than the others, it is difficult to say by just looking around. He has no desk but several tables and chairs. Some tables are without chairs. Some chairs are without tables. The other objects, whether they are books or objets d'art appear completely at home here. Sitting in one chair or another renders a different summary of the room. This becomes easily apparent by walking to different parts of the room which is what 1st-Robe always maneuvers his guests to do.

Conversations begin during the tour of the room and without a hint of direction from 1st-Robe the visitor stops, maybe more than once. When 1st-Robe thinks the visitor has found his comfort zone he takes a seat leaving the visitor feeling free to do the same for his own reasons and without a clue as to the Robe's calculus.

What 1st-Robe wants is to have the visitor settle with the view of the room that suits him. The view, for 1st-Robe, becomes a metaphor for the visitor's mind set. In this manner, the visitor suggests his penchant for heavy, clear, symmetrical or some other habit of mind. Without giving it thought, Officer Gim-bal settles into a chair that is surprisingly comfortable in a room that is the same.

"The Robes Chamber does not place any stipulations on the manner of death for the individual named in the death warrant or the expediency of its enforcement," advises 1st-Robe settling into his chair with undistracted familiarity.

"So I will not have to defend those choices," confers Gim-bal with a similarly relaxed attitude.

"It goes without saying that since it is intended to be limited to the named party or parties," collogues 1st-Robe unperturbed by the attention to redundancy, "it cannot involve anyone else."

"No innocent bystanders," says Gim-bal distracted by a polished pedestal supporting a large glass ball.

"In this case that includes Ali-ce Mim-sey," allows 1st-Robe letting his ease reinforce the gravity of his statement. "But in any case, if a bystander is at risk, you can put off executing the warrant for a more appropriate situation."

"Promptness and discretion are preferred but the details are up to me?" Gim-bal asks.

"Yes, and of course, you will have to wait until the warrant is issued," treats 1st-Robe.

"Has anyone tried to persuade Dr. Mim-sey to withdraw his petition?" Gim-bal asks.

"It is the last formality before any vote," says 1st-Robe.

"It will be especially important in this case," says Gim-bal.

"Dr. Mim-sey, as the petitioner, will be invited to the Robes Chamber and given the chance to withdraw the petition," says 1st-Robe absently nudging the heavy crystal sphere into a roll.

"What about the presumed party to be executed?" Gim-bal asks with a little tension coinciding with the lurch of the heavy globe.

"This is a summary petition," confronts 1ˢᵗ-Robe as the orb slows in its approach to the edge of the table. "The onus is on us to decide the fate of the designated person."

"Dr. Mim-sey's motives could be highly self-serving," predicates Gim-bal with some relief as the polished globe begins its retreat from the edge of its equally polished table.

"We have given that a lot of attention. In this case, with the arguments of Dr. Mim-sey's motives taken into consideration," says 1ˢᵗ-Robe without urgency as the globe returns to the center of the table with increasing speed, "there is still a risk to our civilization that the death warrant addresses."

"So the vote is not based on the opinions, or motives of Dr. Mim-sey," says Gim-bal.

"No matter how self-serving," says 1ˢᵗ-Robe as the globe glides past the center of the pedestal without restraint.

"May I infer that you do not approve of Dr. Mim-sey motives?" Gim-bal asks tensely.

"You may not. We can, and some Robes do, express their opinions along these lines but that is mostly to clear the air," says 1ˢᵗ-Robe as the globe slows to a halt somewhat short of the edge. "Those opinions do not contribute to the vote and are not expressed publicly."

"Is that why the Robe's opinions are published with the votes and the petition?" Gim-bal asks as the globe returns to an invisible center.

"Yes," says 1ˢᵗ-Robe with a smile. "You would be surprised at how much the prospect and the effort to put one's opinions in writing eliminates petty concerns on the part of the Robes."

"It does not seem to elevate the impetus of the petitioner," asserts Gim-bal.

"All petitioners, regardless of the origin of the petition, have an agenda. Dr. Mim-sey, does what they all do," says 1st-Robe without a glance at the sphere as it rolls to the brink again. "He appeals to the highest standards and discounts his own exaggerations."

"My concern is that while I do not know Roy Abbit, I do know the Mim-seys," says Gim-bal guessing that the tabletop is not simply dimpled but curved enough to drive the mass of the ball back and forth.

"You aspire to know what you can or should do?" 1st-Robe asks.

"Yes," says Gim-bal as the globe finds the center of the table again.

"This is not being as patronizing as it sounds but since you do not know the vote, you cannot do anything," states 1st-Robe while Gim-bal studies the framed and mounted Eye-moth above the now settled sphere.

"I forget there is no vote, yet," deposes Gim-bal estimating the Eye-moth to be as big as two hands. "It would be a mistake to prepare, even Ali-ce, for a vote that would execute Roy Abbit."

"Not only do you not know what the vote will be, but if you try to prepare Ali-ce for the vote it will imply that you know something you do not," says 1st-Robe pleased to own an Eye-moth.

"That is not my intention," affirms Gim-bal guessing legal collection of the moth ended centuries ago.

"To be sure," predicates 1st-Robe recalling the Eye-moth's history with cults and icons. "You are looking out for her. But the effect of that is that you appear to know something you do not."

"I suppose, there is a part of her that will discount my insistence that I do not know how the vote will go," asserts Gim-bal aware that the Eye-moth's photosensitive scales may still be viable.

"If you try to console her with what the vote could include, she might think you were telegraphing the vote. I am sure you do not want to do that," says 1st-Robe mindful that the light sensitive scales disguise its very state of life or death.

"But you have not said I cannot associate with the Mim-seys or even Roy Abbit," says Gim-bal noting that, while alive, the Eye-moth can choose its patterns and colors to deliberately or passively reproduce its environment like a camera.

"We are just getting to that part of the conversation," says 1st-Robe. "You can say and do as you wish in all cases."

"You can assume that I will not make my job harder by a deliberate word from me but what if I inadvertently complicate things?" Gim-bal asks.

"Can you think of any vows or maneuvers on your part that would facilitate the events to come?" 1st-Robe asks forgetting the Eye-moth for the moment.

"I can omit revealing my role as executioner until after publication of the petition, vote and warrant," says Gim-bal detecting the image of a lamp and a row of books on the wings of the Eye-moth.

"That is all we ask," enjoins 1st-Robe leaning a little forward in his chair.

"This sounds like the end of the conversation," imparts Gim-bal preparing to stand.

"Unless you have something else on your mind," remarks 1st-Robe rising from his chair.

"Only hypothetical concerns," divulges Gim-bal now standing at eye level with the Eye-moth.

"You will not be allowed to communicate with the Robes anymore until after the vote," says 1st-Robe. "If you need to, you can contact any of our staff at any time and in any manner."

"Will they forward my concerns to the Robes?" Gim-bal asks.

"At some point before the vote all contacts outside the Robe's Chamber will stop," says 1st-Robe noting Gim-bal's profile in the wings of the Eye-moth. "You will not be told when that is."

"What happens if a death warrant is issued?" Gim-bal asks with no thought of the Eye-moth.

"You will be called to the Robes Chamber by a member of our staff," reveals 1st-Robe. "There will be a private ceremony to invest the warrant and then you will be free to execute it."

"The vote and the petition will not be published until after the warrant is enforced?" Gim-bal asks unaware that this part of the conversation has lasted long enough to passively impress his image on the wings of the Eye-moth.

"Exactly," counsels 1st-Robe as he moves toward the door. "If there is no warrant, the vote and the petition will be published without ever mentioning you and your service in reserve."

"In that case, I will find out along with everyone else," says Gim-bal following 1st-Robe.

"That is the way it is done," says 1st-Robe while opening the door.

"Thank you," says Gim-bal before stepping out of the office.

"And you have our thanks as well," says 1st-Robe before closing the door.

1

1 1

1 2 1

1 3 3 1

1 4 6 4 1

1 5 10 10 5 1

1 6 15 20 15 6 1

Chapter 29

Chapter 10111000....

Chapter ACDE

Chapter $(1 + 5x + 10x^2 + 10x^3 + 4x^4)$

Chapter $(a^4b^0 + 5a^3b^1 + 10a^2b^2 + 10a^1b^3 + 4a^0b^4)$

Chapter $2^0 + 2^2 + 2^3 + 2^4$

Episode 85

    The call to board the flight home catches Dr. Mim-sey by surprised. This is not because of the required absence of clocks and watches. It is because he has not finished his drink. This is a drink he knows to be rare and expensive, but cost free and unlimited for him. If Dr. W-abe notices his friend's ambivalence he does not show it.

His only apparent concern is that his passport is handy in case of need. That, he checks while he boards the flight. The two board as if they were together. As if they were together both in purpose and destination.

The cultural penchant for "Don't bother me with the rumble and the sway of getting me there." is not lost on the Banker's transport. The passenger's cabin is gimbaled and essentially circular with some effort to disguise its shape with furnishings, focal points, lighting and accouterments. By longstanding tradition, Flight attendants never enter and exit through the same 'door'. A custom that predates flight, requires that no matter what entrance the attendants use to enter the passenger cabin, it does not serve as the exit for that visit. If you want to speak to the captain of the ship, the attendant exits in direct pursuit of your wish immediately but not through the portal of entrance.

The captain, promptly or not, visits in response to your request. He may or may not enter through the same portal that the the flight attendant used but he will not exit by his portal of entrance. There are no hesitations, missteps or even accidents - so strong is the custom. This much is certain, even in a cabin with only two portals, you will never know which one someone will enter from but you can be sure they will not exit from the entrance-one.

On Clidim, the combination of motion without a sense of motion, direction without a sense of direction and time without a sense of time render a complete but unimposing comfort for a traveler who prefers to travel without imposition. For Dr. W-abe, there is no luxury that is not incidental, as long as it is the luxury that he expects. For Dr. Mim-sey, the occasion should be commensurate with the luxury.

When it is not, he is irredeemably disappointed. One or the other has been shortchanged. Dr. Mim-sey, who treats every conversation as a means to an end is not anticipating disappointment this time. He sees the end of this conversation just as it begins even if Dr. W-abe does not.

"There is nothing more than analogy in the symmetry of the Pascal/Chu triangle," emphasizes W-abe lounging in one of four oversized chairs. "Analogy does not a paradigm shift make."

"Without Mr. Abbit's manuscript or its publication you would be safely and comfortably correct," says Mim-sey sinking into his chair more aware of the lighting and music than the flight taking off. "But he is proposing a calculus."

"That is just a little too generous," says W-abe as the pneumatic chairs perform their job.

"It is more than a little too innocent to dismiss his logical, algebraic and statistical applications as mere analogy," says Mim-sey as the deck, pictures, tables and chairs level with the flight-craft usually to the impercipience of otherwise distracted passengers.

"Do not get in a temper," urges W-abe keying in an order for food and drinks.

"You can give Abbit some credit. Yes, he has a long way to go to achieve a paradigm shift but it looks like the shortest rout to one," submits Mim-sey feeling clueless as to what W-abe has ordered for them.

"I guess I will have to give you that but there is more than a paradigm shift between a black hole, a white hole and a worm hole," says W-abe betting himself on which door will open.

"Is it more than the difference between an expanding universe or a contracting universe?" Mim-sey asks as the door behind him opens bringing a smell of hot savory food.

"I do not think even an analogy can come close to that difference," declares W-abe.

"The mathematical or analogical symmetry between an expanding universe and a contracting universe is complementary, is it not?" Mim-sey asks moving to the dining table.

"Like two angles that sum to ninety degrees? Yes," outlines W-abe.

"If both angles were forty-five degrees, they would be similar," affirms Mim-sey.

"Yes," accepts W-abe taking his seat while the wait service sets the table with food. "This kind of symmetry is also typical of the odd numbered rows of the Pascal/Chew triangle."

"But this kind of symmetry is not typical of the black hole/worm hole/white hole analogy or symmetry?" Mim-sey asks.

"And the difference is cosmic. You might stipulate that there must be a white hole for every black hole," espouses W-abe waiting for the wait service to exit before he picks up his fork. "But you cannot by any stretch of mathematics or analogies hold that every black hole is connected by a worm hole to a white hole."

"But you can stipulate both by mathematics and analogy that the path between a black hole and a white hole is by way of a worm hole," prompts Mim-sey peeking under lids and tops.

"The need is the connection of the worm hole," urges W-abe.

"So look at the even numbered rows of the Pascal/Chu triangle," prods Mim-sey.

"The symmetry is different," explains W-abe plating hot and cold servings for his friend. "There is a subset separating, or maybe connecting, the complementary subsets."

"What do you suppose is the significance of that intervening subset?" Mim-sey asks.

"If we were talking about an expanding/shrinking universe," construes W-abe serving the entre for himself the as well as Mim-sey, "the intervening subsets would provide a transition between the two."

"This same subset could, simplify or facilitate that transition between an expanding universe and a contracting universe," interprets Mim-sey while pouring hot and cold drinks for them both.

"In fact, without such a transition, the theories would compete rather than combine," elucidates W-abe as he starts with a quaff of his drink.

"Theories about an expanding universe would remain distinct from theories of a contracting universe," says Mim-sey as he seasons his hot drink with a sweetener and a taste.

"Just like white holes and black holes with no worm holes to connect them," says W-abe ready to lose himself in his waiting food and drink.

"Without the benefit of such an analogy," declares Mim-sey taking a bite of his favorite sweet red fungus, "mathematicians could work forever in separate camps on expanding and contracting universes, black and white holes or any number of complementary and similar theories without combining their efforts."

"In the most general way possible, by way of consistency and analogy, row three of the Pascal/Chu triangle suggests that a Theory of Everything benefits from eight dimensions not ten," says W-abe while he collects a combination of smokey pate and pickled white fungus before reaching for his favorite blue sauce.

"But row four has the added benefit of six dimensions to connect the micro-dimensions to the macro-dimensions," jubilates Mim-sey alternating between his hot drink and his red fungus. "For those who need the ten dimensions, this is the first row that supports them."

"We have come full circle but not without Mr. Abbit," says W-abe reaching for his drink.

"And consequently not without escaping our resolve to eliminate Mr. Abbit's mathematics, even at his expense," says Mim-sey with an exalted sense of self-satisfaction.

Episode 86

The light in not shining directly into the window but the source is definitely that of sunrise. It cannot be called dark anymore. There is no new light added to the room when the door is opened. If Sister Ali-ce were asleep, like Jenny, she would not have noticed that the door is opening. But Ali-ce is not asleep and her dialogue is finished. Her exchange with the soon to be born Roy Abbit is primitive for obvious reasons but just as obviously focused for its singular purpose. Everything in those few hours before sunrise between Sister Ali-ce and Roy is intended for his grab at survival and his commitment to his grasp. She does not prepare him for shock or desperation. That will take care of itself.

She prepares him for commitment to commitment. A commitment that is utterly selfish. So selfish that the thoughts and wants of the soon to be new 'now' eclipses everything. The old now of sympathetic heartbeats lost forever; the chill of the air; the heat of the light; the vulgar heave of a self stolen from a lifetime of pampered buoyancy; the familiar comfort of closeness without dimension now turning into the terror and insult of being gracelessly pulled from the center of everything into the middle of nowhere; these altogether and separately.

Woe to anyone who cannot respect or redeem the egomania of the infant parted forever from all that has always been his. Ali-ce's work is done and she is not surprised to see Edgar. She waits for Edgar to look away from Jenny's bed and speak.

Under the burden of a restless night at home, despite the hope of getting a little sleep, Edgar says, "Sister Ali-ce, I am not surprised to see you. How long have you been here?"

"A little over two hours, Edgar," reports Ali-ce. "Jenny has been asleep most of the time."

"I am surprised. She is so distressed," says Edgar feeling consoled with the good news.

"You both want this baby. How can you let the doctor talk you into this?" Ali-ce asks coming very close to raising his anxiety even more.

"They cannot hear a heartbeat," falters Edgar looking to see if Jenny is still asleep. "If they do not remove the baby, we could lose Jenny or the chance to have another baby too."

"Is she ill?" Ali-ce asks with a reassuring look for Edgar when he looks back to Ali-ce.

"Not yet. But If we wait she could get very ill very fast," quavers Edgar.

"You cannot lose them both," says Ali-ce with Edgar resisting the urge to check on Jenny.

"We both want children," says Edgar calmed a little by returning to some of the ground that they have thoroughly covered. "Losing this one must not prevent us from trying for another one."

"What if this baby is alive?" Ali-ce asks with more than rhetorical motives.

"A life of hours or days is the only thing that can make this situation worse," says Edgar.

"Being three months premature is dangerous but not without hope," councils Ali-ce

"We would be his murderers," says Edgar choking on the words as well as the thought.

"How can you be to blame?" Ali-ce asks with other villains in mind.

"We would not be held responsible but we would feel responsible," trembles Edgar.

"The nurses want to baptize the baby, if there is any sign of life," broaches Ali-ce.

"They want the baby to get into heaven," says Edgar while mentally exploring something unconsidered.

"That is the least they can do," acknowledges Ali-ce illuminating their motives and options.

"The Sisters might have only a few minutes," quakes Edgar.

"Do you want to ask Jenny?" Ali-ce asks to firm up the possibility.

"We have not decided this before now," shrinks Edgar as the possibility shrivels into impossibility. "I do not think there will be time before the operation."

"By the time she is out of the general anesthetic it maybe too late to ask," says Ali-ce studying Edgar's face.

"I do not want to ask just before the operation," says Edgar.

"She might be consoled to know how much we all care," says Ali-ce renewing her approach with Jenny's own sentiments.

"Jenny and I have made some good friends here," affirms Edgar shoring up his spirits.

"We all care so much for both of you," asserts Ali-ce standing for the first time.

"You have never given up on this baby," acknowledges Edgar turning toward Ali-ce.

"I do not think you or Jenny have given up either," confirms Ali-ce assailing any possible doubts.

"Yes. Baptize the baby if there is a chance," declares Edgar.

"When will they perform the surgery?" Ali-ce asks almost as if she has a new task.

"They are ready now," apprises Edgar as he returns to his original purpose on entering the room. "I asked to be the one to wake her so they can begin the anesthesia."

"I will speak to the nurses," says Ali-ce. "We will wait outside the operating room."

"If it were the delivery room, you could be there with us," says Edgar as Ali-ce turns to the door.

"I am with both of you anyway and always," says Ali-ce as Edgar turns to Jenny's bed.

Episode 87

Perhaps J. Rus-ty Gim-bal is anticipating and resigned to a permanent exodus from the house at Gates Street but, nevertheless, the living room feels completely commodious. There is nothing in the room that you could not get from a Carolina furniture factory or a Virginia furniture store. The everyday estate sale habitue would be charmed and disappointed at the same time.

Charmed at the good sense and good taste of the original owner and disappointed at the lack of anything exotic. The exotic is required not just for the sensational but also to distinguish the shopper as superior to the now failed landlord. Superior for never having owned such a spectacle and for not submitting to the temptation to take possession of the marvels until they could be had at a ridiculous discount.

Furnishing the house or even this room is not a priority. He buys what he wants when he sees it. Money is no object but expensive things escape his interest. It is enough to buy it if he likes it. In one sense, it is like the contemporary 1st-Robe's office. The room does not impose on him. It tends to disappear during a conversation.

But there the similarity ends. There is not one object in any of the rooms that is identical to any of those in 1st-Robe's office except that chairs are chairs and so on.

"I thought you would not be back so soon," says J-R One.

"There is no telling when the Robe's Chamber will vote or how they will vote," says J. R. Gim-bal while he divides an apple into sections.

"Unless there is a call to Investiture for the warrant, you do not have to go back," concludes J-R One taking a few of the apple sections from Rus-ty.

"Even if the vote comes soon, there is a chance I will not have to return on short notice," says J. R. Gim-bal while J-R One drops the apple sections onto the wood chips that more than half fill the terrarium.

"Where is Ali-ce?" J-R One asks as he watches the wood chips quake under the apple slices.

"I have not seen her since our last conversation. No one had been asked to execute the warrant by then," says J. R. Gim-bal as a swarm of acrid ebony millipedes unhurriedly envelop the apples.

"Why did you agree to do it?" J-R One asks as ever more millipedes surface and crowd onto the apple chips.

"I was not situated well to refuse. Who among us can get to Roy Abbit?" J. R. Gim-bal asks while adding some more apple slices to the terrarium of wood chips and the deepening layers of chaotically articulating millipedes.

"No doubt they could not ask Ali-ce and asking one of us is the same as asking you," derives J-R One as nearly a third of the terrarium has become a slowly churning carbon black knot.

"It is Mim-sey's petition. Who else has a professional or familiar relationship with both Ali-ce and her father?" J. R. Gim-bal argues. "Acting in behalf of one is to countervail the other."

"Refusing as a way to avoid conflict would put you in Ali-ce's camp as far as Dr. Mim-sey is concerned," contends J-R One now adding apple sections to a small hand net.

"Can you think of anyone who would read it any other way?" J. R. Gim-bal asks.

"Do you suppose the Robes considered this before asking you to enforce a possible warrant?" J-R One asks.

"I did not ask so I can only speculate," says J. R. Gim-bal parking his knife on a table.

"What do you think she will do when the warrant is published?" J-R One asks.

"If there is a warrant," affirms J. R. Gim-bal while J-R One lowers the net and apple sections into the mass of millipedes, "it will only be published after the fact, after the warrant is executed."

"It will mean the end of this project. Will it not?" J-R One asks as the ink-black melange ruminates.

"We can not hang around for an investigation," says J. R. Gim-bal sitting back in his chair.

"What will Ali-ce Mim-sey do?" J-R One asks with the migrating millipedes crowding into the net.

"That is a separate question," J. R. Gim-bal contends as J-R One lifts the nearly full net.

"She can make a life here, if she wants to," says J-R One emptying the net into a glass bowl with steep sides. "With Mr. Abbit and the rest of us gone, she might be a mystery but not one that points to us."

"She will not be in a hurry to go home under the circumstances," declares J. R. Gim-bal picking at an occasional wood chip.

"Her stay here on Earth will be noticed at home," says J-R One handing Rus-ty the bowl of preoccupied millipedes.

"The only ones that will notice will be the Mim-seys," cites J. R. Gim-bal, "and his supporters."

"The more prominent he becomes the more conspicuous her absence becomes," reminds J-R One as they move to the kitchen and while Rus-ty sets the bowl of shuffling millipedes on the counter.

"Not just her absence from Clidim but her preference for this planet and its Earthlings will be obvious," says J. R. Gim-bal.

"She could become an ally of the opposition that is likely to develop on Clidim," details J-R One.

"She could nurse allies on Clidim with a sympathy for Earth," says J. R. Gim-bal placing a lid on the bowl.

"What can follow from that?" J-R One asks adding water to a double boiler.

"Nothing in terms of Earth," says J. R. Gim-bal retrieving some broccoli from the crisper.

"It could take generations to form a coalition on Clidim that has the ability to bring their idea of justice to Earth," says J-R One as Rus-ty cuts the broccoli florets from their stems.

"Yes, initially she might blunt her father's success," moralizes J. R. Gim-bal layering the double boiler with broccoli florets, "but she might also exaggerate the divisions that could fragment our world culture."

"What do we do now?" J-R One asks.

"We wait for the orders to end our presence here on Earth and execute our exit," narrates J. R. Gim-bal stratifying the percolating double boiler with a mahogany colored fungus. "Our companies will get new CEO's from these Earthly ranks and no one will know about our alternative lives and covert missions."

"That sounds more like the answer to, 'What do we do then?'?" J-R One asks.

"Oh, yes. We wait to hear from The Robe's Chamber, Dr. Mim-sey and Ali-ce Mim-sey and try to satisfy all of them," offers J. R. Gim-bal lifting the bowl of milling millipedes.

"It is good there are three of you," opines J-R One lifting the lid from the double boiler.

"For the time being," says J. R. Gim-bal adding the teeming mass of millipedes to the steamer, "the three of us can start making enemies with the principals of the companies that I, uh, we direct."

"You want them to think about retiring us?" J-R One asks returning the lid to the steamer.

"When we get the word to go, we want to be able to quit or be fired," concludes J. R. Gim-bal taking his seat at the kitchen table. "We are not going to be here to pick up paychecks but we do not want to appear to disappear."

ENDING CHAPTERS

1

1 1

1 2 1

1 3 3 1

1 4 6 4 1

1 5 10 10 5 1

1 6 15 20 15 6 1

Chapter 30

Chapter 01111000....

Chapter BCDE

Chapter $(1 + 5x + 10x^2 + 10x^3 + 5x^4)$

Chapter $(a^4b^0 + 5a^3b^1 + 10a^2b^2 + 10a^1b^3 + 5a^0b^4)$

Chapter $2^1 + 2^2 + 2^3 + 2^4$

Episode 88

Because W-abe and Mim-sey take the Banker's flight every year to the Director's Retreat, they have a sense of how long the trip should take.

Direct flights between the resort and the building-port always seem to take the same amount of time. The mental clock begins with the acceleration of the takeoff and the clock stops with the deceleration of arrival. Dr. Mim-sey always wants to verify his estimate but does not always get to. Dr. W-abe never wants to verify his estimate. He knows whether the duration of the flight is more or less the same as before.

He also knows that Dr. Mim-sey will study his watch when he gets it back and he will compare his watch to any clock he encounters immediately after the trip. But the effort does more to confuse the issue than to resolve it. Mim-sey has to estimate the time in the Banker's Lounge and their travel speed.

He has to decide whether or not his watch has been reset when it does not agree with the clocks he can find or whether the local clocks are just off. Dr. W-abe watches Dr. Mim-sey go through this almost every year. There are trips when Dr. Mim-sey does not go through the flight-time ritual.

W-abe can see it on Dr. Mim-sey's face. If deceleration begins too soon, they are making a side trip to drop off or pick up something. It will not be a passenger, there is never any indication of where the stop is made and there is rarely more than one side trip.

As soon as Dr. Mim-sey decides this is not a nonstop trip the distraction of timing the flight ends. Dr. Mim-sey is able to discuss things that are best suited to an undivided attention. This time the clues of the side trip are there so Dr. W-abe knows how this conversation will go.

"When do we begin planning Mr. Abbit's end?" W-abe asks pushing his empty plate away.

"We are past the planning and the end has been set it in motion," says Mim-sey not yet finished eating.

"What is your daughter going to do?" W-abe asks lingering with his cold drink.

"She will not know before I get her back here on Clidim. Not even then, if I can avoid it," says Mim-sey.

"What makes you think she knows much, if anything?" W-abe asks refilling his glass.

"On the one hand she will not leave Mr. Abbit if she thinks he is in imminent jeopardy," says Mim-sey. "She has her suspicions but she cannot confirm them without leaving Earth to do it."

"No offense, but she cannot suspect the true potential of Mr. Abbit's manuscript," says W-abe as Mim-sey ignores his nearly empty plate and contemplates his drink.

"Mathematics is not her field," manifests Mim-sey while reaching for the chilled pitcher.

"What keeps Mr. Abbit from finding out the importance of his manuscript?" W-abe asks as Mim-sey fill a tumbler.

"He is his own worst enemy there. He is phobic about promoting his own work," affirms Mim-sey moving along with W-abe to bigger chairs and smaller tables. "Abbit does not want his work stolen by someone who could turn it into someone else's money."

"That could be anyone who is a better writer and a trained mathematician," says W-abe.

"Exactly," asserts Mim-sey as the wait service clears and polishes the dining table. "But the first book on the topic is still the seminal book even if it is badly done."

"Mr. Abbit would get acknowledgment and a little money, at least," professes W-abe.

"And, no offense again, most mathematicians would not condescend to read his work," proffers Mim-sey feeling for his pocket watch out of habit.

"Abbit has no credentials," predicates W-abe trying to not smile at Mim-sey's absent minded search for his watch. "It would take less manipulation than sheer murder to keep things that way."

"He may live thirty or forty more years. Even if my daughter were willing to cultivate his friendship for that long," says Mim-sey guessing he has time to finish his drink, "she would not do it at the expense of his sole achievement. One that would barely raise him above ignominy and a petty life."

"Replacing Mr. Abbit's keeper periodically would only risk discovery of his merit and our presence," says W-abe judging that Mim-sey is getting anxious about his beverage nearing its end and keying in an order for fresh drinks.

"Besides, the need for a continued or permanent presence on Earth is gone," tenders Mim-sey now wondering if he should hurry the rest of his drink. "We have learned from my daughter's experience, and others, how to come and go without revealing ourselves."

"And I suppose Earth's technology makes it easier to survey its knowledge base now?" W-abe asks keying in an order for hors d'oeuvres as well.

"That and the secret stuff will not go undiscovered," proffers Mim-sey looking more relaxed. "The advances in math and physics will be indicated by the kinds of experiments that get funded."

"I suppose significant weapons will be limited to Earth's surface or near it," says W-abe satisfied that Mim-sey is not concerned about the duration of the flight, at least for the moment.

"Communications and travel will develop separately," says Mim-sey setting the empty glass on his drink table. "Which means the means of delivering weapons will be addressed well after the fact."

"Besides, we have the advantage of knowing about them without them knowing about us," proposes W-abe before draining his glass.

"And by eliminating our establishment on Earth," propounds Mim-sey wondering what is delaying the wait service, "we have a better chance of remaining undiscovered."

"When will our permanent presence on Earth end?" W-abe asks undistracted and unperturbed by Mim-sey's perturbation and distraction.

"We will start that as soon as Ali-ce gets back here," declares Mim-sey.

"You are keeping her in the dark about that too?" W-abe asks as Mim-sey begins to stir.

"I do not want her to figure out whether or not we are going to leave Mr. Abbit alone," says Mim-sey looking at his empty glass and feeling his watch pocket, again, for his accustomed timepiece.

"By the time she does, she will not have the benefit of any of our current resources on Earth," says W-abe.

"And without significant resources at her disposal she cannot frustrate the termination of Mr. Abbit," says Mim-sey while he is beginning to suspect the reason for the absented wait service.

"The only loose end is the murder of Mr. Abbit," says W-abe as he takes the wait service's cue by standing.

"At the very least, the executioner does not need to know why Mr. Abbit must die," says Mim-sey now standing with the reluctance and resignation that comes with the end of the flight.

"But you still have to contain any notes or copies of his work," predicates W-abe. "It would be a mistake to count on his work going undiscovered forever."

"That is the weak link," carps Mim-sey as he retreats to the oversized chair he started the flight with. "If his notes or a manuscript survives so does his danger."

"This means that Mr. Abbit will have to meet his end before he begins his research for his manuscript," rallies W-abe while he settles into his own pneumatically compensating chair.

"His library records will tell us when that was," says Mim-sey feeling a little less than happy.

Episode 89

The hospital is equipped and staffed like anyone would expect. That much makes it look like a hospital. But beyond that it does not look like one. The windows are too big. Rooms are too small and too few. Hallways are too large. In fact, all along all the hallways there are ersatz rooms for equipment, beds, supplies and office spaces for interviews and nurse stations. That is because this building's use as a hospital is just its latest evolution. This is originally the home of a merchant. In the centuries that followed, it grew with the business from a house into a warehouse cum home. Needs made it into a fortress and finally a palace.

This is a pedestrian palace that money made. It has no role in history, politics or religion to give it a pedigree. There is a little local history but it is hard to find. It is a hospital out of necessity. Nobody who knows the needs of a hospital thinks this is where to put one. But the possibility of building a new hospital for the neighborhood is out of the question. So this palace serves as a hospital instead of one built for the purpose.

The requisite quiet is easily displaced by any conversation or activity. Quiet becomes the hum of appropriate noises rather than the rattle of inappropriate noises. Equipment that is operating to a hospital purpose is quiet. Repairing or setting up that equipment is noise. Conversation involving hospital operations is 'quiet'. All the other conversations are noise. Nurse Sister James and Nurse Sister Fiona are having a quiet conversation at the door of the operating room.

"Can you see anything?" Asks Nurse Sister James.

"No. But the operating light is in position," narrates Nurse Sister Fiona accommodating her colleague without giving up her disadvantaged vantage point. "They are ready to begin."

"This is so sad," relates Nurse Sister James. "Sister Ali-ce is praying in such earnest."

"If there is any sign of life we can baptize the baby but that is just to prepare for the funeral," ascribes Nurse Sister Fiona.

"I hope we can tell Mrs. Abbit we were able to baptize the baby when she comes out of the anesthesia," instances Nurse Sister James. "It will be the only good news we can give her."

"I think they are resigned to losing the baby," says Nurse Sister Fiona into the glass.

"Everyone is except Sister Ali-ce," says Nurse Sister James resigned to the relay system.

"If she were speaking out loud, she would be talking to the baby even now," states Nurse Sister Fiona looking away from the door long enough to focus on Sister Ali-ce.

"Almost certainly, the way she is praying," laments Nurse Sister James looking on as well.

"I wish there were something that could be done for premature births," deplores Nurse Sister Fiona returning to her vigil and the circle of shoulders obscuring Mrs. Abbit.

"The problem is usually heart and lungs," details Nurse Sister James trying to prepare herself for the almost certain grief. "They are just not strong enough, yet."

"And any infection stresses the heart so," says Nurse Sister Fiona with resignation.

"This baby will be spared the stress of a natural birth but that may not help," protests Nurse Sister James trying to leverage hope against expectation.

"An incubator, some extra oxygen and sister Ali-ce is the baby's only hope," says Nurse Sister Fiona while studying the surgeon's choreographed shuffle for exceptions and expectations.

"Do you hear that?" Nurse Sister James asks without the benefit of any kind of view.

"Is that the baby?" Nurse Sister Fiona asks wondering where the cry is coming from.

"The baby came out crying," responds Nurse Sister James with her ear to the door.

"And loudly too," declares Nurse Sister Fiona while rejoicing in her own certainty.

"Almost like a scream," extols Sister Ali-ce with a certainty of her own.

"Maybe there is hope," enthuses Nurse Sister James as she replaces hope with conviction.

"There has always been hope," stirs Sister Ali-ce convincing herself as if for the first time.

"Perhaps the baby will live even more than a few hours," rouses Nurse Sister James.

"It sounds good," says Nurse Sister Fiona commingling both the facts and hopes.

"I can tell the Abbits the baby is alive," says Sister Ali-ce feeling restored with confidence.

"The baby will not be able to leave the hospital for weeks. Perhaps a few months," delivers Nurse Sister James tempering her joy and hopes with reality again.

"That is not bad news, now," says Sister Ali-ce anticipating her next prophylactic obligations.

"Here they come," proclaims Nurse Sister Fiona relinquishing her post at the door.

Episode 90

Even thought it is early April, the trees are heavy with ice. It could be the last cold weather event of the winter for this part of Virginia. A walk to the shuttle means lots of dripping water and some falling ice.

This does not promise a comfortable walk but the view is spectacular. It is another thing Gim-bal will miss once he leaves Earth. Suburban life and winter temperatures are just another way his home and Clidim is so far removed from Earth. It is this contemplation and this view that is interrupted by the news.

"Your 'Notice to Appear' has arrived," says J-R Two studying a handheld device.

"Is Ali-ce still with her parents?" J. R. Gim-bal asks while staring at the trees perhaps for the last time.

"Yes, and Roy Abbit is still at Number Two Meeting Street," volunteers J-R Two.

"I am being called back to the Robe's Chamber," declares J. R. Gim-bal hoping he has anticipated everything and knowing there is no such thing. "That means they have issued a death warrant."

"Do you have to go to the Investiture?" J-R Two asks while looking for some advantage to the situation.

"I do not have the authority to kill anyone until I have that piece of paper in my hand," says J. R. Gim-bal as he plumbs J-R Two's vain attempt at optimizing the process.

"Are you coming back here after the investiture?" J-R Two asks genuinely.

"I expect to go to Charleston first," articulates J. R. Gim-bal as he answers far ahead of J-R Two's questions. "Tell Ali-ce I will need her place on Wappoo Creek."

"She will want to know when her place at Wappoo Creek is available," says J-R Two.

"Tell her to park here and take a commercial flight to Charleston, if that is where she wants to go. Make sure she knows this before she leaves her parents, she may have to adjust her schedule," states J. R. Gim-bal.

"Will you need to know her travel plans for any, uh, reason?" J-R Two asks.

"No, I expect to see both her father and Roy Abbit before I return to this house," says J. R. Gim-bal. "So there is a chance I will see her too but she does not have to change her plans for me."

"Are you going to Glasscastle before returning here?" J-R Two asks as he retreats to second tier details.

"I need to spread some hate and discontent. I want to be able to quit, if I have to, when I get back," exhorts J. R. Gim-bal.

"What else?" J-R Two asks as he polls third tier assignments.

"Have you put that bucket with beef liver and beer in my shuttle's freezer?" J. R. Gim-bal asks.

"Yes," says J-R Two as they step off the porch and walk toward the ice clad trees with their branches drooping from the icy burden and shedding water like fountains. A portrait of grief formed by the unnatural drape of the branches and the raining melt. All this punctuated by random, petulant shocks of falling ice as if the agony is still insufficiently manifest.

"Is that everything?" J-R Two asks.

"You and J-R One have the 'grief and acrimony detail' for the other companies. We need to be able to quit or be fired once we get notice to leave Earth," details J. R. Gim-bal redundantly.

"You are pretty sure that is coming?" Asks J-R Two with similar redundancy.

"There will be no need for us to be here, on Earth, after Mim-sey's petition is published. The mission will change from one of collecting information to the containment or subjugation of Earth," speculates J. R. Gim-bal with an eye to a certain, if not very distant, future. "We cannot have anyone permanently here under those circumstances."

"What about Ali-ce?" Asks J-R Two cluelessly.

"We will always be able to come and go according to the perceived needs of the time," declares J. R. Gim-bal dismissively. "She can come and go according to her circumstances as well."

"I suppose as factions grow and as Clidim's culture fragments," embroiders J-R Two sidestepping a puddle, "there will probably be more expeditions to Earth than ever."

"You may be right," submits J. R. Gim-bal. "Ali-ce may be more important than ever."

"Perhaps," underscores J-R Two as they approach the shuttle and arrive at the end of any room for conversation.

1

1 1

1 2 1

1 3 3 1

1 4 6 4 1

1 5 10 10 5 1

1 6 15 20 15 6 1

Chapter 31

Chapter 11111000....

Chapter ABCDE

Chapter $(1 + 5x + 10x^2 + 10x^3 + 5x^4 + x^5)$

Chapter $(a^5b^0 + 5a^4b^1 + 10a^3b^2 + 10a^2b^3 + 5a^1b^4 + a^0b^5)$

Chapter $2^0 + 2^1 + 2^2 + 2^3 + 2^4$

Episode 91

Dr. Mim-sey does not like being so far away from his office and his '-burb home. It is little consolation that it is his second favorite home of the three he owns. The third one is supposed to keep his wife happy. He tells her to stay there when he needs her to and this time he wants to talk to Ali-ce alone.

Ali-ce is spending the day with her mother. That has made her late. Dr. Mim-sey does not like it when anyone is late. He takes it personally, even with family. But it is all right, this time. Mim-sey is not planning on being pleasant. Neither is he planning on dissembling with his daughter.

He is so sure she knows what this is about she could almost miss this visit with him. But Mim-sey wants to be sure of himself, even if it is at her expense. It does not matter what the inconvenience is for her. He does not care what the cost is to her sympathies for Earth or Roy Abbit.

Dr. Mim-sey is throwing crickets in the pond. He pays his staff to do this but when he has the opportunity, he likes to do it himself. The crickets cannot swim. That is part of the pleasure. He hopes the fish are hungry. That completes his tableau of terror.

The fish are large enough to eat the crickets in one bite but they always seem to masticate the poor things first. The fish have no teeth but their mouths are bony and strong enough to break the cricket into pieces before eating them. This is where Ali-ce finds him.

"Father," voices Ali-ce as she enters her father's spacious garden atrium.

"Father?" Mim-sey asks still concentrating on his fish pond. "Are we going to use English?"

For the time being," declares Ali-ce as her father slowly turns to face her.

"Does that include me?" Mim-sey asks while he is aware of the missing exchange of greetings.

"I want both of us to use this language," asserts Ali-ce hoping to softening the tone a little.

"All right for the moment," vents Mim-sey. "But I am your father and I will do as I wish."

"As long as you wish this conversation to continue then it will be in English," says Ali-ce.

"Is this an homage for your Earth friend Abbit?" Mim-sey asks.

"Among other things, yes," says Ali-ce without reducing any of the figurative or literal space between them.

"You are angry," says Mim-sey feigning his interest in a handy plant, "but I do not know why."

"You do know," says Ali-ce without needing her father's body language to confirm the lie.

"And you do too. I guess," says Mim-sey. "I have to agree to the termination of Roy Abbit."

"He was alive when you called me home," says Ali-ce stepping to the edge of the pond and the brink of the real conversation.

"Abbit became incidental after his manuscript was written. He could have lived as he wanted as long as he was not doing anything that we considered dangerous," ventures Mim-sey as he moves to sit on a textured and carved, spessartine, orange, bench that forms a conversation size semicircle.

"But the manuscript is the source of danger," hazards Ali-ce.

"Exactly," states Mim-sey while hoping Ali-ce is smart enough not to sit or stand so close that he has to move again. "So he has to die at some point in his time line before he begins his manuscript."

"Without producing a manuscript there is no need to contain Roy Abbit or his work," speculates Ali-ce as she settles at the opposite end of the crescent shaped bench and facing her father.

"He has to die because of what he has done not what he might do," utters Mim-sey.

"Does he have to die to remove his manuscript from his history?" Ali-ce asks.

"When he dies early enough in his time line his manuscript and notes will, uh, evaporate," broaches Mim-sey buoyed by the opportunity to reveal his cleverly effective solution.

"Evaporate?" Ali-ce asks turning away, a little, to focus on the meaning of the word.

"The way it is explained to me ...," says Mim-sey still self-absorbed and warming in his own cleverness, "The matter and energy that were involved with Mr. Abbit become affected by other things rather than by him."

"A pencil he picks up when he is alive will not be touched by him because he does not live to use it," substantiates Ali-ce while returning her focus to her father and what he has done.

"So that pencil takes a different path in history. It may gather dust in an apartment that goes unrented. It may be found and taken up by someone or drift off to a different history molecule by molecule, atom by atom, beyond any perception," says Mim-sey fascinated by the irreversible simplicity.

"Do his notes become blank pages or disappear?" Ali-ce asks doubtfully.

"Ink and paper, he uses, never meet in history so they degrade fairly quickly," communicates Mim-sey while mentally hugging himself with his thorough originality. "The history of the atoms and molecules do not have the inertia they would have with the benefit of a consistent history."

"They stop going with the stream of history and start going against it," articulates Ali-ce.

"History no longer supports the ink and paper in association," concedes Mim-sey with a little effort at patience for the topic and Ali-ce. "They go their separate ways in a new history."

"What about the people who knew him?" Ali-ce asks.

"The people who knew him while he was alive and up to his death know, or come to believe that he died," vouches Mim-sey.

"What about the people who knew him in college? What if he dies before then?" Ali-ce asks.

"You mean suppose he dies, say, at age ten, what do his college classmates experience?" Mim-sey asks as if he had not considered this particular aspect of the dispersion.

"Yes," asserts Ali-ce while hoping for her father to hesitate.

"Most of them lose track of him. After they notice he is absent from their experience they attribute his absence to faulty memory or strange dreams," says Mim-sey pleased with the opportunity to present another aspect of his thoroughness. "They remember him but they cannot place him in time. Any temporal records will dead-end and rapidly deteriorate While they last, they will show where he was but not where he is."

"The food he did not eat?" Ali-ce asks while giving up on winning this one and standing up.

"It gets eaten by someone else, or not," says Mim-sey without concern and standing as well.

"The apartment he did not rent?" Ali-ce asks belaboring the point while exiting the atrium.

"It gets rented by someone else," deigns Mim-sey dismissively while ushering her from behind.

"The clothes he did not buy?" Ali-ce asks as she crosses his study with finality in her voice.

"They get bought by someone else," asserts Mim-sey with a matching finality while passing his desk.

"And your copy of his manuscript?" Ali-ce asks while focusing on the front door as if to leave.

"It should feel lighter by now. Soon I will not be able read it or turn the pages without them falling apart," maintains Mim-sey with a glance back at his desk.

"Like a really old book?" Ali-ce asks as if trying to comprehend.

"Exactly, would you like to see it?" Mim-sey asks with a new blush of conceit.

"You have the manuscript here?" Ali-ce asks.

"The remoteness of my house serves its safekeeping. Here...? IT IS..." Mim-sey murmurs.

"What is the matter?" Ali-ce asks mustering a perfunctory gesture of concern.

"It is in perfect condition," chokes Mim-sey with surprise.

"Perhaps it has not been long enough," shams Ali-ce twisting the verbal knife.

"But he has been dead more than 50 years," whines Mim-sey completely absorbed by the fear of failure.

"Then perhaps Roy Abbit is not dead enough," says Ali-ce without regard to being heard.

Episode 92

Roy can see it from his attic apartment window. The low mound is grey and probably four or five miles away. People tell him it is Fort Sumpter. The huge white house that raises this window to its view is right at the end of Meeting Street in Charleston. It is next to the battery which is the last piece of land on the peninsula at the confluence of two great rivers.

Living and working in South Carolina is the fallout from a corporate sell off. Living and working south of Broad Street in Charleston is the real surprise to Roy. That they need their grocery store and that a store needs its cashier is the smallest part of the story.

As Florida's population eclipses New York City's, Miami's megalomania eclipses theirs. Homes and apartments are more expensive than comparable ones in New York or California and rents are equal to the mortgages. A cashier's job in Miami no longer pays the rent anywhere there. A cashier's job is the latter day newspaper boy's route. It is a job you get while you are still living at home.

On a more down to Earth level, Roy thinks Ali-ce will like this neighborhood. He wants her to see the big antebellum homes, not to mention the beautiful streets and gardens. His room in the large white house is one of five in the attic. The fifth room has 'plunder' which is what the landlady calls the stored furniture that fills it. It only costs him a week and a half of his month's wages.

This means that Ali-ce will not have to offer him any of her loose diamonds for his needs and he will not have to accept any. It also means that, when Ali-ce visits, she can stay there too, if she wishes. The conversation has not gotten to that point yet.

"What did you find in Rome?" Roy asks holding the big iron gate open for Ali-ce.

"Most of what I found was in Verona," admits Ali-ce.

"So what did you find in Verona?" Roy asks facing the brick behemoth across the street.

"As you might guess, what I did not find is what is most telling," says Ali-ce turning left.

"If you will tell me, I will not ask anymore, "What was, uh, not in Verona?"" Roy asks.

"I could not fine many records that showed you were born in August," affirms Ali-ce.

"But my birthday has been August 13 every year of my life," says Roy looking at Ali-ce.

"Your records that remain intact all show that August 13 birth date," declares Ali-ce.

"So why does my birth certificate say May 15?" Roy asks looking toward the battery.

"You will begin to understand when I remind you that I said 'the records that remain intact'," says Ali-ce looking alternately at the big white house and the iron fence surrounding it.

"I expect some records to be lost or degraded. What is the significance?" Roy asks.

"The records that are lost are lost a lot of ways but the records that are merely degraded are significant," offers Ali-ce turning left with Roy toward East Bay Street and the Cooper River.

"Someone tried to change them?" Roy asks as they walk along the eastern edge of the house.

"Someone tried to change them but not how you think," divulges Ali-ce .

"Do you think your father had someone tamper with the records?" Roy asks.

"No. The records showing an August birth date were aging faster than the same papers they were filed with," imparts Ali-ce at the railing that separates East Bay Street from the river.

"Which means someone substituted a forged document," remarks Roy as they turn South.

"No, it means that some of the papers are on a different time line," protests Ali-ce.

"Even if that were possible, how would anyone know?" Roy asks.

"Normally, things age in predictable ways because of their composition. Things can oxidize or decompose at different rates," says Ali-ce joining Roy with a look over the railing and the river. "But identical things degrade or decompose similarly under the same circumstances."

"So all of the papers in a file change color and fade similarly because the environment for them is the same," judges Roy as they resume walking south with the setting sun on their right.

"Exactly, but the same things on different historical paths do not age similarly even under identical circumstances," contends Ali-ce while looking ahead to the southern limit of the battery.

"How is that possible?" Roy asks still watching the river as they walk.

"Things, or artifacts, on the same time line age along with everything else. They follow the same time line. Heat, cold, oxygen, light, everything interacts in the context of the current circumstances," says Ali-ce while Roy bounces his left hand along the top of the iron railing.

"These things and the interactions change over time," says Roy.

"But things that are on a different time line interact out of context and usually without the benefit of contingent circumstances," claims Ali-ce looking west down a path bisecting the park.

"Then the contingent circumstances must represent some kind of inertia, one that usually slows down degradation," subscribes Roy as he waits to resume the stroll either south or west.

"Things that are on a different time line do not have the benefit of this inertia and so they degrade faster," espouses Ali-ce while continuing south toward Murray Boulevard.

"So how do you know that this is why my 'August' birth records are aging differently?" Roy asks as he now walks on Ali-ce's right for no particular reason except for the change.

"I would only know after more time and more aging. At least for the records generated on the occasion of your August birthday," advances Ali-ce.

"So is that what we are going to do?" Roy asks crowding left to let someone pass.

"That is not what I did," asserts Ali-ce pleased at the excuse for closeness.

"You went back in time to see how the records were aging?" Roy asks looking at Ali-ce.

"I went back in time to see your parents," says Ali-ce as they near the end of East Bay Street.

"You did? How were they? How was I?" Roy asks walking faster without realizing it.

"Your parents were fine but you were in the greatest danger of your life," says Ali-ce.

"How old was I? What happened?" Roy asks looking at the approaching corner.

"You were not born yet," affirms Ali-ce. "I was not able to prevent your abortion."

"But I survived," acknowledges Roy thinking he should be more confused than he is.

"Not without changing your birthday by three months," says Ali-ce absorbed by the facts.

"Was this because of your father?" Roy asks.

"I am pretty sure it was," deposes Ali-ce descending a small set of stairs and into a sense of guilt.

"So I can plan on another attempt at murdering me?" Roy asks failing at the joke.

"I think I can stop even that," vouches Ali-ce recovering her composure and resolve.

"What makes you think so?" Roy asks slowing his pace to match Ali-ce's.

"It was his choice for this attempt and that tells me a lot. It tells me that if there are any more attempts, they will have to be later than this one," predicates Ali-ce as she focuses more on her words than the Battery. "The murder was less about you and more about your manuscript."

"What makes you think so?" Roy asks while cultivating the hope.

"Such an early attempt in your history might have been intended to have the maximum effect on artifacts connected to you," says Ali-ce as they approach a path across the battery.

"By dying fifty years before my manuscript, its degradation is faster than if I died only a few years before my manuscript," guesses Roy hoping he is in step with Ali-ce and her thinking.

"Because the historic path is much farther removed," acknowledges Ali-ce looking north.

"So your father is more interested in removing my manuscript than me," broaches Roy.

"He will kill two birds with one stone no matter how or when he does it," says Ali-ce. "It is just that earlier is better than later."

"The faster my manuscript degrades, the fewer who see it. It keeps my work out of the hands of people who can use it," conveys Roy.

"The people who use it are the threat my father fears," says Ali-ce.

Episode 93

Without any mention of its history, Officer Gim-bal is directed to the same small library where Dr. W-abe saw Dr. Mim-sey's original petition. The indexing activity is done. Now there sits, on the shelves, a bound copy of the petition with its indexes. On one of the tables are also two copies of the warrant and an embossed, personalized, luxuriously bound copy of the Ritual of Investiture for the warrant.

Gim-bal listens as the clerk explains that this ritual is written for this investiture. The Robes memorize the ritual to conduct the investiture and the attendees follow the investiture with their personal copies. Then the clerk tells Officer Gim-bal to compare the two copies of the warrant. On one copy he sees his name and authority to execute the warrant. He sees, clearly written, who is sentenced and the description of the offense. On his copy of the warrant, he sees that this information is redacted.

The clerk indicates that only a redacted copy of the warrant is to be published. The same is true for the published version of the petition and index. It is redacted and the public will not see any information that identifies Earth, Roy Abbit or any living Clidim principal. Gim-bal takes his version of the warrant and follows the clerk to the Robe's Chamber.

Twelve bells sound a twelve note Bodumnloi melody for the Bemyem word *piu* for 'protect, defend, keep.' On the last note the Preceptor's Chorus announces, "*Piu.*"

The twelve Robes stand. Room lights dim by half and the ruby globe in front of 1st-Robe glows red with its own illumination.

"*Are we fully formed?*" 1st-Robe projects.

"*We are,*" reflect all in unison.

"*How are we formed?*" 010nd-Robe projects.

"*We are among the Twelve,*" reflect all in unison.

"*Why are we formed?*" 110rd-Robe projects.

"*For Justice,*" reflect all in unison.

"*Justice for whom?*" 001th-Robe projects.

"*Justice for all,*" reflect all in unison.

Eleven Bells sound the melody for *ta* (get, have, obtain). On the last note the Preceptor's Chorus harmonizes, "*Piu ta.*" The aquamarine globe in front of 010nd-Robe glows light blue.

"*How have we justice for all?*" 101th-Robe projects.

"*By notice and decree for all,*" reflect all in unison.

"*How have we Justice?*" 011th-Robe projects.

"*By purchase and consent,*" reflect all in unison.

Ten Bells sound the melody for *na* (each, every). The Preceptor's Chorus sings, "*Piu ta na.*" 110rd-Robe's sapphire globe glows purple.

"*By whose consent?*" 111th-Robe projects.

"*By consent of all,*" reflect all in unison.

"*Can consent be had for less?*" 0001th-Robe projects.

"*No consent for less,*" reflect all in unison.

Nine Bells sound the melody for *ci* (cast out). On the last note the Preceptor's Chorus sings, "*Piu ta na ci.*" The heliodor globe in front of 001th-Robe glows green-yellow.

"*By what purchase?*" 1001th-Robe projects.

"*By death,*" reflect all in unison.

"*Can justice be had for less?*" 0101th-Robe projects.

"*No Justice for less,*" reflect all in unison.

Eight Bells sound the melody for *im* (according to, by). On the last note the Preceptor's Chorus intones, "*Piu ta na ci. Im.*" The 101th-Robe's diamond globe glows white with its own illumination.

"*So reflect all?*" 1101th-Robe projects.

"*So say all,*" reflect all in unison.

"*Have we the petition?*" 0011th-Robe projects.

"*The petition we have,*" reflect all in unison.

Seven Bells sound the melody for *he* (example, case, instance). On the last note the Preceptor's Chorus sings, "*Piu ta na ci. Im he.*" The orthoclase globe in front of 011th-Robe glows light grey.

"*Names the Principals to the petition?*" 1st-Robe projects.

"*The petition names the Principals,*" reflect all in unison.

"*What Principals are named in the petition?*" 010nd-Robe projects.

"*The Principals are redacted,*" reflect all in unison.

Six Bells sound the melody for *si* (wipe, clean). On the last note the Preceptor's Chorus modulates, "*Piu ta na ci. Im he si.*" The fluorite globe in front of 111th-Robe glows lavender with its own illumination.

"*Names the petition home and hearth of the principals?*" 110rd-Robe projects.

"*The home and hearth of the principals are named in the petition,*" reflect all in unison.

"*What are the home and hearth of the principals?*" 001th-Robe projects.

"*The home and hearth of the principals are redacted,*" reflect all.

Five Bells sound the melody for *vi* (send out, issue). On the last note the Preceptor's Chorus chants, "*Piu ta na ci. Im he si vi.*" The demantoid globe in front of 0001th-Robe glows leaf green.

"*Have we the warrant?*" 101th-Robe projects.

"*The warrant we have,*" reflect all in unison.

"*Names it the heirs to the warrant?*" 011th-Robe projects.

"*The warrant names the heirs,*" reflect all in unison.

Four Bells sound the melody for *ii* (so be it). On the last note the Preceptor's Chorus vocalize, "*Piu ta na ci. Im he si vi. Ii.*" The spessartine globe in front of 1001th-Robe glows orange with its own illumination.

"*What names are heir to the warrant?*" 111th-Robe projects.

"*The names are redacted,*" reflect all in unison.

"*Names the Executor to the warrant?*" 0001th-Robe projects.

"*The warrant names the Executor,*" reflect all in unison.

Three Bells sound the melody for *i* (be advised). On the last note the Preceptor's Chorus trolls, "*Piu ta na ci. Im he si vi. Ii I.*" The iolite globe in front of 0101th-Robe glows violet.

"*What name is the Executor to the warrant?*" 1001th-Robe projects.

"*The name is redacted,*" reflect all in unison.

"*Know we, here formed, all the principals, names, homes and hearths?*" 0101th-Robe projects.

"*Principals, names, homes and hearth are known,*" reflect all.

Two Bells sound the melody for *li* (all rise, stand). The rhodolite globe in front of 1101th-Robe glows pink with its own illumination.

"*Know we, here formed, the names that are heir to the warrant?*" 1101th-Robe projects.

"*The heirs to the warrant are known to those here formed,*" reflect all in unison.

"*Know we, here formed, the Executor to the warrant?*" 0011th-Robe projects.

"*The Executor is known to those here formed,*" reflect all in unison.

"*Knows the Executor the heirs to the warrant?*" 1st-Robe projects.

"*The Executor knows the heirs to the warrant,*" reflect all.

"*Let justice be served,*" 010nd-Robe projects while applying the seal (*gea*) to the warrant.

One Bell sounds for *yi* (done) and signals the exit of the Robes. The topaz globe in front of 0011th-Robe glows golden yellow.

"*Let justice be served,*" reflect all in unison.

The globes provide the sole illumination. The room empties with a swarm of red, light blue and purple; pools of green-yellow, white and violet; shards of pink, light grey and lavender; slashes of leaf green, orange and golden-yellow.

All exit with the melodies of "*Piu ta na ci. Imhe si vi. Ii I.*" coursing through the ears and hearts of all those present.

They issue out with the agitation of 'protect, defend, keep;' commiserate with 'get, have, obtain, each and every;' recommit to 'cast out, according to;' unite with 'by, example, case, instance;' rededicate to 'wipe clean, send out, issue;' cohort with 'so be it, be advised' all cantillating in the minds of the dispersing witnesses.

Everyone is similarly committed to protecting the redacted information as long as it remains redacted. Not just for their lifetimes but even at the expense of their lives as tradition requires.

Officer Gim-bal departs with all the others but with his own resolution:

Enemies of our world *(i)* be advised. We stand ready *(piu)* to protect, to defend and to keep our world uncompromised by you. You and your threat will be *(ci)* cast out. You and your poison will be *(si)*, wiped and cleaned from existence. You and your terror will be *(vi)* sent out and issued from hiding and resources. According to *(im)* this petition and this warrant, in this example *(he)*, in this case and this instance you and your threat, your poison, your terror will be purged. The innocent, each and everyone *(na)* will get *(ta)* and have relief from your assault. We will obtain peace and obtain justice by execution and warrant.

So be it *(ii)*. "*Piu ta na ci. Im he si vi. Ii I.*"

1

1 1

1 2 1

1 3 3 1

1 4 6 4 1

1 5 10 10 5 1

1 6 15 20 15 6 1

Chapter 32

Chapter 000001000....

Chapter F

Chapter $(1 + 6x + 10x^2 + 10x^3 + 5x^4 + x^5)$

Chapter $(a^5b^0 + 6a^4b^1 + 10a^3b^2 + 10a^2b^3 + 5a^1b^4 + a^0b^5)$

Chapter 2^5

94.

 Ali-ce's shuttle to her ship is on the coast near her father's office, so, like it or not, they travel together that far. The Lev-Train is the fastest means there. That is why Ali-ce likes it. It is also the smoothest ride possible. That is why Dr. Mim-sey prefers it.

This Lev-Train is like any significant ship. Tons of sculpted, metal impregnated glass. Its metals are arranged to produce and manage the magnetic fields that levitate the train and propel it at nearly the speed of sound. Only the cost of noise abatement prevents it from going faster. Accommodations are generous and the smoothness of the ride insures that there is no sense of traveling. Lights and bells give notice of acceleration and deceleration to advise the passengers to take their seats, if they wish, but furnishings and architecture do not signal any utility except for comfort. Most of the passengers travel in groups to take advantage of lower costs. Dr. Mim-sey thinks of them as 'those who reduce the cost of his suite'.

He is outfitted for dining, sleeping and lounging. Each option is sufficiently separated by facings and drapes so that each area of his car is virtually a separate room. It compares very well with parts of his home in the building-burbs. He normally has a lev-car to himself when he travels. But this time Ali-ce is with him.

"You can go back and save Roy Abbit," blusters Mim-sey without sincerity. It is not too late."

"I only have to find out when he dies and stop your agents," proposes Ali-ce to tin ears.

"And every time you stop my agents I have an entire timeline to choose from," vaunts Mim-sey while trying to decide if he would be more comfortable in one of the other chairs.

"The advantage will always be yours," protests Ali-ce without conceding anything.

"More than ever. Our permanent establishments have been called back," clarifies Mim-sey. "You can go back to Earth and stay, if you want, but you will be completely without any of our resources."

"A few diamonds go a long way on Earth," disputes Ali-ce as she walks to a chair and settles in.

"You will only make the job difficult. You will never make it impossible," says Mim-sey.

"But I can make it unnecessary," agitates Ali-ce ready to confront her father, finally.

"What makes you think so?" Mim-sey asks wondering if the ottoman should be a little closer to his chair.

"Mr. Abbit becomes irrelevant if his book is published before he dies," illustrates Ali-ce.

"What is the difference?" Mim-sey asks more concerned with arranging his throw for comfort and effect.

"In that case, when he dies, his published books do not evaporate," explains Ali-ce without hurrying to discommode her father's assumptions. Only his notes and his manuscript face that fate."

"The books become the publisher's history," analyzes Mim-sey adjusting his table light for the trip. "Roy Abbit, as a source, can only be explained as a fiction as time goes on."

"And you are not going to collect all the loose published copies," formulates Ali-ce as she loosens another brick still without her father noticing. "This time it is you who will not have our resources on the Earth."

"What makes you think that if his manuscript gets published that very many copies will be sold?" Mim-sey asks barely concerned with what he considers a closed subject.

"Ordinarily, you would be right. Ordinarily, demand drives the market," presses Ali-ce.

"Who is going to want the book?" Mim-sey asks feeling like he is repeating himself.

"If you are thinking of informed readers, you are right. The book is dead before it is published," collogues Ali-ce handing her father his own argument on a platter.

"Thank you. And the publisher destroys the unsold copies without any help from me," says Mim-sey while thinking how this conversation could make this trip a lot longer than necessary.

"But what if someone, perhaps someone like me, finds a publisher and buys his first printing before it is printed," says Ali-ce as she probes her father's comprehension or more precisely his lack of appreciation of what is coming.

"Mr. Abbit makes fifty-five hundred dollars," says Mim-sey without concern for the math.

"And his benefactor has 20,000 copies of a book that no one wants to buy in the first place," acknowledges Ali-ce while peeling the onion all the more relentlessly.

"Are you an informed reader of one?" Mim-sey asks challenging her role as benefactor.

"Not even informed. I still do not know why his manuscript is important," conveys Ali-ce.

"So, you are angry because you cannot save Mr. Abbit or his book," remarks Mim-sey.

"I am angry because it is you who tried to destroy him and his achievement," says Ali-ce.

"Tried? Succeeded. You have just described the very pillars of my success," says Mim-sey feeling grateful for the obvious end of this conversation and the prospect of a different one or even silence.

"And now, even though it breaks my heart, I must knock those pillars down," imparts Ali-ce.

"I do not think you can," says Mim-sey without the least urge to test his hopes for quietude.

"It is already done," says Ali-ce while watching her father. "There are 20,000 libraries with a copy of his book and millions of informed readers with notice that those libraries have that book."

"Those books can be recovered," comments Mim-sey feeling provoked by another empty gesture.

"Not ahead of the demand for more copies and not without the cooperation of those libraries," conveys Ali-ce as she sheds some light on the substance of her father's predicament.

"You have anticipated only part of my plan," divulges Mim-sey unwilling to concede any flaw in his scheme. "I can kill him as many times as I want before he publishes."

"But you cannot kill his informed readers, their journal articles and class lectures," imparts Ali-ce while driving a wedge deeper into the potential failure of her father's best hopes.

"It is not necessary to kill him often," reveals Mim-sey. "I just have to kill him early."

"Do you think I have not anticipated that?" Ali-ce asks as she resists the urge to raise her voice.

"Mr. Abbit was in his fifties when he finished his manuscript. I have fifty years to choose from," discloses Mim-sey while hoping the train is not going as slow as it seems.

"Before you get too comfortable with your options, I have something for you," confronts Ali-ce as she turns to the penultimate task in her plans for her father.

"And it is framed too. What is it?" Mim-sey asks wondering what it is supposed to mean.

"It is a botanical souvenir," says Ali-ce while thinking it will be the first and only thing he ever throws away in his lifetime. "Earthlians have the custom of drying and framing flowers and leaves."

"And these leaves?" Mim-sey asks without a clue.

"From a forest in Scotland," continues Ali-ce as she looks for a sign of recognition.

"With all you have to do, you troubled yourself ...?" Mim-sey asks as he stalls for the point.

"They are from Birnam Wood. Do you recognize it?" Ali-ce asks deliberately avoiding a sarcastic tone.

"Should I?" Mim-sey asks hoping for a quick explanation and an end of this conversation.

"I sent you a copy of William Shakespeare's *Macbeth*," contends Ali-ce while feeling very much like her father's daughter.

"Oh, I vaguely remember," deflects Mim-sey taking the bait.

"Do you?" Ali-ce asks thanking the witches.

"In this play, Malcolm suggested that the troops carry branches from Birnam Wood to disguise their number. But what does the siege of Macbeth's castle have to do with me?" Mim-sey asks while wondering if Ali-ce has plans for a siege of the Robe's Chamber and trying not to smile.

"Macbeth felt safe because those troops were born of woman and so could not touch him," responds Ali-ce as she studies her father's face for the light of cognition.

"But his confidence was misplaced because Macduff was among those troops and could claim the exception because he was born by 'cesarian section'," warrants Mim-sey recalling that turn of the story.

"Mr. Abbit was born by cesarian section," resumes Ali-ce as she connects the last of the dots.

"A coincidence," says Mim-sey trying to escape his daughter's trap the moment he sees it.

"But Roy Abbit was not being born, he was being aborted," says Ali-ce as she presses the point.

"Yet another coincidence," ripostes Mim-sey while mentally fleeing from the conclusion.

"Except, I have found his birth records. He was born the first time, full term in August," conveys Ali-ce outlining the putative history.

"When was the abortion?" Mim-sey asks as he blinks no longer able to escape the facts.

"Three months earlier in May," imparts Ali-ce as her father swallows before dropping his jaw without parting his lips.

"The abortion had something to do with me?" Mim-sey asks by way of denial.

"The abortion had everything to do with you," asserts Ali-ce by way of accusation.

"And that is how we lost Mr. Abbit?" Mim-sey asks as he settles for winning ugly.

"I am here to tell you that we have not lost Mr. Abbit," says Ali-ce without blinking.

"How do you know?" Mim-sey asks openly fearing the answer.

"I was there with the priest who was there to accept the body of the baby for burial," prevails Ali-ce. "Contrary to expectation he was born alive. I was there and he was baptized."

"So is this souvenir a token of my defeat?" Mim-sey asks as he lets the artifact fall in his lap.

"Keep it as a reminder of what any similar effort faces," says Ali-ce blinking finally.

Episode 95

Battery Park is close-by but Roy spends only a little time there. The park is ringed with a two lane street that separates it from an elevated walk on one side. It could have been an enviable park if cars were never invented.

If you look over the side, you do not see a beach. So you do not think you are looking at the Atlantic Ocean. This is where the Ashley and Cooper Rivers meet the Atlantic. Where is the Charles River? Should this be Coopertown or Ashleytown?

The oak trees and the linear oyster shell paths converging on a capstan from the U.S.S. Maine are ordinary. The smallness, the featurelessness, the breeze, the shade and the unlikely remoteness make it a pleasant and comfortable place night and day. With the right fence and a Wright house the park would convert to a nice residence.

But the idea is only amusing as a burlesque of money and power. For both Roy and Ali-ce, who enjoy anonymity more than money, even this burlesque is out of the question. Ali-ce, who chooses among the worlds with privacy and purpose, would feel like a museum bug with an all too large pin and an all too large label with an all too large word summing her up for an all too large, gawking and awe seeking crowd that is herded by in amazing numbers.

For Roy it would be an emptiness that exceeds the scale of Ali-ce's several worlds. Millions of light years might as well replace the fence. More rooms than you can use might as well be more galaxies than you can use. All the people you do not know on the other side of the fence are no different from all the people you do not know across the galaxy.

Where is the comfort or bliss in that? Comfort and bliss for Roy and Ali-ce is in the ease of their mutual company and the ease of conversation that makes even pivotal topics approachable.

"So my birthday is the Ides of May not the Ides of August?" Asks Roy pulling at the gate.

"May 15 or August 13, we could celebrate your birthday twice a year," parleys Ali-ce.

"If I were forty years younger," says Roy closing the iron gate, "I would like that idea."

"How about celebrating your birthday for the ninety-three days between the birthdays?" Ali-ce asks.

"Maybe," says Roy turning to the clapboard house, "if it did not remind me of my murder, my first murder."

"I do not think you will have to worry about that anymore," imparts Ali-ce.

"How is that?" Roy asks while he pulls open one of the double front doors.

"My father is not going to kill you for sport. Without a reason to do it, he will not," advises Ali-ce entering with Roy into the dark oak clad vestibule dimly lit by a small yellowing chandelier.

"I do not like the option of going back a few years and not writing my manuscript. It is all I have," declares Roy waving a friendly hello to the landlady sitting, as always, in the sitting room.

"If you did, your manuscript and notes would begin a new history - one that would resemble a history after your death," apprises Ali-ce waiving a friendly acknowledgment too.

"Even if I am not dead?" Roy asks as they ascend the wide staircase to the second floor.

"If you go back and start a new history, one that does not include writing the manuscript, the history that preserves and promotes your manuscript and notes will be lost," advises Ali-ce marveling at a stairwell that is as wide as a lev-train cabin.

"Without the history that sustains my manuscript and notes, what do they do? Evaporate?" Roy asks marveling that the stair does not squeak and groan after so many decades.

"It is more like aging," delineates Ali-ce. "But your notes and manuscript age faster."

"The effect is the same whether I were murdered or even died of old age," utters Roy.

"The difference is the longer you live to preserve and promote your manuscript and notes the longer it takes them to age after you are gone," says Ali-ce nearing the second floor landing.

"And the longer your father stares danger in the face," says Roy taking Ali-ce by the hand.

"As long as preserving and promoting your ideas is limited to you, you are the solution to the problem," says Ali-ce letting Roy steer her to the left down a wide hallway with many doors.

"Publishing a book is no solution. If it does not sell, we are back in the here and now," says Roy oblivious of the impressively wide hallway that serves the second floor guest rooms.

"There is another option and I have taken it," declares Ali-ce still hand in hand with Roy.

"Can I start looking forward to my next birthday?" Roy asks with a smile.

"And many more," delivers Ali-ce hoping for Roy's approval and the sense bliss that comes with being published. "I have put copies of your book in 20,000 libraries worldwide."

"That just makes me, my death more important" says Roy before he can think any further.

"Actually, it does just the opposite," states Ali-ce steering Roy more to the point as he steers her to yet another stairway. "Now there are 20,000 historic paths for your book."

It is a lot more work to murder 20,001 people," says Roy reviving his sense of humor.

"And no advantage to murdering you," vents Ali-ce with relief.

"Who knows what will become of the information in my book. People will read and use the information," chimes in Roy enjoying the closeness that this narrower stairway imposes.

"Other books will be written and published. Technologies and theories will emerge," broaches Ali-ce while recalling that his third floor room is as close and cozy as its stairway.

"But this will take a long time," says Roy a little disappointed, "perhaps several centuries."

"Is this bad?" Ali-ce asks stepping into the common room serving the floor and wondering at Roy's disappointment.

"I do not have that many years left," says Roy closing the door to the common bathroom.

"We could take a hop into the future," renders Ali-ce taking Roy's hand again.

"Who is going to cash the checks?" Roy asks crossing the common room to his door.

"The royalties can go to a corporation. They can pay you to write the next book whatever your name is," pictures Ali-ce. "They do not have to know you are the same person."

"They will not even consider the possibility that I am the same person," interprets Roy.

"There will be all kinds of ways for you to collect the money," relates Ali-ce.

"Will I live longer?" Roy asks leading Ali-ce into the cozy single room he calls his.

"I do not think so," admits Ali-ce looking around at the garret beams and the garish fabrics that serve every purpose but harmony. "You, uh, we will just live in different centuries."

"I will be that much harder to for your father to find," agitates Roy.

"I have already told him he does not need to bother," says Ali-ce with approval and ease while closing the door to Roy's room and the light of the common room.

"Well, there is nothing left to do," remarks Roy sitting on the bed and taking off his shoes.

"We do have a birthday to celebrate," enthuses Ali-ce hopping into the center of the bed.

"How long should we celebrate?" Roy asks with genuine interest in the question and undressing as if the smallness of the room presupposes it.

"How about 59 days?" Ali-ce asks as she luxuriates in a bed that fills half the room almost naturally.

"I like that idea - one day for each of my years. What will the landlady think?" Asks Roy without supposing that she had thought things out long ago.

"I am already looking forward to next year," delights Ali-ce while stretching for the light with the freedom of familiarity and the freedom from clothes.

"I am already looking forward to now," says Roy as he squirms into Ali-ce's embrace with a freedom that matches hers.

"Um," strums Ali-ce in the accustomed comfort and warmth of bedstead, arms and legs.

"Um," hums Roy with an urge to mutuality that completes his will to surrender.

Episode 96

Dr. Mim-sey has discovered something new about Dr. W-abe. He can get intoxicated enough to become vulnerable. Maybe that is how he got married. This is an unseen side of Dr. W-abe. Perhaps that is why no one ever sees him drink this much. It takes years off his life. He is like a little boy.

Dr. Mim-sey wonders if Dr. W-abe is beginning to recognize the kind of power the 'post petition' Dr. Mim-sey will have. Could Dr. W-abe be working on a place close to the new top? Dr. Mim-sey considers the possibility that Dr. W-abe wants to be there when Mim-sey gets there.

Maybe Dr. W-abe wants to quit being an engineer and become Dr. Mim-sey's right-hand man. He will not be free to drink so much. What will the Mim-sey public think?

Dr. W-abe, for his part is not planning on ever being sober again. The helplessness of intoxication could not disguise or compare to the powerlessness he feels at the moment. His hope of influencing Dr. Mim-sey one way or another has gone from small to nonexistent. Since he is only able to watch, the less he sees the better.

So he watches Dr. Mim-sey's self-absorbed, self-indulgence in food and drinks. He sees himself indulging Dr. Mim-sey and waiting the little bit of time for Dr. Mim-sey to aspire for even more. More, at the moment, means moving the party across the building to Dr. W-abe's home.

Dr. Mim-sey as a gesture of accommodation that is just short of condescension goes along with Dr. W-abe's suggestion to see the batball field on the way. The matrix elevator makes it easy. The prospect of Dr. W-abe's whim evaporating, almost immediately, makes the detour an insignificantly small price to pay. Dr. Mim-sey can afford to be that generous as long as he is not being taken for granted.

The matrix elevator takes less than two minutes to get to the river. To Dr. Mim-sey's private appreciation Dr. W-abe is already agitated. He can count on spending less time at the battery than it took to get there. But against Dr. Mim-sey's self-serving hopes, the battery is not empty.

They see Officer Gim-bal and Gim-bal sees them. With predictable unselfishness and courtesy, Dr. W-abe acknowledges Gim-bal by approaching him. Once Officer Gim-bal takes up the conversation, Dr. W-abe, true to form, drifts off telling Dr. Mim-sey to meet him upstairs.

"I am going to have to do something about this lighting," says Mim-sey thumping a light pole.

"It is just old," apprises Gim-bal while he deliberately opposes Mim-sey's opinion. "All those years ago, when they were put in, this was a lot of light and very expensive."

"Real estate is up more than 800% in the last twenty-five years. Street lights are more than twice as bright these days," says Mim-sey enlarging on his superficial command of the issue.

"We are probably the only two individuals who have been in this park since its rededication," interposes Gim-bal with the intention of changing the subject.

"Things are going to change around here," says Mim-sey casting about for a new topic.

"My assistants, in Lynchburg, tell me that you want to see me, personally," says Gim-bal.

"I just left my daughter at her shuttle down the coast," says Mim-sey nonresponsively.

"I plan on leaving after this," says Gim-bal. "Maybe I can see her on Earth."

"What is in the bucket?" Mim-sey asks while focusing on the only new object in view.

"It is beef liver and some beer," entices Gim-bal as he deliberately omits specifics.

"What is a beef?" Mim-sey asks while compulsively examining this new idea.

"A beef has four legs and weighs more than ten Earthlians," clues Gim-bal with few details.

"What do beefs do?" Asks Mim-sey as he stares into the blackness of the bucket.

"They are raised for food. One beef, along with bread and vegetables, will make six thousand meals," says Gim-bal.

"How do you eat those livers and beer?" Asks Mim-sey as he wobbles over the bucket.

"That I do not know. I am chumming with these," says Gim-bal.

"I know what that is. But here?" Mim-sey asks as he approaches the antique railing.

"Are there fish in this river?"

"It seems so but they are hard to see in this light," says Gim-bal without prospect in his voice.

"Can I chum some?" Mim-sey asks while looking for an opportunity to beat Gim-bal at his own game.

"Okay, an Earthlean sailor in New Orleans taught me what to look for," says Gim-bal as he deliberately keeps the details brief. "Use this fork and throw a liver in where the light is good."

"How will they know to jump for the liver?" Mim-sey asks still missing the point.

"They will not jump. You have to drop it in the water, about six feet out," says Gim-bal.

"How will I see anything? The water is so muddy," says Mim-sey to the almost featureless water.

"You can tell when they are swimming just below the surface. They disturb the surface of the water," prompts Gim-bal.

"I do not see anything," squawks Mim-sey while casting bait impatiently and about to give up on the fish.

"Throw another liver in. Just do not do it too soon. We do not want to run out of liver," gushes Gim-bal as he cultivates Mim-sey's interest.

"While we are at it," says Mim-sey as he finally recalls through his intoxicated haze why he wanted to see Officer Gim-bal. "I want to tell you my petition in the Robe's Chamber is coming to a vote very soon."

"It must have taken years to get there," says Gim-bal

"Not that long," says Mim-sey with some glee, "It is a Summary petition. And now it is coming to a vote."

"You need me to submit new testimony?" Gim-bal asks.

"That is not why I have you here. After the vote I will not have any time for our project," proclaims Mim-sey with a terseness coming from the decision having been made a long time ago.

"Who is going to replace you?" Gim-bal asks as if to miss the point.

"Can I throw another liver in?" Mim-sey asks while occupied with his own agenda.

"Go ahead. Watch the spot where you throw it in. The fish has dark smooth skin, almost black, so it will be darker than the water," says Gim-bal. "If it breaks the surface, you will see it."

"You remember that the petition identifies Roy Abbit and Earth as a threat to our world," prompts Mim-sey elated by his megalomania as it resonates with his monomania.

"How many of the Robes are going to believe you?" Gim-bal asks.

"I do not know but even if they refuse to consider it, it is a million pages of official information published by the Robe's Chamber," gloats Mim-sey. "Our world will explode."

"The legends of Earth will become the legions of Earth overnight," says Gim-bal.

"Exactly," proclaims Mim-sey feeling pleased that Gim-bal is finally beginning to appreciate his prescience. "Just imagine the terror from the top to the bottom of our civilization."

"Do you think that is a dangerous thing to provoke?" Gim-bal asks.

"Let me throw in another liver," enthuses Mim-sey while expecting to provoke a dangerous fish as well. "This mutual myth could go on forever without any more help from me."

"What about the panic?" Gim-bal asks without specifying the panic he has in mind.

"I have that figured out. There will be a lull between the publication of the petition and the riots," schools Mim-sey. "It is during that interval that I will promise protection and safety for Clidim."

"What makes you think the public will believe you?" Gim-bal asks cultivating Mim-sey's ego.

"I have two, actually three, arguments for that. One, the Earth has not attacked yet. I will say that it is mostly out of their fear. Everyone on Clidim will understand that word. Two, I will point to my followers as a sizeable segment to draw our defenses from," states Mim-sey directly to the muddy water.

"More liver?" Gim-bal asks prompting Mim-sey. "What is the third argument?"

"I, as the most knowledgeable, am the one to lead this world not only to safety but also to preemptive victory," grits Mim-sey while tensing with his own self-esteem.

"I see a fish. Do you see how big that one is?" Gim-bal asks with a hand on Mim-sey's shoulder.

"Where? Let me throw in more liver," says Mim-sey without taking his eyes off the water.

"There," springs Gim-bal without taking his hand off Mim-sey's shoulder. "Right where you threw it."

"That is a fish?" Mim-sey challenges feeling awed and frightened at the same time. "It looks like a tree."

"It is bigger than I expected," grunts Gim-bal while grabbing Mim-sey's clothes behind his knees and pushing on his shoulder. "Here is another piece of liver," Gim-bal heaves. Heaves in every sense of the word.

Mim-sey howls with the diamond hard glass railing digging into his waist, "Quit pushing."

"I must push," hectors Gim-bal to the bottom of Mim-sey's jerking feet.

"I cannot hold on. Help me." screams Mim-sey as he hits the water.

"More Liver?" Gim-bal asks into the turbulence of fish and flailing limbs.

"*Logeo.*" screams Mim-sey risking his breath futilely.

"I think I see more than a few of those big fishes," hammers Gim-bal. As if Mim-sey would like to know. As if he is listening. As if he can still hear.

"*Juozam. Logeo,*" screams Mim-sey with his bloody face barely breaking the surface of the thrashing water and scarcely interrupting the silence of the empty riverfront.

"There is no liver left, except yours," harasses Gim-bal galvanizing his commitment to his Robes warrant.

"*Juozam....,*" screams Mim-sey now the designated threat to the safety of Clidim's civilization and the heir to the Robes' Summary Warrant.

"I cannot hear you Dr. Mim-sey. Maybe you will have something to say when you surface. I am waiting but I cannot wait forever. *Piu ta na ci. Im he si vi. Ii I,*" chides Gim-bal to a storm of mahogany stained water and heaving fish.

1

1 1

1 2 1

1 3 3 1

1 4 6 4 1

1 5 10 10 5 1

1 6 15 20 15 6 1

Chapter 33

Chapter 100001000....

Chapter AF

Chapter $(1 + 6x + 11x^2 + 10x^3 + 5x^4 + x^5)$

Chapter $(a^5b^0 + 6a^4b^1 + 11a^3b^2 + 10a^2b^3 + 5a^1b^4 + a^0b^5)$

Chapter $2^0 + 2^5$

Episode 97

Junior Commander Poon is as bored anyone would expect a traffic cop to be. But he is nowhere nearly as lonely as anyone would expect. He, here, for the only time in his life does not consider himself lonely. This is Clidim's busiest intersection, whether coming or going, and everyone is a stranger with other concerns greater than any for himself, Jr. Com. Poon.

Most of his subordinates serve with him longer than anyone he has known as co-workers or neighbors on Clidim. Most of them become friendly and many become friends. His friends return to Clidim at the first opportunity but his long distance friendships take years to fade. New friendships come to him across the horizon of time and circumstances.

That is what keeps him on the job. He will not do anything to risk his position not even a promotion. So he pays close attention to every ship that leaves or enters the 244,000 mile diameter bubble that divides Clidim's inner space form the rest of the universe.

The demotion that comes with failing to support the protocols of crossing the threshold of Clidim's inner space does not just put his position at risk, it threatens his personal universe. Not until now did he suspect that the heroic defense of protocol could risk his happiness as much as incompetence.

If his abilities became known he could become promoted back to Clidim. The planet of glitter and superlatives may not promise more boredom than this outpost provides. But the utter and permanent emptiness of the billions of strangers that populate Clidim is waiting for him on the next rung of the ladder.

With a firm grip on his affection for the status quo, Jr. Com. Poon is anticipating the arrival of Officer Gim-bal. Officer Gim-bal, for his part, is not concerned with 'Jerkin' Poon's motives. He only needs to be sure that his plans will go his way. This is intended to be his first and last meeting with Jr. Com. Poon. Otherwise the next visit will be with a warrant.

"This is not a hotel," declares Ensign Gen-Sjen without looking at his commander. "It is an insult. Why way out here? How long will we have to put up with this? Who would pay for such a thing? Why? This is a troop outpost. There are no luxuries. It is strictly self-service and sparse. No hotel staff. We have rations not cuisine."

"'Hotel Services' is just a technical term. It applies to our accommodations for civilians who are not under arrest. This time, they are in transit from who knows where to Clidim," advises Jr. Commander Poon looking steadily at his junior officer. "They are going home. Just like we would like to. So we will treat them like we would like to be treated."

"Not like troops?" Ensign Gen-Sjen asks as he looks at his captain.

"Not any more than we have to. Make requests instead of orders. Make their accommodations as contiguous and self-sufficient as possible so that they do not have to be told that anything is 'off limits'. Schedule activities and events rather than place limits on access to food, water and toilets. Create a 'night' that promotes sleep and inactivity," articulates Jr. Com. Poon without looking away. "We can use the 'night time' to restore housekeeping and supplies that are used during the day."

"So we work around the clock as usual?" Ensign Gen-Sjen asks.

"Almost. Take two-thirds of our manpower and divide it by 125. That will make nine troops to serve the needs of each 'guest'. Three troops for each shift of the clock for each guest. They will eat and sleep during the guests' night.

Our guests will understand that and hopefully not impose on them too much. The guest troops that are not directly assisting guests will restore the guests' spaces that are vacated in the cycle of activities, events and sleeping.

The remaining third of our manpower will assume all the duties of station operation - two shifts on, one shift off. Meals on the double shifts for operations so they can get as much sleep as possible," instructs Jr. Commander Poon while leaning back in his chair.

"How long will we have to do this?" Ensign Gen-Sjen asks.

"I do not know, yet," responds Jr. Commander Poon blinking slowly. "I will ask Gim-bal. But the better we are at keeping the guests from getting restless the longer we can last and the easier it will be."

"I will get the ball rolling, now, before our guests arrive. We can use the rehearsal to work the bugs out of the routine How long until the first group arrives?" Ensign Gen-Sjen asks as he moves to the door.

"Only two and a half days but they will not all arrive at once. Keep that under your hat," prescribes Jr. Commander Poon as he gets up to leave too. "You need to operate as if they were all here. The problems will show up sooner and get fixed sooner as well. Get going. I am going to see Officer Gim-bal now. Initiate the changes by this time tomorrow."

Episode 98

It has been almost five hours since Roy punched out from an uneventful day at work and the sun is setting on the postcard setting of Charleston and the Battery. Roy and Ali-ce are sitting on the east side of the bandstand enjoying the advent of cooler temperatures and the prospect of going to their attic room that is cooling with the eastern sea breeze. They are not alone and do not expect to be. The Battery is full of people with the same motives, if not circumstances. The difference for Roy and Ali-ce is, as always, in the choices they have made. The bandstand, even though it is at the very center of the Battery, is the least popular spot in town at the moment. The tourists want to get as close to the shore as they can. The bandstand is as far from that as space and geometry permit. The perimeter streets harbor cars and their drivers as far from the bandstand as parking and cruising permit. Tailgating, car stereos and their air-conditioning complement the attractions of the perimeter. Except for an occasional pedestrian taking a short cut to or from the perimeter, Roy and Ali-ce can count on and enjoy a comfortable isolation with a certain certainty.

"Ali-ce, I am afraid," relates Roy with his gaze fixed on Ali-ce while she takes advantage of the elevation to study the horizon ahead, "that our bandstand mountain top is losing its charm."

"I do not detect a sense of fear and cannot guess what it has to do with charm even familiar charm," quibbles Ali-ce with a smile and taking a comfortable focus on Roy's face.

"I suppose you are smiling at the joke but 'afraid' means 'I think' along with the idea that the thought is not a welcomed one," clarifies Roy in the glow of Ali-ce's admiration.

"I am smiling because when my thoughts turn to you, I always smile. Did I make a joke? Was it a good one?" Ali-ce asks hoping for a generous compliment.

"Just between us it was," underscores Roy undeterred by Ali-ce's maneuvers around his point. "What I am trying to say is that you seem preoccupied, maybe even worried."

"Oh. My distraction is growing lately. It has been two months since I have heard from the company," illuminates Ali-ce no longer able to hide her inner concerns. "That means no news from my family and no connection to the company in the usual sense of purposeful work."

"You are not afraid you are being fired. Are You?" Roy asks.

"It is a little early to start thinking about that. But the shut down of our role, here on Earth, that my father promised is well underway. Almost seventy of the lower echelon workers have left the company under a variety of pretexts," discloses Ali-ce with the vain hope that her fears would sound unreasonable. "Some have found alternate work back in Lynchburg. Others have been fired. Some have quit or gone on vacation without returning to work."

"You have already said that you can stay no matter what. So what is the concern?" Roy asks with an effort to minimize the potential in her worries.

"I am concerned that things are going a little too fast," admits Ali-ce restating her doubts.

"What were you expecting?" Roy asks with the hope of finding a flaw somewhere else.

"I imagined that my father would show up at the house in Lynchburg and shuttle the one hundred or so Clidimites home through Lynchburg at about three or four each month during the summer months and six to seven a month during the winter months with the whole operation taking two years," illustrates Ali-ce while resting her gaze on the horizon again.

"I suppose the lower the number of launches over two years would serve the need to be inconspicuous. But why do, or did you expect smaller numbers to leave in the summer? It seems like more people could be cut in the summer," explains Roy hoping he had found the fatal flaw.

"Launches in Summer are a lot more dangerous. Earth is closer to the sun," specifies Ali-ce almost as if she was reading from a flight manual. "That, along with the large moon of Earth, makes our kind of launch calculations more difficult and optimum opportunities are fewer."

"Is that what worries you?" Roy asks without doubting her pilot skills for such a hazards.

"I am 'afraid' as you say that my worries are larger than that. Gim-bal will have all 125 Clidimites out of here in eight months. Two months are already gone. My father is not only not here, no one has seen or heard from him for three months," renews Ali-ce with her old fears.

"It could be a little too soon to worry about your father. He must have a lot to do," defends Roy hoping to be done with the subject for a while, maybe longer.

Ali-ce smiles again without letting the thought escape her lips that it might be a little too late for her father and maybe for Roy too.

Episode 99

Landing in Lynchburg three hours later than he is used to and six hours later than he wants to means its summer for the '-burg. The dogs are up later. TV's are on later and porch lights are off later. But little by little night conveniently envelopes Gates Street. With the fabric of a moonless darkness and a passing train, J. Rus-ty Gim-bal sets down among the trees between his house and the train tracks.

Officer Gim-bal is left to his disembarking routine without distraction while his assistants guarantee his uninterrupted focus. J-R One is waiting outside his boss' ship and J-R Two is sitting in the glow of a TV watching the street through his livingroom window without attracting attention. Gim-bal is not surprised or disappointed when he heaves a large dark case through the hatch and feels it taken from his hands even before he crosses the threshold of his ship.

He knows in the short trek through the trees that J-R Two will close the livingroom curtains and take his position at the back door of the house. There, he will wait for Gim-bal and J-R One to cross the threshold of the house before turning off the back porch light.

"This is different. You have never come back with anything you could not leave on the ship without telling us," declares J-R One while ambling along the carpet of pine needles that comprise the directionless path. "I thought we were best served by complete mutual confidence."

"This grave object is the warrant," murmers Officer Gim-bal while focusing on the figurative and literal darkness ahead. "I have to keep it until all the names in it have been served."

"The requisite description of the offence and just a few names cannot require such a heavy volume," rallies J-R One hopefully as light blue lights blink on at their feet.

"You anticipate the solution to your own conundrum," says Gim-bal while ignoring the puddles of light and his assistant's vain hopes. "There are more than a few names in this instrument."

"The news is that the Mim-sey petition and the warrant are redacted. So I do not have to open this tome to know I will not learn anything by my effort. Why carry it here and I suppose everywhere you go?" J-R One asks as they step into the glow of the back porch light.

"You are getting distracted by the details," says Gim-bal. "For several thousand years, the law for executing warrants has involved possession of the warrant. This is the first warrant that has been so heavily redacted. This just seems to makes the unwieldily warrant superfluous."

"One concern remains. With a large number of redacted names, Is the benefit of the warrant, as a reference, compromised?" J-R One asks as he reaches the porch. "You must need to consult it in order to execute the warrant. Specifically, if you must execute a huge number of individuals, how are you going to be sure you have served everyone named in an edited warrant?"

"My answer may not seem to be one, at first, but before I am through, you will see that what I am saying is true," says Gim-bal. "When I picked up my copy of the warrant, I was shown an unredacted copy. I looked at the list of names and knew by the time I finished that I knew all the names and locations of all the individuals subject to the warrant."

"It follows then that while I feel like I cannot guess who the subjects of the warrant are, I must already know who they are," adduces J-R One as he pass into the house behind Gim-bal.

"You, I and J-R Two know who they are. Some you will recognize right away. Some you will resist acknowledging until the end. One or perhaps two of whom you will refuse to believe ever," persists Gim-bal while settling into his usual chair in his only house on this planet.

"Should it happen, in our lifetimes, that the unredacted warrant is published, as the law requires, you will see that your disbelief was wrong and the warrant has been served as it is written."

"So the only thing that keeps us from being identically informed, even now, is our individual inability overcome our individual fears and suspicions," J-R One says almost to himself.

"Yes," pledges Gim-bal as comfortable with his opinion as he is with his house and chair.

"Am I to presume that this is the end of the conversation and that we can take up another topic?" J-R Two asks while easing forward in his chair in anticipation of the next issue.

"I would be surprised if you took any longer than you did to address our concerns with the repatriation of our Clidimites to our planet," posits J-R Gim-bal with a glance at J-R One.

"We have a lot to digest and I do not see how more discussion will serve it," presses J-R Two while settling back into his chair in a pretense of unhurried deliberation.

"Neither do I. Have there been any surprises?" Gim-bal asks as a signal to continue.

"In the three months since Mim-sey initiated the exodus and only two months since he disappeared, we have moved forty-seven of our astronauts off the planet. The only surprises are which individuals do not want to go," evinces J-R Two hoping his contribution was significant.

"I will let you tell me the details in the order you choose," offers Gim-bal by way of encouragement. "I presume that will serve efficiency and an economy of information."

"The ones that are co-operative and willing have been encouraged to quit their putative jobs. Some of those are employed in Lynchburg to wait their turn to leave. This makes their exit less conspicuous, if not suspicious. The ones that do not want to go have been fired and transported first. The lack of the usual trail of employment or unemployment, for that matter, is consistent with individual efforts to shed the problems of being fired," says J-R Two.

"Anyone who cares about the matter will expect them to resurface once their employment trail resumes," prompts Gim-bal in order to endorse J-R Two's logic and move things along.

"When that doesn't happen, all of us will be gone," urges J-R One.

"Everyone cannot be seen to gather at or associate themselves with Lynchburg. That will connect them with this house since I am ultimately their CEO," proposes Gim-bal.

"The attrition of the company ranks will be spread out along an eight month arc and a myriad of circumstances," tenders J-R Two. "They only need access to Lynchburg. They can go around the world, if they have to disappear. Then they can leave through this house."

"There is no particular pattern to the individuals and their corporate responsibilities to suggest anything suspicious?" Gim-bal quizzes.

"Where there is no cohort there is no conspiracy," says J-R One in acknowledgment.

"I have never been troubled by the number or frequency of launches from those trees out there. Am I proven right so far?" Gim-bal asks not yet ready to let go of his concerns.

"As long as we can choose when we launch or land that will never be a problem. As long as there are no changes in routines that cloak us we will not attract attention. If there are no neighborhood concerns about how many individuals come and go from the house we will not get attention. As long as our coming and going does not attract attention because of manifestations on radar, radio and television electronics, no one will look for us or at us," says J-R Two.

"So there have been no significant changes in any of the arenas that overlaps us?" Gim-bal asks almost completely convinced that his assistants have been as thorough as he had hoped.

"No changes in neighborhood activities including the train schedules and cargo. No changes in airport routines including equipment, activity and employees. No changes on a pretext of weather, economics or seasonal events," rallies J-R One hoping this finish the topic.

"That sounds like more than two J-R's can guarantee," challenges Gim-bal.

"At this point, we have the assistance of those who are anticipating repatriation," says J-R One while he reaches for an apple to bait his dinner. "We are down to seventy-eight now. As their number decreases, the need for their services decreases too,"

"We are in effect closing the door behind ourselves," says Gim-bal as he gets up and goes to the kitchen.

1

1 1

1 2 1

1 3 3 1

1 4 6 4 1

1 5 10 10 5 1

1 6 15 20 15 6 1

Chapter 34

Chapter 010001000....

Chapter BF

Chapter $(1 + 6x + 12x^2 + 10x^3 + 5x^4 + x^5)$

Chapter $(a^5b^0 + 6a^4b^1 + 12a^3b^2 + 10a^2b^3 + 5a^1b^4 + a^0b^5)$

Chapter $2^1 + 2^5$

Episode 100

It is the most pristine space on the station but Officer Gim-bal find the ruse creepy even though it is intended as a courtesy. It is Jr. Com. Poon's stateroom. No clutter, no dust, no mementoes, but only superficially, pristine. All evidence of anyone's use of the room is gone.

Officer Gim-bal is supposed to think this room is exclusively his for the few minutes he is here. But he is sure that if he opens one drawer, one door, one book he will find evidence that this is Poon's stateroom. Vainly, he waits as uncomfortable sitting as he is standing. He alternates between sitting and standing hoping that one will be better than the other.

But he knows better. Even on the largest station, the space is limited. There are always more people, things and things to do than space provides. Double, triple duty for everything and everybody. Three people to a bed. One for each shift. The numbers go up for tables, seats, toilets and showers. The exception, of course is the Commander's Stateroom. But not a complete exception. Gim-bal is only the other soul to share Jr. Com. Poon's Stateroom, lately. So he waits, alternating between impatience and the consolation that this will all be over sooner or eventually.

"I am sorry for the delay," announces Ensign Gen-Sjen as he knocks on the doorjamb. "The Commander will see you now."

"I hope I have not kept you waiting too long Officer Gim-bal," adds Jr. Com. Poon as he enters his room.

"Not at all," offers Gim-bal. "Clidim should be as efficient."

"I am not sure how to take that. But time is not to be spared up here. So can we get down to business?" Jr. Com. Poon.

"I am always glad to suspend protocol," advises Gim-bal as he pushes the schedule across the table toward Jr. Com. Poon. "Here is the book. It describes the schedules and data you will need for three consecutive visits. A little more than forty astronauts with each visit...,"

"That is not consistent with the 125 for each visit that I have prepared for. Why should I set aside resources for that many, if the real number for each visit is 66% less?" Jr. Com. Poon asks.

"Slow down just a little. You do not have to prove anything to me. I appreciate your pace, if not haste," discloses Gim-bal without looking at the book in Jr. Com. Poon's hands. "Each of the forty or so will be met with about 80 family members arriving from Clidim, hence the 125."

"We are not going to be able to quarantine 375 visitors on this schedule," quibbles Jr. Com. Poon as he puts the schedule down.

The relatives will be quarantined on the planet," states Gim-bal as he turns a little toward the door and away from Jr. Com. Poon. "The astronauts will be quarantined at their destination."

"You distinguish the astronauts' destination as someplace other than Clidim," carps Jr. Com. Poon as he moves toward the door.

"Nothing in this schedule says anything about quarantine locations since that does not involve your station. Everyone, including you, assumed it would be here or Clidim," corrects Gim-bal as he moves still closer to the door. "Make sure everyone keeps assuming that very thing."

"Obviously, we are our own quarantine station," concludes Jr. Com. Poon as a note to himself. "So our personnel who are planning to leave about the time of the visits will not be able to leave as planned."

"It is only a small delay. All of the other astronauts have come through quarantine without vectors. Including myself," coaxes Gim-bal.

"I assume the astronauts will have their orders by the time they arrive here and the family is visiting because they know, at least, that their astronaut kin will not be returning home, yet," says Jr. Com. Poon.

"Their orders will not come through you or anyone on your station. You will not know any more than the families or astronauts," says Gim-bal. "Tell the crew not to pry or to collect information in this regard."

"I assume that includes me," says Jr. Com. Poon almost to himself.

"You guess correctly," says Gim-bal with Jr. Com. Poon following him out of his stateroom.

Episode 101

It is not dark yet. It does not need to be. Roy and Ali-ce know their room will be cool enough in this dusk to expect it to be comfortable. It will be dark soon. But Roy and Ali-ce are often occupied enough with their own concerns or each other to notice when it does get dark outside.

Sometimes, however, the darkness can be metaphorical. Ali-ce finds a courier delivered package on Roy's bed. She feigns amusement, if not surprise, to disguise her dread. Could this be a herald of the darkness she fears. This is one time her association with Roy Abbit has worked to her disadvantage. This time she has no address for conventional correspondence except Roy's.

"Who is it from? Open it," prompts Roy.

"It is from the boss, or at least the publisher," says Ali-ce.

"Is it an assignment? Did you write something? Where do you get the time?" Roy asks while Ali-ce opens the envelope.

"It not from the publisher. It is from Gim-bal. He has resigned. I have purchased his house for $120,000.00," remarks Ali-ce relieved at the surprise but not completely.

"I know you never have to worry about money. Not as long as you have that fistful of rocks. But where did you get $120,000.00?" Roy asks sitting on the bed to change the subject.

"He just dumped the money into his account to substantiate the transaction," says Ali-ce.

"Why does he want you to have the house?" Roy asks.

"He is leaving and he does not want the property to appear abandoned - on paper at least - by him at least," comments Ali-ce barely aware of Roy and his beckoning.

"Is he in trouble?" Roy asks.

"Not legal trouble," imparts Ali-ce as she turns to join Roy on the bed. "He or his assistants have been working on getting fired so he could leave the company."

"What is he going to do now?" Roy asks.

"He is not looking for a job. This is his exit plan," says Ali-ce holding her papers on her lap. "He left two months ago. If he is coming back, he cannot get here in less than four months."

"We only have to wait and see what he does when he gets back," says Roy.

"I do not think he is coming back," concedes Ali-ce as she stands and walks to the dresser before turning to look back at Roy. "He has turned the house over to me. His assistants will be gone soon. My father's promise to remove our establishment from this planet is almost done."

"What is left?" Roy asks sitting up and thinking about getting up.

"I suppose there are eighty or so of us left to evacuate including his household assistants," grants Ali-ce. "I suppose I need to quit or be fired but I do not have to leave Earth."

"If they tell you to leave, you will have to," says Roy as he stands and faces Ali-ce.

"If I do not leave," says Ali-ce. "I will be arrested, if and when I return home."

"They are not going to send anyone after you?" Roy asks as he walks up to Ali-ce.

"I cannot be extradited with the help of any law enforcement on this planet. Gim-bal is the only law enforcement officer from our planet and he is not here. He will settle for executing me rather than escorting me anywhere," confesses Ali-ce as she puts her head on Roy's shoulder.

"That is not good," says Roy as he puts his arms around her. "I cannot let him do that."

"I have not been asked or told to leave Earth. I probably will not. I am probably more important here now that he and everyone else is gone," asserts Ali-ce returning his embrace.

"Then we do not have anything to worry about," contends Roy ready to move on.

"Not exactly, Father's exodus may not have been the only reason for Gim-bal's return to Clidim," continues Ali-ce while pushing away from Roy and his effort to change the subject. "If the Robe's Chamber has published a warrant, he would be the one to execute it."

"Your father is not going to have **you** arrested or executed," argues Roy.

"But if the vote went his way, **you** could be arrested and executed," claims Ali-ce.

"We have four months to find a place to hide," defends Roy.

"Then he has as many months as he wants to find us and kill us. You according to the warrant and me for obstructing justice," says Ali-ce.

"Is obstructing justice a capital offense?" Roy asks in disbelief.

"If I am obstructing the execution of a capital warrant, it is," says Ali-ce.

"What are we going to do. I can surrender to Gim-bal," declares Roy as Ali-ce turns to face him again. "That will spare you. You have to let me do that much."

"I was not considering that and I will not," says Ali-ce while she steps around Roy and over to the bed. "We have four months before he gets back. We must get married right away."

"How is that going to stop or slow down my execution?" Roy asks.

"It will make my new house community property. You will inherit it," delivers Ali-ce sitting on the bed. "You will need it or want it, if you or both of us survive what is next."

"What is next? He will not have any trouble landing next to your ship at that house and finding us in it," relates Roy as he takes off his shirt.

"We do not have to go to Lynchburg to get married. Just as soon as we are married, we are leaving for Clidim. By the time he gets here, we will be gone by three or four months. It will take him three months to get back to Clidim just like us. That will give us three or four months to find out what we can about the vote on my father's petition. Maybe we can file petitions of our own. If you are elected for execution, maybe we can get it set aside. If we cannot do that, we can petition for a pardon. He cannot do anything while the petitions are pending," says Ali-ce.

"That buys us some more time but it does not sound easy," says Roy pulling off his shoes.

"It is a lot harder than it sounds. You and Earth are not supposed to exist. No one has to prove that you do to vote for your execution. But for a pardon we will need to show that you exist before they will vote on a pardon. The House that issues a pardon cannot hang its reputation on phantoms and legends without paying a political price. As soon as you and/or Earth exists, a constituency will form around you," says Ali-ce reaching for the light switch.

"Well, let us get going and get married and, uh, get going to Clidim," says Roy.

Episode 102.

Mer-cy Mim-sey lives in the Chi-cago Valley and has ever since her family bought it for her as a wedding gift. Early in their marriage, before her husband, Ma-x, is elected a Doctor of Civil Service, she and Ma-x visit the valley as often as they can. He promises to build her a house here and keeps it within a few years of becoming Dr. Mim-sey. Over the years, she lives here more and more and he visits more and more. Almost no one calls the valley the Chi-cago Valley except the Mim-seys.

When Ali-ce tells her mother that she has found an entire city called Chicago and they both shrieked with laughter. Chi-cago with the hyphen and in Bemyem means 'Let us go shopping'. Somehow it increases the delight of visiting her mother in Chi-cago and deepens the sympathy when she says she misses Chi-cago.

Building a house and setting up housekeeping in that remote valley meant buying and bring back something with every trip out of the valley. Every trip began with the words, 'let us go shopping'.

So the Mim-seys come to call their valley Chi-cago. Mer-cy is being visited by her brother Hu-go. He is not a Mim-sey and it is his painful task to prepare Mer-cy for the awful reality that she too is no longer a Mim-sey.

"You are never here this time of the year, Hu-go," fusses Mer-cy.

"I cannot stay long. But Doctor, uh, Ma-x has been gone a month now. I cannot leave you alone, up here, under the circumstances," grumbles Hu-go before taking a careful sip of tea.

"I have gone longer than a month at a time without the comfort and the company of Ma-x," huffs Mer-cy while pouring herself a cup of tea.

"I bet you call or he calls every few days. You both know what each other is doing and where you both are. No one goes much more than a few days without that much. I bet you and Ali-ce connect at least once a month. Most keep in touch from one waking hour to the next, " imputes Hu-go drinking freely now that he is sure his tea is not too hot.

"Yes. But just because I do not know where he is and what he is doing now does not mean he is dead or gone," argues Mer-cy looking at her teacup without picking it up. "He could be on his way to Earth."

"He could not board a ship without someone knowing it and no one has a record of that. It is a misdemeanor to conceal that information in an official investigation. It does not mean jail time but it could wreck someone's career," enjoins Hu-go setting his teacup down. "No one makes enough money and no one wants to make half of what he is making now for the rest of his life."

"I will not hear from Ali-ce until she moves into Gim-bal's house," judges Mer-cy finally picking up her teacup. "If she is already on her way here, it will be at least three months."

"It pains me to tell you this way. While I cannot tell you what I know about Ma-x's petition, I can tell you that Ali-ce has nothing to do with enforcing the warrant," says Hu-go constrained by his office against his sympathies.

"Is that as clear as things get?" Mer-cy asks while setting down her teacup.

"I cannot say anymore and I cannot stay any longer," asserts Hu-go without getting up.

"No one has a better brother. I will know that as clearly as ever, by the time the fog rises from this mountain valley tomorrow. I am just as sure what you have told me will come into focus. Neither of us can ignore the fog of your office for the time being. But I can count on that lifting too," contends Mer-cy as she pours some hot tea into Hugo's teacup to warm it.

"I hope that in the clear light of the near future you will not shrink from the view," persists Hu-go before picking up his teacup and taking a deep sip.

"And I hope a lifetime of affection between a brother and sister will serve us both," pleads Mer-cy before taking a deep sip of her tea as well.

"It is a hope that would have been lost on our younger years," entreats Hu-go as he stands and waits for Mer-cy to usher him to the door. "Back then we were less aware of our familiar affection and how complex life really is."

"So you think we will get through this intact?" Mer-cy asks while standing. She drops her napkin on her chair and walks with Hu-go toward the vestibule.

"There is a chance. Perhaps we can be grateful that we are not as young as we used to be. I will see you for the recess months," pledges Hu-go before he turns to leave the house.

1

1 1

1 2 1

1 3 3 1

1 4 6 4 1

1 5 10 10 5 1

1 6 15 20 15 6 1

Chapter 35

Chapter 110001000....

Chapter ABF

Chapter $(1 + 6x + 12x^2 + 11x^3 + 5x^4 + x^5)$

Chapter $(a^5b^0 + 6a^4b^1 + 12a^3b^2 + 11a^2b^3 + 5a^1b^4 + a^0b^5)$

Chapter $2^0 + 2^1 + 2^5$)

Episode 103

Jr. Commander Poon is not unhappy with the way the first visit has gone and he is not unhappy with the way the second visit is going. But he is not happy either. The numbers are right. The supplies are sufficient. Replenishment is en route and on schedule. No problems have surfaced with either visit.

That is part of the problem. As aggravating and unanticipated as problems are, it is peculiar when there are none. Officer Gim-bal is with this group too but that cannot explain the perverse perfection of this operation.

Also absent is the usual information. Where are the astronauts coming from and where are they going? Why not do all this on Clidim? If these are family and friends of the astronauts, why are they so glum. Why are ships that can travel at speeds beyond the speed of light supplied with three months of provisions? How far will they travel like that?

"Any problems with the second visit?" Jr. Com. Poon asks.

"None," answers Ensign Gen-Sjen. "But you already know that."

"I thought you would find something," says Jr. Com. Poon spinning a pen just above his desk thanks to the peculiarities of ersatz gravity.

"The last thing Officer Gim-bal said was no prying or collecting information," says Ensign Gen-Sjen who is most unimpressed with e-grav.

"Has Officer Gim-bal said anything or asked for anything? I can reasonably expect to know that much," responds Jr. Com. Poon.

"Nothing," comments Ensign Gen-Sjen who finds e-grav counterintuitive. "I thought you might have seen him from time to time."

"Not since he left this book," says Jr. Com. Poon who finds e-grav endlessly fascinating. "I have had full and uninterrupted use of my stateroom since."

"Have you been in the visitor spaces?" Ensign Gen-Sjen asks with some expectation of what the answer will be.

"Just to pass through," concedes Jr. Com. Poon who compares e-grav to a kind of magnetism that includes nonferrous materials. "The operations and activities are going exactly according to his book."

"Is that the one he left here about four months ago?" Ensign Gen-Sjen asks grateful, at least, that e-grav is not as crude as static electricity.

"That is another problem," asserts Jr. Com. Poon who finally pushes his pen to a rest onto his desk. "No typos. No deviation from the published schedule. What kind of staff does he have? It is no small one."

"You mean all the ships coming and going are on time and with the supplies and services described in that book?" Ensign Gen-Sjen asks.

"The ships named, the captains named, the manifests specified, arriving and leaving on time," conveys Jr. Com. Poon.

"That has done a lot to prevent problems," says Ensign Gen-Sjen. "We have enough to do without the added work of problems and delays."

"I guess I should not kick the tires on a gifted car," fumes Jr. Com. Poon considering spinning his pen on the wall. "But it just speaks to something more than a one man operation, a lot, a huge lot more."

"Has anyone said anything about where they are coming from or where they are going?" Ensign Gen-Sjen asks without a glance at the acrobatic pen.

"No," fusses Jr. Com. Poon while putting his pen in a drawer. "But it would not make any sense for them to go back where they came from."

"It would, if they want to take advantage of the time shift that goes with traveling above the speed of light," grants Ensign Gen-Sjen.

"They cannot go too far into the future or the past without losing connections and support with Clidim," explains Jr. Com. Poon.

If they go too far into the future, who is going to be here for communications and repatriation," illustrates Ensign Gen-Sjen following his captain's lead.

"I suppose. They could comeback to a technology that does not interface with theirs. They could come back to a Clidim that does not know or believe who they are," suggests Jr. Com. Poon. "They could be destroyed before or shortly after landing on the surface."

"If they go too far into the past. There will be no technology to interface with. Even if we found out where they went, they could be on a planet with no technological interface. They could be banging rocks together. They could not plan on living long enough for technology to catch up," judges Ensign Gen-Sjen.

"When did you last eat something that did not agree with you?" Jr. Com. Poon asks while checking his desktop for any other floaters.

"I cannot remember. Why?" Ensign Gen-Sjen asks.

"Only one Watch Officer per shift can tour the visitor spaces. Is there anyone you cannot depend on?" Jr. Com. Poon asks. "Someone who is likely to shirk his duties, sleep and generally screw up?"

"Noncom Earwax," informs Ensign Gen-Sjen.

"That cannot be his name," objects Jr. Com. Poon.

"It is Err-wick. But you know what we do to names," remarks Ensign Gen-Sjen checking his mental list of names for some of the best ones.

"I am afraid I do. Put 'Earwax' on roving watches that he can screw up. He cannot be accused of paying attention so he will not be presumed to be collecting information. Put yourself and myself on the other two roving watches. That way we will have all watch shifts for visitor's spaces covered," prescribes Jr. Com. Poon. "Both of us will be called often enough to duties that our coverage will obviously be spotty."

"That should keep you and I from looking like we are doing little more than marginally following required procedures. Now about your indigestion. If you are not called off watch conveniently, hit the nearest toilet and turn on the air scrubbers. No one will question how long you are in there."

"Write down any words you do not recognize - phonetic is fine. Make sure you write it down. Make sure you have your notes secured before you open the door. I do not want you to be seen putting things in your pockets. I do not want you to be seen touching your pockets - zipper and belt are okay."

"Do not go off duty without coming to my stateroom. If I am not here, have the watch officer page me. I want you to personally hand me anything you have in writing. I do not want that stuff to turn up anywhere. I will dispose of anything that will connect you to your information. Any questions?" Jr. Com. Poon asks as he gets up from his desk and thereby signaling the end of the dialogue.

"If I think of any, I will not ask anyone but you," responds Ensign Gen-Sjen as he stands in preparation to be dismissed.

"That only leaves the two of us to screw this up. I have this roving watch. Post the watch schedule as usual," declares Jr. Com. Poon preparing to leave. "That should give 'Earwax' time to see it. Make sure he does not miss a watch. If you have to make a substitution you know what I want."

"That would be...," says Ensign Gen-Sjen.

"That is okay. I am scared of what I will hear. Dismissed," barks Jr. Com. Poon lifting one foot slightly to feel the pull of the e-grav.

Episode 104

A honeymoon among the stars would be the envy of anyone but a honeymoon among a myriad of galaxies is beyond anyone's ability to dream. Ali-ce, for her part is undistracted by the grandeur. She is looking ahead to mounting danger. She can count on the danger of this trip being eclipsed by the perils of events on Clidim. The remoteness of her mother's home offers some prospect of relief.

Roy, for his part, is uncomprehending since the galaxies look like stars and the stars look wildly similar. He is equally uncomprehending of the danger they both face. His sense of his mortality, in any form, has escaped the one form that has any meaning that of reality.

The unreality of any mortal hazzard for Ali-ce is girded in denial and his admiration for everything about her. Not the least of which is her prescience guiding them through not just the lovely woods but the dark and deep ones too. The worst connotations of sleep remain miles away, even light-years if such a thing could be comprehended. But peace is tentative.

The threat of reality dawns on Roy when he considers their hasty marriage. The selfish side of him is grateful for the speed and lack of pomp. The selfless side of him is pained to think of the sacrifice Ali-ce has made. Surely, she wanted all the trappings. As many friends and family as she could manage. The best venue, beautiful clothes, food and accommodations. As the details mount in his mind, Roy retreats deeper into his own thoughts.

Perhaps he can recruit the help of Mer-cy. Perhaps things can be resolved to have a wedding on Clidim. At the very least a real wedding will be a priority when they return to Earth. They both take refuge in the uneventful aspects of the trip. Roy has never been in space so Ali-ce is thoroughly occupied with the most important elements and minutia too.

"How do you get from here to there?" Roy asks with his face in the freezer drawer.

"Well that depends on what is 'here' and what is 'there'," smiles Ali-ce.

"I understand that when we are near Earth, and I expect it is the same for Clidim, there are all kinds of details that make a successful take-off and landing possible. But out here you could go in any direction. How do you know when you are going in the right direction?" Roy asks while trying to open the wrapper of a frozen fruit stick with the usual initial frustration.

"Oh. You know how a flashlight helps you get through a dark stretch?" Ali-ce asks.

"Yeah. But I know something about the room or yard I am trying to navigate," says Roy.

"Well mostly the navigation computer and I know something about the neighborhood we are in," says Ali-ce reaching for Roy's unyielding fruit stick with routine resignation. "Then there is something you have done with your flashlight all your life without thinking about it."

"I cannot guess what that is," says Roy

"You follow the beam," says Ali-ce with a smile that supplies a metaphor for the literal.

"The same is true for those who drive at night," says Roy. "But I do not see any headlights on this ship and even though it feels like we are standing still, I do not think we are waiting all those hours and days for our high beams to reflect back from this galaxy or that."

"Well, that part is a little different. We send light, or photons ahead of us and we chase after them. Light likes to travel in a straight line. We aim the light in the direction we want to go and follow it," says Ali-ce. "The computer determines how well we follow the photons and where we are. We can send out photons as often as we want and keep track of where we are going."

"You make it sound simple. I am sure you are doing that for me. But I have heard that galaxies and black holes bend light. So it sometimes, if not often, follows a curved path. What about that?" Roy asks.

"Yes. That makes the flight path a little more complicated. We just follow the curves. We do not insist on, so called, straight lines. Besides there is a bonus. This tendency to curve alerts us to black holes and galaxies that are difficult to see any other way," relates Ali-ce while taking a bladder of hot tea out of the microwave. "We can travel faster than we could any other way because the photons show us where these obstacles are whether they are on the map or not."

"But, at least some of the light you send out has to be reflected back off of something," supposes Roy recognizing, for the first time, why Ali-ce never takes her tea with sugar.

"The photons are projected from three spots on the hull. They travel almost parallel to each other but converge eventually. When they converge some are scattered back. If we are directly behind the point of convergence the reflected photons will land exactly between the three spots on the hull. The photons that land outside the center spot tell us how to adjust our trajectory," reveals Ali-ce.

"I guess it is like hanging a weighted string between the legs of a tripod. If all the legs are the same length the string will be centered between the legs," grants Roy. "So when the light bends on your photon-tripod it is like having one leg longer or shorter than the others."

"That is a good analogy but why did you wait so long to ask about this? The trip is almost over," adds Ali-ce while she watches Roy take another drought of ice cold fruit juice.

"Mostly because the trip is almost over. I have been living from moment to moment essentially. The incidentals have been more real than the ultimate end of the trip. The end is not so abstract anymore. How we got here primarily involved what got us here - the ship. Knowing more and more about how this ship supports our existence, how it preserves us, used to tell me how we got here," concedes Roy as he mashes the frozen fruit stick into a green slush.

"Well, we are not there yet. We will have to slow down to an almost comprehensible speed," advises Ali-ce. "Just to make sure this ship supports our existence, we will have to spend the last twenty-eight hours of the flight in the flight suits and they do not have intercoms."

"Why?" Roy asks before sucking a few milliliters of green slush out of its wrapper.

"We cannot talk while in the suits but it is not so you cannot ask all those questions. The real reason for wearing the suits is that they provide protection against most equipment failures. The suits will become our seatbelts and roll bars. Also there is no need to keep the rest of the ship livable. We will have all the necessary comforts within the suits," asserts Ali-ce while testing the bladder's temperature with her finger.

"Why no communication between the suits? What if you get injured or killed?" Roy asks.

The last thirty hours are on automatic pilot. If anyone needs to help us or save us we will be detected, intercepted and helped by resources that exceed yours and mine. We will be traveling below the speed of light so our telemetry will be way ahead of us," says Ali-ce. "Clidim will be waiting and ready, especially if we need help."

"Okay, why no intercom for the suits?" Roy asks.

"The soft-suit is the flight suit. It provides the backup atmosphere and the ship provides the protection of the hard-suit. The hard-suit allows the pilot to work outside the ship. The ship is intended for one person - the pilot. No one planed on two people wearing both suits at the same time," replies Ali-ce before taking more than a mere sip of warm tea.

"So both of us wearing suits is just redundant?" Roy asks.

"With both of us in suits, the one and only pilot seat becomes redundant," says Ali-ce.

Episode 105

Anyone who wants to know what is going on overnight at the towers and spaces that serve the Robes needs only to observe the lights. Actually, it is the lack of light that is telling. Lights go on following the Qu-ad's. The Qu-adrants move through the towers and spaces cleaning their designated areas. Each area is a 'quadrant' identified with a binary number.

Of the millions of quadrants, some get cleaned every few hours some get cleaned every night. There are some spaces that get cleaned every decade. All of them get cleaned on schedule. One would suppose it is just a 'dust and polish detail' given that the towers and spaces have seen nothing but polished and dust free surfaces for centuries.

While that is the most conspicuous accomplishment of the Qu-ad's, it is the least of their duties. They must not know who is who or what is what. Papers are unread. Books are unopened. Screens are doused without a glance. Dead bodies are rare but their disposition is flawless and efficient.

Spaces that are occupied are circumvented until shift instructions give them access. The longer Qu-ad's can go without communicating among themselves on what is seen or heard the longer they keep their jobs. Most Qu-ads have been Qu-ads for generations.

The transportation hubs are lit on the approach of traffic and doused as it subsides. Elevators sigh into place as they are approached. Escalators ease into operation in anticipation of use and quietly stop when riders disembark. R. A. T.s blink into service on a moments notice. Quadrant lights and services are accessed with a key card that validates schedules and services. Special details are qualified before the transaction and assigned according to temporal circumstances.

Immersed in this silent, around the clock, efficiency, 1st-Robe's train pulls up to the tower platform. He disembarks and proceeds to his office without inconvenience. The person he expects to be waiting for him is there with details that he hopes will be very nearly the last. 1st-Robe does not stand on ceremony. Governor Err-wick does not stand. His rank does not require it.

"Hello Governor," bids 1st-Robe as he enters his office. "I presume and hope that your presence here and on time means that plans have gone as planned. I presume similarly that you are no longer a Noncom."

"You assume correctly," chats Err-wick with a nod of recognition and without a hint of subordination. "My surveillance ended with my quarantine and return to Clidim. Noncom Earwax will generate a paper history down here and eventually disappear."

"I know what you have to say is not good," coaxes 1st-Robe without disclosing what he knows. "Please tell me it is not as bad as I fear."

"We have lost 128 but without a liquidation," advises Err-wick leading the 1st-Robe.

"How did we lose them and why 128?" 1st-Robe asks leaving the Governor to his game.

"Officer Gim-bal's solution, which is his prerogative, was not to execute the Clidimites on Earth. It was his judgement that 125 deaths in a necessarily short period of time would become conspicuous. Local law enforcement may even prevent him from completing his task," observes Err-wick with an undisguised tension. "At the same time the ever shortening list of Clidimites would likely suspect the fate of their colleagues and in a futile effort of self-preservation would make his job almost impossible."

"That was the majority opinion against the vote for the 125. But the purpose of the warrant was to eliminate Dr. Mim-sey's effort to officially establish the existence of Earth. The return of the 125 en mass would do that despite the execution of Ma-x Mim-sey. The celebrity of the 125 would make Ma-x Mim-sey a martyr. The liquidation of the 125 could only be concealed as long as the warrant and petition remains redacted," divulges 1st-Robe as an intern enters with a foamed glass carafe and matching cups . "The effort to publish an unredacted warrant and petition is already mounting."

"Officer Gim-bal has capitalized on a number of problems that already have a history with the Earth project," declares Err-wick as 1st-Robe pours drinks for two. "First Gim-bal got them off the planet. He promised, the ones that did not want to leave, a family reunion and an immediate return to Earth.

The others that feared premature death were promised no compromise of their lives and comforts. And in the tradition of Clidim, anyone and everyone who dies away from Clidim will be returned to Clidim and their families. The 125 consented and met with family members to console their fears and substantiate the agreement."

"I can see how this is going. Diffused across the lifetimes of the 125, the urban myth of Earth will not reify with respect to them. The families will not promote the myth on the promise that their loved ones will return in the course of their careers. That is a longstanding protocol," maintains 1st Robe while reaching for his cup. "Today's version of the myth will not migrate into reality."

"Their individual lives and stories as explorers will proceed on the path that they embarked on without the interruption of Ma-x Mim-sey's petition. His effort to insinuate the myth of Earth into the lives of the astronauts will not get traction or recruit new believers. There will be those who will be offended by his attempt," continues Err-wick. "The perennial believers who do not need evidence of Earth will still insist on the existence of Earth but will not find new substantiation."

"I can see how temporary removal of the 125 from Earth will solve a lot of problems. But he does not seem to have solved the problem of the Clidimite's continued existence on Earth," says 1st-Robe. "At the very least some of those 125 lives and lifetimes could substantiate the existence of Clidimites on Earth; or at least the existence of aliens on Earth from somewhere."

"That problem has been solved by exploiting the other primary problem of the Earth project. We have been diligent in our comings and goings not to let the problem of time shifts interfere with our best efforts. What Officer Gim-bal has engineered is a return to Earth by the 125 but into pre-technological Earth history. Living or dead Clidimites will not be detectable as aliens.

They will live out their lives and expire naturally without the means to identify even their remains as alien," conveys Err-wick without acknowledging 1st-Robe's insult by picking up his cup first. "They will have the consolation of being spared execution and of living out their lives immersed in the cultures with which they are already find familiar and interesting."

"We will not have the intimate connection we have had but perhaps periodic visits will be sufficient," assures 1st Robe satisfied that he has harpooned the Govenor's sense of official superiority with his teacup gambit. "Without repopulating Earth with Clidimites the periodic visits will stop with the removal of the remains of the last of our explorers. I guess the exile explains the difference between 125 and 128."

"Yes," reveals Err-wick. "While Officer Gim-bal and his two assistants were not designated in the warrant for execution, they went with the 125 back to Earth and into its pretechnological history."

"It is humbling to think he would do that," deplores 1st-Robe. "I do not suppose he was just being thorough though."

"I do not suppose so either," construes Err-wick. "Perhaps he was rewarding his two assistants for their service. I presume they will live lives that will be as fulfilling as any in modern Earth. After all, their exile was voluntary. I also suppose that Officer Gim-bal was not seeking retirement. His self-exile was his covenant with the families. His self-sacrifice cements our covenant and our obligation to those exiles."

$$1$$
$$1\ 1$$
$$1\ 2\ 1$$
$$1\ 3\ 3\ 1$$
$$1\ 4\ 6\ 4\ 1$$
$$1\ 5\ 10\ 10\ 5\ 1$$
$$1\ 6\ 15\ 20\ 15\ 6\ 1$$

Chapter 36

Chapter 001001000....

Chapter CF

Chapter $(1 + 6x + 13x^2 + 11x^3 + 5x^4 + x^5)$

Chapter $(a^5b^0 + 6a^4b^1 + 13a^3b^2 + 11a^2b^3 + 5a^1b^4 + a^0b^5)$

Chapter $2^2 + 2^5$

Episode 106

The Viyebiucla, also known as the 'forensic bus', started out a hundred years ago with the best of intentions. It stopped growing at 500 million tons by the time it was fifty years old. Its original purpose was to rescue the crew and cargo of disabled space freighters, also glass, that could be as large as 150 million tons.

It is the largest piece of glass circling Clidim. Its worth is justified on the lives it saves or could save. Before the Viyebiucla, the fate of a disabled freighter is that of stranding forever in the deepfreeze of space or as a lethal fireball plowing into one of Clidim's largest oceans, fortunes and skills permitting. The putative priority of the forensic bus is to save the crew of doomed freighters. Then, circumstances permitting, the Viyebiucla rehabilitates the freighter so it can continue its service including delivering its jeopardized cargo. Freighters that cannot be saved are stripped of their cargo and broken into pieces. The pieces are sold to glass smelters or disposed of by steering them into an ocean.

The insurance industry eventually twists the mission of the Viyebiucla into its present role. Rescuing the crew is now a legal convenience to establish salvage rights for an abandoned vessel. In order to be saved, the crew must abandon its ship. The Viyebiucla takes ownership of the abandoned ship and its cargo. Now both are theirs to sell. The Viyebiucla, as salvager, make more money than the insurance premiums could generate. The Viyebiucla makes its stockholders very happy. Over time, stockholders for the forensic bus have evolved into trust funds held by insurance companies since they typically outlive its stockholders.

The pedestrian insurance company stockholder never touches the trust funds generated by the Viyebiucla but makes his money from the millions of small premiums that eclipse policy values by scores of thousands of times. An insurance company that pays as much as five percent of its holdings for insurance claims faces a stockholder's revolt in the form of stock sell-offs and falling stock values.

The principals of the company are quickly removed by the stockholders and are replaced with directors who propose to be more responsible with the stockholder's profits. Successful directors are rewarded for their efforts by those stockholders. All of this is almost academic for Captain Ham-pton.

He is not yet afraid of losing his job for misfeasance. As always, he is focused on efficiency and productivity. But that little tiny ship in the middle of the air-dock puts him as far from E 'n' P as he can get. The specter of misfeasance is threatening to materialize.

"Why is this piece of bottle glass tying up my air-dock?" Prime Salvor Lo-thar asks peering over a railing at the little ship 225 feet below.

"It is not bottle glass and while it is only a few kilo-tons it is every bit as valuable as any 150 million ton hull you have waiting to sit here," vents Captain Ham-pton resting his lower back against the same railing.

"Why not dismember this chip on the cold-dock?" Prime Salvor Lo-thar asks while turning to rest his back on the railing as well.

"We cannot get the hatch open and the spectrum of organics suggests that there might be someone alive or recently dead onboard," advises Captain Ham-pton.

"We can say it was dead when we cracked the hull," fusses Prime Salvor Lo-thar. "It will not be the first frozen steak we have ever found."

"This 12 billion light-year wide freezer works against us," warns Captain Ham-pton. "It could preserve evidence that the passenger was alive when the hull was opened. You know where that would land us."

"In 100 years no one has won that gambit and more than a few have tried," beefs Prime Salvor Lo-thar still hoping for a quick solution.

"This time we have to be more than a little careful. This ship is owned by a niece of our 1st-Robe. Her whereabouts are unknown like her father's. There is more than just the smell of life or a former life on that ship. The telemetry, possibly recorded, is English," warns Captain Ham-pton hoping to turn the corner on Prime Salvor Lo-tar's opposition.

"We speak English because it is an upper echelon language. That astronaut comes from similar circumstances. The occupant could have hijacked the Mim-sey ship and killed the 1st-Robe's niece," suggests Prime Salvor Lo-thar. "We could be doing the 1st-Robe a favor by delivering the remains, however recently deceased, of the perpetrator."

"No one would question the demise of such a dangerous individual. But this astronaut does not seem to know how to operate the intercom. We do not even know what gender the astronaut chooses just by listening to, let us say, 'him' and he does not know what button to push to connect us," lectures Captain Ham-pton. "We will have to meet this astronaut face to face and alive for more than merely grammatical reasons."

Now that he is on Captain Ham-pton's side of the issue Prime Salvor Lo-thar asks, "Does the occupant know he has been rescued?"

"I do not think so. Reduce the light by two-thirds. Open the vessel. But do not enter it right away. If the astronaut is a murderer the victim may be onboard. The perpetrator may feel the need to continue his career of desperation and I do not want to lose anyone here or provoke his worst fears," charges Captain Ham-pton.

"If the astronaut is suited up, there may not be a noticeable environmental difference with an open capsule. There will be no change in atmosphere, orientation, light or ersatz gravity. If the occupant is not suited up or at least does not have a helmet on, then he might notice a change in atmosphere," construes Prime Salvor Lo-thar. "If we wait long enough, the astronaut might investigate the breach to the air-deck and emerge on his own."

"I will have someone in a hard suit waiting 100 yards away," instructs Captain Ham-pton exploiting the momentum of the dialogue. "I do not think the astronaut will be carrying or planning on using a weapon that could threaten our confederate with that kind of protection."

"What good will that do?" Prime Salvor Lo-thar asks myopically.

"His ship has a similar hard-suit. Even if the astronaut is wearing the soft-suit, the confederate will look familiar. If the astronaut is unarmed, we will have our confederate take off his helmet. This will, further, disarm the astronaut's worst fears whatever they are," continues Captain Ham-pton with growing confidence. "The astronaut may take off his helmet too. This will telegraph and enlarge the sense of safety. I will use the air-deck speakers to begin a dialog with the astronaut."

"What if the astronaut does not come out?" Prime Salvor Lo-thar asks narrowing his focus on and widening his commitment to the strategy.

"The astronaut could be on the defensive or as it seems so far - he is unaware of being rescued," argues Captain Ham-pton.

"So we send the confederate into the ship to meet and greet?" Prime Salvor Lo-thar asks as he takes an increasing ownership of the plan.

"The confederate will have a portable speaker, microphone and camera. They will be our eyes and ears. My voice will greet the astronaut. The confederate will keep his helmet on for safety and to maintain the impression that the voice is coming from the occupant of the suit.

If the astronaut threatens the confederate, he can drop the portables and bug out. That way we will still have communication," relates Captain Ham-pton with one last remaining task before executing the mini-mission.

"What if he flees his ship?" Prime Salvor Lo-thar asks.

"The confederate can follow him and stop at the hatch. The confederate can then block the hatch, so no one can get back in," relates Cap-tain Ham-pton. "Once he is outside the ship, we can begin some kind of dialogue. Our first task is to establish two-way communication."

"One last thing before we get the ball rolling - Who is going to be the confederate?" Prime Salvor Lo-thar asks without a clue. Captain Ham-pton supplies that clue with a slight smile.

"Oh. I will get suited up and take my position, uh, 100 yards from the breach," says Prime Salvor Lo-thar without alternatives of his own.

"I will have the watch officer announce the special detail," says Captain Ham-ton every bit the Captain of his 500 million ton ship.

Episode 107

As the largest maneuverable ship serving the 122,000 miles of space above the surface Clidim the Viyebiucla has its advantages. The forensic bus does not have to shuttle back and forth from Clidim to take advantage of skilled workers, special environments and heavy equipment that otherwise exists only on the planet's surface. The Viyebiucla has almost everything it needs. If profitability requires it, whatever a 150 million ton freighter needs it gets. If it needs a one and half million ton ablative coating to get through Clidim's atmosphere, the Viyebiucla can lay it on. Emergency medical attention is fundamental and postmortem services save lives by containing vectors and exculpating suspicions.

Ultimately, on Clidim, there is neither celebrity nor celebrity-trial waiting for an astronaut. He or she can count on going home or to jail in complete anonymity as soon after touchdown as routine circumstances permit. The same rules apply to Roy Abbit. He is by definition and circumstances an astronaut.

Problematically, there is no record of his birth or training on Clidim. But since he is on the Mim-sey ship, he is by law already on Clidim. So the question becomes, "Is there any way to limit his ability to come and go anywhere that is ordinarily available to anyone on Clidim?"

Similarly problematic is when and how did Ali-ce Mim-sey die? What was his role? Is he a hero or a villain? Examiner Hu-me and 0011th-Robe are on the Viyebiucla to settle these questions and a few more.

Where Roy Abbit goes when he leaves the Viyebiucla depends on the answers to these questions. Profitability requires that Roy Abbit leaves the ship as soon as possible, one way or another.

"No offence, but I thought the Robes would not send its, uh, most junior judge," carps Ex. Hu-me a little proud of his tact given the insult.

"It is easer for the other Robes to take over my calendar. The higher up the food chain you go the harder it gets," chats 0011th-Robe not willing to dignify the slight. "Besides I will be back on Clidim pretty soon."

"But this involves 1st-Robe's niece," clarifies Ex. Hu-me with an implied slight.

"Precisely, the Robe's Court must send someone. I am as far from the appearance of impropriety as the Court can get," defines 0011th-Robe allowing Ex. Hu-me to step into a trap of his own making.

"You do not think anyone would suggest...," Ex. Hu-me begins.

"It is a little late to go on the defensive Ex. Hu-me," 0011th-Robe imposes almost embarrassed at taking advantage of the advantage so easily.

"I will leave you to defend the considerable dignity of the Robe's Court. I only have to determine if there is evidence of a crime here," counsels Ex. Hu-me exploiting condescension in order to take back the advantage.

"There may be evidence of a crime but it will have to be specific to Roy Abbit," declares 0011th-Robe dragging Ex. Hu-me back to the issue at hand. "That will get him off this ship and that is why we are here."

"I think we can both agree that there is no motive to ascribe to Roy Abbit. He has no pilot skills. It would not serve him any purpose to off the only person who possess the skills to land the ship and him safely," states Ex. Hu-me as his effort not to make the Robe's case for her.

"He could count on being intercepted and rescued just as he is now," utters 0011th-Robe readdressing Roy Abbit's culpability.

"That is not only a dangerous maneuver there is no evidence that he planned on such a thing," renders Ex. Hu-me ambiguously. "There is no record of Abbit or Mim-sey making statements to that effect."

"There is evidence of a four month old fetus. Perhaps he did not want the child or the marriage," relates 0011th-Robe unambiguously.

"They were newly weds. It does not seem that they were at a point in the marriage where either one would want anything but a life together. Again there is no record of statements that would suggest that they did not want their marriage or the child," recites Ex. Hu-me.

Enlarging on the absurd as a heuristic fallacy, 0011th-Robe quibbles, "He says he has no idea how or when Ali-ce died."

"Neither do I. They were in the slowdown loop around a black hole. They were traveling above the speed of light. Time as we know it was standing still. The rate of slowing around the black hole brought the time line of the ship into correspondence with ours. Their clock and calendar came into correspondence by the time they slowed down to the speed of light," explains Ex. Hu-me in an effort to defend the obviously absurd.

"But time for Roy Abbit and Ali-ce Mim-sey was normal to them," stresses 0011th-Robe suggesting a relative time frame for Roy Abbit.

"They cannot say anymore about what happened in our time frame than what we can say about what happened in their time frame," acknowledges Ex. Hu-me not willing to extrapolate and reconcile a second time frame.

"What then was the manner of death?" 0011th-Robe asks feeling free to move on to other potential issues of culpability.

"As nearly as we can tell, both Roy Abbit and Ali-ce Mim-sey were subject to the same environment. The environment around the black hole had various levels of radiation from the longest to the shortest frequencies. Unfortunately, the hard suit provided resonant cavities of various fixed sizes. Some of the cavities focused some of the energy like a microwave oven or an antenna. The water, fat and bone molecules of Ali-ce's body vibrated at different speeds producing lethal heat in some cases and blood clots in others. We have never seen such extensive internal trauma," says Ex. Hu-me.

"They came too close to the black hole?" 0011th-Robe asks.

"Not exactly. If it was as simple as that, both Roy Abbit and Ali-ce Mim-sey would have been affected equally by gravity, heat, or radiation, lethal or not," instructs Ex. Hu-me.

"So it was just the special circumstances of that suit and that flight path. Is it likely that Roy Abbit maneuvered Ali-ce Mim-sey into that suit with the purpose of saving himself at her expense?" 0011th-Robe asks.

"His morphology is better suited to the soft-suit," judges Ex. Hu-me. "It is unlikely that he would know about this unique danger of the hard-suit in the environment of a black hole, if we did not."

"Differences in time frame not withstanding, How could she have died without his knowing until his rescue hours later?" 0011th-Robe asks.

"They were not floating around the cockpit during the black hole maneuver. They were belted into separate niches. This prevented injury from uncontrolled movement, either theirs or the ship's. It also allowed them to plug into ships life support too," maintains Ex. Hu-me.

"Oh. I remember. Even though we are weightless we still have our normal mass and inertia. If I hit the wall at a seemingly slow speed, I would still have the inertia and mass of 172 pounds," observes 0011th-Robe granting the exculpating circumstances offered by Ex. Hu-me. "The impact would be the same as bouncing around in normal gravity. I suppose."

"The ship's environmental support also served to scrub the heat and gasses that killed Ali-ce Mim-sey. Pressures above one atmosphere were vented out of the ship. Her air supply and temperature was normalized with no outward apparent damage to the her suit," comments Ex. Hu-me reinforcing Abbit's innocence further.

"Why did Roy Abbit not notice her lack of communication or movement?" 0011th-Robe asks. "Would that not be alarming?"

"Perhaps if he had gone longer without rescue, he would have left his niche, certainly out of concern for Ali-ce," offers Ex. Hu-me. "But we found him and took him off the ship not long after he discovered her fate."

"His fortuitous rescue preserved the 'crime scene' and gave him an alibi. Do you find that suspicious?" 0011th-Robe asks.

"Not in the least. It is standard procedure to program a distress signal before a dangerous maneuver allowing time to turn it off once the maneuver is safely executed and before it is broadcast," holds Ex. Hu-me.

"I have one last concern. Now that Roy Abbit knows his wife is dead. What is his reaction to the news?" 0011th-Robe asks.

"He is fortunate to have been accessible to medical attention. He is very distressed. He is sometimes unable to talk or focus. He is sometimes angry. He sometimes thinks she can be revived. He blames himself for not knowing when she was injured. He thinks he could have given her aid and called for help sooner, if he knew more about being a pilot," depicts Ex.Hu-me, "He feels like they would not have made the trip, if it were not for him. He is physically and psychologically helpless for the time being."

"Well. This, along with what we already have, presents no case for any crime," asserts 0011th-Robe satisfied that she will not be contradicted.

"You make it sound like this meeting is over," objects Ex. Hu-me. "We have not been here for fifteen minutes."

"My interns have just informed me that Roy Abbit is on my shuttle along with your notes on the case," declares 0011th-Robe.

"You said to prepare for 120 hours of meetings," says Ex. Hu-me.

"I...?" 0011th-Robe asks. "I have never spoken to you before this."

"But what about the meetings?" Ex. Hu-me asks.

"If you were told to expect a ten minute meeting you would have prepared ten minutes of information. Am I correct?" 0011th-Robe asks.

"This is not the only thing I have to do," protests Ex. Hu-me.

"Fortunately, I will not be keeping you from your, uh, duties. You have no reason to detain Ali-ce Mim-sey's ship or her remains. So you have plenty to do, no doubt," proposes 0011th-Robe as a formal courtesy.

"Everyone knows you were here," objects Ex. Hu-me desperately, "Everyone knows Roy Abbit was here. You cannot deny that."

"Why would I. Anyone who cares is welcome to know that I came all the way out here for a ten or fifteen minute meeting. The conclusion can only be that nothing of substance was found. There was nothing to keep me here. I left with your notes as a courtesy to acknowledge your diligence," says 0011th-Robe.

"I shuttled a passenger to the planet's surface, a further courtesy for a traveler who similarly had no reason to be delayed. While you, with all you have to do, will routinely return the heroic remains of an astronaut and her ship to Clidim. I suggest that you do that with a minimum of delays or there will be a lot of concerned citizens who will wonder what you do have to do," reasons 0011th-Robe.

"I can reconstruct my notes," answers Ex. Hu-me. "I can maintain that this is a cover-up to ferry an alien from Earth to Clidim."

"Who will not suppose that those ad hoc notes were constructed to support the myth of Earth. One with a sweep that is far too broad and costly to justify the trespass of someone as unimposing as Roy Abbit. You will be taken as someone who is trying to substantiate a centuries old myth on the all too brief careers of Ma-x and Ali-ce Mim-sey, if not yours," rejoins 0011th-Robe testing Ex. Hu-me's sense of self-preservation. "You will make yourself a cultist - one of many. Just another one."

"You will forgive me If I cannot spare the time to personally see you to your shuttle. Perhaps one of your interns can do you the service," taunts Ex. Hu-me bereft of both courage and authority.

"Do not discomfort yourself. I have no intention of imposing on you or your staff any longer," conveys 0011th-Robe with no apology for her junior rank. "I will not keep my interns waiting any longer either."

Episode 108

Roy Abbit is leaving the forensic bus on the Robe's ship. He approaches it on an air deck that makes his transport look quite small. He has no way of knowing if this is the same deck that he last saw Ali-ce's ship. By the time it occurs to him to ask, he is not in the company of anyone who can answer the question.

Finding himself in the company of friendly people who can not answer his questions had a custodial ring to it. He considers the various forms of custody and suppose that if he were under arrest he would be in a cell, alone. He supposes that he would be shackled when he was not in a cell. He expects to eat alone or perhaps among other prisoners. But the absence of those measures leads Roy Abbit to suppose this is a custody of a different sort.

His questions about the 0011th-Robe are just as disconcerting. She is a judge. But what kind of judge travels more than 100,000 miles into space to escort the exhausted and broken former husband of an astronaut who probably spent more time on Earth and in its libraries than on Clidim and in its public institutions. She is not a fugitive and he is not a criminal. Or so he has been told. Bunk space on a freighter and a taxi can get him to any appointment with this 0011th-Robe. Roy Abbit is able to determine that this Robe is not the pilot of this ship, perhaps not even the captain. Most disconcerting of all is that this judge is not the captain of Roy's fate either. Neither the judge nor any of her assistants could say where Roy is to go when the ship lands.

This, Roy guesses, explains the judge's preoccupation and absence. Perhaps Ali-ce's haste to return leaves them unprepared for an alien guest. They are busy making arrangements that Ali-ce would have made had she lived. Again, Roy finds himself weeping.

"I am not comfortable imposing on Mer-cy Mim-sey like this," admits 0011th-Robe as she and 1st-Robe enter their shared stateroom.

"She has insisted that she cannot have it any other way," vouches 1st-Robe while adjusting the cabin's ersatz gravity for a relaxed evening and the night's sleep.

"I suppose it is consolation and solace even under these circumstances to welcome her new son-in-law as someone who knows and loves her daughter perhaps as much as she does," concedes 0011th-Robe.

"So she says and so I must believe," grants 1st-Robe. "Besides Ali-ce will be buried there in Chi-cago and Roy Abbit must be there for that."

"I guess it eases the burden on both of them to spend some time with each other out from under the gravity of the funeral," offers 0011th-Robe taking a seat on the edge of her bunk.

"Like it or not, one of our burdens is eased at the same time. The Robe's Conclave is coming in a couple of months," acknowledges 1st-Robe. "We have to keep Roy Abbit out of circulation and occupied until then."

"We cannot guarantee his safety except by keeping him out of the reach individuals who hold him responsible for Ali-ce's death. How we do it is subject to scrutiny and exploitation. There is also the danger from those who are convinced he is an Earthling.

He will be the target of any individual that wants the glory of killing an alien. Dead or alive he would galvanize the cult of Earth's legend quicker and deeper than Ma-x Mim-sey's petition. But how is Roy Abbit's anonymity going to be served by the Robe's Conclave?" 0011th-Robe asks.

"First of all, the mystery of Roy Abbit is just like the mystery Ma-x Mim-sey. It will exhaust itself in the popular culture over time. The quicker the better. The less fuel the quicker. It does not matter whether the mysteries are promoted or ignored," says 1st-Robe. "Time makes the mysteries a dead weight that fewer like to lift. The dwindling number of fans always seek someone in authority to endorse the mysteries."

"This Robe's Conclave will be the largest assembly of Governors in almost 900 years. Without the benefit of this conclave, their endorsement of a legend or mystery has to be won on an individual basis, Governor by Governor," says 1st-Robe. "There never has been a majority of Governors who publicly endorses some version the Legend of Earth."

"I propose that the Robe's Court pronounce the innocence of Roy Abbit in the death of the astronaut, Ali-ce Mim-sey, at this conclave. Further that she was a treasure beyond price," says 1st-Robe. "His naivete looms larger than his inculpability. He has no skill or resource that could have saved her or served her. He is just her friend and eligible for marriage on Clidim had she lived and wished it so."

"The Robe's Conclave will invite any and all Governors to speak to the contrary. I doubt that very many Governors will contradict the Robe's Court and associate themselves with the Legend of Earth," says 1st-Robe while putting two drink bladders in the microwave. "Finally, I will invite Roy Abbit to acquit himself as the Ambassador of an inimical Earth and to confirm his naive relationship with Ali-ce Mim-sey."

"He will represent himself almost as you see him now. He is a cipher. No one will want to use him to promote the legend of Earth," predicates 1st-Robe. "As long as he does not break any laws, his DNA will not be made public. The temporal and parochial memory and history of Ali-ce Mim-sey will be eventually divorced from the timeless and catholic Legend of Earth. This will disarm M-ax Mim-sey's petition," says 1st-Robe.

"Dr. Mim-sey will be viewed as a fanatic and his petition will be treated as a huge fabrication that only occasionally connects with reality. How and where it connects with reality will be debated without resolution. His bombshell will fizzle," says 1st-Robe.

"What about the warrant?" 0011th-Robe asks.

"That will be published last. The fate of those named, of course, is not included with the warrant. The disappearance of Ma-x Mim-sey will be his presumed but unconfirmed execution. The 125 astronauts, also named will be presumed misdemeanors and known to have continued their chosen roles as astronauts without further blemish.

"The astronauts will be honored and buried on Clidim as they die individually and over time. The eventual erection of 128 grave crystals will eliminate any theory of a colony on Earth. The theory of an Earth that could support such a colony will loose its substance at the same time," allows 1st-Robe.

"How was your meeting with Captain Ham-pton?" 0011th-Robe asks. "You can say it is none of my business, now, but I may need to know sometime in the future since I am the putative heavy at this juncture."

"I cannot say how we will cross that bridge, or if we will get to it. Who knows. You might be 1st-Robe by then. In the meantime, it will not hurt to have you know," says 1st-Robe.

"Captain Hamp-ton was as courteous and friendly as his phony, casual demeanor would allow. I do not think he liked subordinating his rank to mine. We really stepped on his toes."

"He could resent that for a long time," says 0011th-Robe with the blinking yellow light quietly signaling the end of the microwave's assignment.

"Just my being there was an insult to his sovereignty," admits 1st-Robe handing a hot drink to 0011th-Robe. "Removing documents and Roy Abbit would emasculate him with respect to everyone under him. He was preoccupied with the neutered elephant in the room."

"He thought you were there to drop the blade on his symbols of power?" 0011th-Robe asks before taking a careful sip of her drink.

"He was cheated out of the consolation of a lower echelon lackey hacking away at his masculine identity," says 1st-Robe while testing his nightcap for a drinkable temperature.

"That could get him a lot of sympathy from the those who do not have the rank or nerve to bring up such a painful topic. But what about all his superiors who can, uh, rub his nose in it?" 0011th-Robe asks.

"I told him that I did not come all the way out there to look down my nose at him. I was there to give him the courtesy that he deserves. I said I have a little piece of paper with me that will scare the pants off anyone who cannot otherwise be intimidated. Just put it in a drawer somewhere. And if anyone, regardless of rank, gives him a sniggling hard time about this, he could take out that piece of paper and hold it far enough away that they have to bend over his desk to read it. After they recognize what it is, they will realize their posture. He will not have to say a word and no one will ever have another word to say on the subject," charges 1st-Robe before taking a sip of his warm drink.

"You said that to him?" 0011th-Robe asks.

"I had to expose his suspicions and get past them," chats 1st-Robe.

"What was the 'little piece of paper'?" 0011th-Robe asks.

"It was a subpoena from the Robe's Court," discloses 1st-Robe.

"Served by the 1st-Robe and apparently without the unnecessary knowledge of anyone except the captain himself," collogues 0011th-Robe before taking a tentative drink.

"The Examiner's office takes the heat and the hit for handing over a chunk of ship's sovereignty," conveys 1st-Robe while squeezing some of his drink into the vent straw. "The captain can defend Ex. Hu-me with faint praise while everyone else steps back out of the spotlight."

"I know my part is to keep the Robe's role as low profiled as possible. But you could still use the Court's craft and crew. Could you not?" 0011th-Robe asks now able to take a full drought of her drink.

"The Court is in recess until after the Conclave. If we used their resources to transport Roy Abbit to my sister's house it would be conspicuous," explains 1st-Robe. "He can travel with me at my expense. No one can assert that he is using Court's resources and privileges for which he is not eligible."

"At the same time, no one can threaten him without threatening you. No one can intrude upon him without intruding upon you," treats 0011th-Robe. "You travel without drawing attention but you have the full benefit of your office. I like your treatment of this situation, if I may presume to say so."

"I appreciate your courtesy. By the time you become the 1st-Robe, if I may presume, you will have earned every courtesy the office merits. Now, we only have to see to the comforts and recovery of my sister and Roy Abbit - at least for the time being," concedes 1st-Robe before turning his thoughts to his drink and the protection of his all too independent sister.

$$1$$
$$1\ 1$$
$$1\ 2\ 1$$
$$1\ 3\ 3\ 1$$
$$1\ 4\ 6\ 4\ 1$$
$$1\ 5\ 10\ 10\ 5\ 1$$
$$1\ 6\ 15\ 20\ 15\ 6\ 1$$

Chapter 37

Chapter 101001000....

Chapter ACF

Chapter $(1 + 6x + 13x^2 + 12x^3 + 5x^4 + x^5)$

Chapter $(a^5b^0 + 6a^4b^1 + 13a^3b^2 + 12a^2b^3 + 5a^1b^4 + a^0b^5)$

Chapter $2^0 + 2^2 + 2^5$

Episode 109.

Older architectural glass has the disadvantage of being hugely unpopular. Besides being monotonously and predictably as transparent as air, it has those green edges that bespeak low rent. For the last few hundred years of its production, it is used as door thresholds, window sills and surfacing for the walls and halls of public buildings.

As public money permits, the oldest architectural glass is replaced with the more popular manufactured versions of jadeite, lapis lazuli, malachite, and opal. The final insult, though not intended, leaves the old, substantial, glass to be found only in the toilets of public buildings. Thus stigmatized, even the most valuable examples of the glass are shunned.

Latter day generations mistakenly assume that the glass that covers the walls and floors; that graces the thresholds and sills; that sculpts the urinals, toilets, sinks and friezes of those privies is manufactured expressly for the lowly environment and by implication is itself lowly. This low-end utilitarian reputation ushers old style architectural glass out of production and into recycling.

Mer-cy Mim-sey, for her part, adores old architectural glass. That green glow on a thick edge provides constant visual interest to underappreciated window sills, thresholds and door jams. So what if the windows are two or three inches thick. As brick exteriors, it only needs sandblasting to clean and render it discretely translucent. Drapes, screens and tapestries keep environmental light from imposing on the living spaces. Pavers, gates, steps, sculptures, columns, and architectural fragments find new and glorious life in the rambling acres of Mim-sey's gardens.

Roy Abbit for his part finds that the disparaged crystal transparency with its green edges is a comfortable reminder of Earth. It is a relief from the kaleidoscopic storm of ruby, sapphire, emerald, garnet, agate, and onyx manufactures that cover the walls, ceilings and floors of every room and exterior of every home and building on Clidim. Roy considers himself fortunate to have seen only a little of that brittle and icy icing on his way through So-Wemtolo and on to Chi-cago.

"More tea?" Mer-cy asks while prying the top off of a bottle.

"Please. I have never had any 'tea' like this," grants Roy.

"We make it many ways, this is my favorite," conveys Mer-cy.

"I hope it does not fret you that everything is so new to me," imparts Roy as he holds out his glass for a refill.

"Not at all. It has a way of renewing things for me. How do you like your rooms?" Mer-cy queries pouring the rest of the bottle into her glass.

"I took to them rather quickly," relates Roy while studying the foam topping his tea.

"My compliments. I have never found new spaces so livable."

"There is a reason for that. Those were Ali-ce's rooms," discloses Mer-cy before taking a sip through the foam in her glass.

"I did not see anything that I would call hers," admits Roy.

"When my brother asked me to let you stay here, I moved her things to another part of the house," explains Mer-cy.

"I am sorry, I feel like such an imposition," pleads Roy.

"Not at all. No one knew her like you and I," coaxes Mer-cy. "I only changed the gender of the fabrics and accessories. My thinking was that her choice of furnishings and placement would be familiar."

"You are right," says Roy. "I feel like I have lived here for years."

"That is my earnest wish," apprises Mer-cy

"It must have been painful to move her things and know she was gone," frets Roy while twisting his glass in tiny increments.

"I cried a lot," confesses Mer-cy taking note of Roy's interesting habit. "But I also enjoyed setting up her new rooms. I just did it sooner rather than later. I wanted to do it for you too. I promised myself that I would not tell you but I have never practiced deception."

"What do you mean you did it sooner rather than later?" Roy asks before picking up his glass for a sip. "Is this a custom?"

"Not in the formal sense. But keeping her things means cleaning them regularly and enjoying them without grieving. There is no virtue in leaving them untouched," judges Mer-cy while lifting up her glass. "Anyone who sees her rooms would wonder about my feelings for her, if her things were infested with bugs and covered with dust."

"I am afraid I have to agree with you there. But would you have waited until after her funeral?" Roy asks while back to twisting his glass.

"Perhaps. But it gave me much relief from the rigors of the funeral," discloses Mer-cy while setting her empty glass on the table.

"Your courage honored Ali-ce like nothing else," offers Roy. "I do not suppose I was the only one drawing on your strength at the time."

"Both she and you had a part in that," deflects Mer-cy.

"I cannot go back to Earth until after the Robe's Conclave but I have to go back," prevails Roy. "I cannot be your guest forever."

"As far as I am concerned you can. But you are like Ali-ce and she is like me. You have to make your own way whether it is here or there. I would appreciate one favor, however," begs Mer-cy while prying at the top of a bottle.

"Of course," says Roy before he finishes his tea.

"Please stay here until you can visit Ali-ce's new rooms with peace and perhaps some joy," craves Mer-cy while pouring two more drinks.

"But I cannot say when that would be," says Roy.

"Do not worry about that," says Mer-cy. "You will be doing me the favor of giving me the same kind support you say I gave you."

Episode 110

If you go back far enough any pedigree suffers and the Robe's pedigree goes back farther than most. The current site of the Robe's Conclave goes back almost three thousand years. Its location, 2,100 miles northwest of So-Wemtolo, is neither convenient nor prestigious.

The Robe's Court appropriated a malachite mining pit that had fallen into disuse with the advent of the more economical manufactured malachite. What was convenient was seating for eight thousand and the aesthetic of wall to wall malachite.

The slope of the pit has been worked into seating, vomitorium and promenades. This allows the easy coming and going of all of the sitting Governors that rule Clidim from year to year. At the inception no one could be sure how many Governors would rule Clidim from one epoch to the next but the hope was that it would be something less than the pit's capacity for awhile. No one supposes that in less than a thousand years most of the Governors would simply loose interest in exercising their privilege of attending the Robe's Conclave in its three year cycles.

The 837th Rob's Conclave interrupts that trend with record attendance. The first redacted petition and summary warrant in history has piqued interest in this Conclave. Then there is the coincidence of Ma-x Mim-sey's disappearance and Ali-ce Mim-sey's Memorial service.

This is enough to galvanize the Governor's interest for two days but the parochial demands of their constituencies, not to mention the limits of a Governor's ability to exploit the drama for their consumption, means even this conclave faces a timely end.

"Thank you, Governor, for your question but we cannot respond to questions about the existence or non-existence of Earth. Could you, please restate your question with respect to the Mim-sey petition?" Robe's Speaker asks through the clamor of Governors demanding recognition.

"Is the Mim-sey petition the single largest collection of proof of Earth or not?" Governor En-os asks while raising his hand in a bid for calm.

"It may to be the largest assemblage of references to an Earth," advises Robe's Speaker to a quieting crowd. "But there is a substantial difference between the assertion of and the proof of an existing Earth."

"In either case, Is that why Ma-x Mim-sey was marked for execution?" Governor En-os asks as the master of the floor for the moment.

"There is no law with respect to the assertion of or the proof of the existence of an Earth. No one can be charged with a crime, capital or otherwise for associating themselves with or promoting a legend or proof of the existence of an Earth," apprises Robe's Speaker.

"You have not answered my question, Mister Speaker," says Governor En-os raising his hand to quiet the mumble of endorsement. "Dr. Mim-sey has not been seen or heard from for six months, now. His disappearance coincides with the only Summary Warrant in Robe's history."

"With respect Governor, Is there a question?" Robe's Speaker asks.

"Was a Summary Warrant issued? Who was named and what were the crimes?" Governor En-os asks looking left and right for strategic support.

"I presume your first question was just foundation since it is widely known that a Summary Warrant was issued by the Robe's Court. The names and trespasses have been redacted for the time being. As you know, this is done for many reasons. Not the least of which is to preempt speculation on what is a crime and what is not.

"We do not want a 'witch hunt' as the legends of Earth have so agonizingly illustrated," says Robe's Speaker to a sea of drawn faces. "We do not want afficionados of the Legends of Earth to suffer any retribution because of the Mim-sey petition. We do not want to promote either conduct or laws that restrict anyone's association with or assertion of such an innocuous legend."

"Clearly at least one of those named in the Warrant has been found to have committed a capital crime. Otherwise, there would have been no warrant. At the same time, it is unlikely that all of the 126, named there in, have committed capital crimes. Those names were resident to the Mim-sey petition as well as all of the Robes and others. The Robe's Court did not name any perpetrators of a crime that were not native to the petition," says Robe's Speaker

"It addressed the guilt or innocence of the parties named in the Mim-sey petition only. As an element of the petition, their part in any crime, however minor requires adjudication. The Robe's Court does not have the jurisdiction to indite anyone not named in a petition under consideration by the Robe's Court," discourses Robe's Speaker.

"How, then, do you explain the presence and existence of Roy Abbit here, now, on Clidim? Is he and Earth under your jurisdiction?" Governor En-os asks with all eyes on him.

"Explaining his presence would not answer your question, Governor. Being here does not make him or anyone and Earthlian or address the Court's jurisdiction," says Robe's Speaker to a rising murmur. "He was named in the Mim-sey petition but that does not make him a criminal or a potential criminal. The same can be said of Ma-x Mim-sey's daughter, Ali-ce, and many others named in the petition."

"Without a crime there is no jurisdiction. He can claim to be from Earth. That is not a crime. His DNA and anatomy are personal until it is relevant to a crime. Even if he looked like the bipedal lizard with a hemi-penis that is often asserted to be typical of Earthlians, that would not prove that he is from Earth or that there is an Earth," says Robe's Speaker.

"As he is now, or as a bipedal lizard, this does not prove the existence of Earth. All anyone can say is that he is from 'elsewhere' and that 'elsewhere' could be on Clidim or another planet. Neither of which would prove the existence of Earth," drubs Robe's Speaker.

"Should he wish to return, How does Roy Abbit get back to 'Elsewhere'?" Governor En-os asks amid a crow of consensus.

"I can tell, by the tone of your question, that you have grown weary of your own pursuit. But before you are overtaken with exhaustion, please allow me to respond," remarks Robe's Speaker into a sea of shocked silence. "He is not an astronaut and has no way of communicating where he came from. While he knows nothing of Clidim, he knows little more about anywhere else. He has committed no crimes," clips Robe's Speaker.

"He is free to stay or go as he wishes. He can go anywhere our astronauts go; cabin space permitting. While we have less than thirteen thousand astronauts across the universe, that is more access to the universe than he can avail himself of in one lifetime. Even if he came from here, the same can be said of Clidim. It is unlikely he can find that 'elsewhere' even here in the years he has left, " says Robe's Speaker.

"Your concern for my vitality is acknowledged. Civility does not permit a challenge to your sincerity, even if I had such a question. Please allow me to anticipate your comforts and permit me to make one last request," says Governor En-os casting right and left for allies.

"You may...," says Robe's Speaker unbullied.

"Please present Roy Abbit for acquittal by...," interjects Governor En-os with the clamor of his cohort of Governors.

"Roy Abbit, please stand," declares Robes Speaker. "We have no crimes or misdemeanors to attribute to you. In any event, should we need to, we can find you, here, elsewhere or anywhere."

"You can stay here or return to your Earth as you wish," says Robe's Speaker placating the fringes. "In any event, we cannot protect you from harm, retribution or any trespasses against you here or on any 'Earth'."

"Before the Robes, the Governors, and the population of Clidim can extend the benefit of our law to you, you must acquit yourself as worthy of that benefit. Then, any harm or trespass that is visited upon you can and will be prosecuted to the extent that the law applies and to the extent that your benefactors can muster in your behalf. What do you have to say that will earn you the respect of the citizens of Clidim and the benefit of our laws?" Robe's Speaker asks

"Mister Speaker I have no property or prestige to recommend my worthiness. I have lost everything. So I will be brief," says Roy in the sinking drone of Governors preparing to listen. "I was the friend and husband of the Astronaut Ali-ce Mim-sey. "I suppose that there may be resentment of my being here and sharing in the affection expressed for her in the memorial service that opened this conclave. I am here without the invitation or approval of many of those who prize her memory."

"But the memories of her are all that I have. My promise to be brief prevents me from dwelling on even those. I can, before you have lost the will to listen, tell you about her favorite sentiment. She quoted it often; sometimes, when she thought I would not hear her," says Roy. "Perhaps she was trying to rise above her fears for herself, for Clidim, perhaps for me. On other occasions, She would laugh and mock at the gravity of her fears with the same words. By the time of our wedding, we were so accustomed to the sentiment that she included these words in our marriage vows."

Perhaps to assuage the disappointment of not having any of you with her at that prized occasion. Perhaps it was in preparation for our wedding on Clidim that she edited these words from Shakespear's Romeo and Juliet into Bemyem.

I began with: Xu weimkeha:

"Hietu, xo iekemn rie jimiu weiu toeyo me xo ta ie dangzai ye xolimn ximclimn,

Noimnjemn ta wem woa cleimn jeim teiu tiye. Pemn cle xu de kocle slohie lae dimjo laeya ,"

Then she continued with: Di heimgea xu dumn:

"Tiepae xu xo kemn zam peim bie ke juyimn rie yana wemnwo ta wemnslam"

"Pemn yemse iem'ie yaxim rloxae geimn hemn me laeca jeimqumn pemnkea dim jae zou gemnjae

Az zamkeim geimn sloyo heim'ie dumn weiu yuo ximnnge ke slumnpie ta loi cliu di eyimn."

Episode 111

Perhaps he saw it. Perhaps he turned to take his seat at the time. No one else will admit to missing it. They bowed - all twelve. There is not anyone who will suggest that it was mere courtesy. It is a full, respectful, spontaneous and unanimous bow.

While history does not record a Robe bowing to anyone there is no doubt as to what it means. Any insult or injury visited on Roy Abbit would be felt by the Robe's themselves. Who would dare that?

Roy does see the look of pride in Mer-cy Mim-sey's face. For a second, all he can see is Ali-ce's face in hers. Just as quickly, the realization of her loss overwhelms him.

He is grateful for the chance to sit down. He is grateful for the immediate exit of the Robe's and with it the end of the Robe's Conclave. His exit in the company of Mer-cy Mim-sey is not missed either.

If there is any doubt about what the Robe's signaled, it is removed with her comfortable association with Roy Abbit. Anything but the highest respect for Mer-cy Mim-sey and her friend is out of the question.

Perhaps because Roy Abbit appeared so tired, Mer-cy elects to take the sluice train back to the Chi-chago valley. It is famously slow and the last of its kind. It takes four hours to travel 500 miles and it needs a push-car to get it started. The sluice train rides in a half-round trough with water pumped in under the train. The pushcar forces the engineered hull over the water. The hull is engineered to maximize turbulence. The turbulence creates supersonic cavities of steam. The steam forces the train forward. Dropping the pushcar reduces the weight of the train and it accelerates more. The acceleration increases turbulence. The increased turbulence creates more supersonic bubbles of steam forcing the train forward, faster. Creating a plume of 'steam' that rivals the clouds.

For almost 100 years, it is the fastest smoothest ride on the planet. Unfortunately, it is not the cheapest. Magnetic levitation eventually eclipses cavity levitation in speed and tonnage. Ridership determines the rest of the story, almost.

The Chi-cago valley is on the edge of Planet Park. Each year, millions of riders submit to paying the premium to ride the train and see the park. They do not see much. All the cycads look the same. The cycads hide the exotic animals. The rivers are too cold to swim in - too rocky and swift for the millions to boat on. The hotels are too expensive to stay in for more than a day or two. All of this makes it all the more impressive to say you have been there - thus the imperative to go there. Being able to say you have been there makes the trip worth it for the rest of your life.

Mer-cy Mim-sey does not hesitate to pay the cost or take the time. She needs the privacy of her own car and the services of Succor Geon. Succor Geon would not refuse any request of Mer-cy or her brother. But he does not have the first clue on how to medicate Roy Abbit.

"He is resting," conveys Sr. Geon as he enters Mer-cy's room.

"Can we do any more for him?" Mer-cy asks turning to Sr. Geon.

"No. But not just because of his alien physiology. If we knew as much about his physiology as ours, there would still be very little more that could be done. Perhaps we could make him more comfortable. Perhaps we could keep him alive a little longer," imparts Sr. Geon while taking a seat. "For appearance sake, we could hurry him to a top notch facility and have a team of Succors hover over him.

They would all be mindful of what we are now. He is having a heart attack. It is irreversible. Without certain monitoring equipment, the time of death will only be approximate. The precise moment will depend on his brain activity, uh the lack of it."

"We are 500 miles from Chi-cago. Is he suffering?" Mer-cy asks.

"We will not move the train for the time being," says Sr. Geon. "I do not think he knows he is dying. So his anxiety is less than expected."

"There is no reason he should be alone. Is there?" Mer-cy asks.

"You can sit with him, if you want to. But do not face him," stresses Sr. Geon while looking toward the door to Roy's room.

"Why?" Mer-cy asks rising and facing Roy's room.

"You will recognize, telepathically, that his interior monologue is much more stoic and pacific than his physical appearance would have you believe," says Sr.Geon. "His facial expressions could distress you but if you look at his hands you will see they are relaxed more often than not."

"Can he communicate?" Mer-cy asks stepping toward Roy's room.

"I have not intruded but I suppose that he has enough experience with Ali-ce that something of a dialogue is possible," advises Sr. Geon.

"Then he should not be alone. Just yet," insists Mer-cy.

"I can wait out here. You can call me if you need me," invites Sr. Geon while watching Mer-cy enter Roy's room.

"*Roy, how are you feeling?*" Mercy projects looking at the floor.

"*I feel very weak but I am not tired,*" reflects Roy lying motionless with his eyes open. "*I do not understand that.*"

"*Perhaps, it is the medication. I can ask the Su... Doctor,*" projects Mer-cy sitting next to a window with a view of the setting sun.

"*I will not second guess him on the med's. You can mention it when you see him,*" reflects Roy without blinking "*Just make sure he knows I am not asking him to change anything.*"

"*That will be a first,*" projects Mer-cy looking at a piece of floor claimed by a patch of sunlight. "*Everybody wants to change something.*"

"*I would like a little light. Does it have to be so dark?*" Roy reflects with his eyes open but motionless and unblinking.

"*I suppose it is the sedative. There are sedatives that make your eyes very sensitive or insensitive to light. Can you tolerate the lack of light a little while longer?*" Mer-cy projects without betraying the fullness of the afternoon light.

"*As long is it is not much longer. I guess,*" Roy reflects.

"*It will be a little more than four hours before we get to Chi-chago,*" projects Mer-cy without indicating that the train has not even left the station. "*I am sure he will want to change to a photophilic sedative by then.*"

"*I feel so dizzy. Is that the sedative?*" Roy reflects.

"*I can ask,*" projects Mer-cy. "*Maybe the doctor has something.*"

"*Wait. I do not want you to leave, yet. I want to apologize for ruining your day,*"reflects Roy with his chest slowly sinking.

"*How can you say that?*" Mer-cy projects.

"*You had to call a doctor. I must have frightened you and delayed getting home,*" reflects Roy as Mer-cy pulls the blind across the window and the setting sun. "*I am sure you had better things to think about.*"

"*We were only separated for less than an hour. Not long enough to even delay the train,*" projects Mer-cy. "*Besides, I had so little to be concerned about that I had time to reflect on Ali-ce's memorial service. Do you know how easy it is to be proud of her and proud of you?*"

"*All I did was sit and listen but it was beautiful and consoling. You are right about the pride for Ali-ce,*" reflects Roy.

"*What about your eulogy today?*" Mer-cy projects.

"*Oh. I do not think o that as a proper eulogy,*" reflects Roy. "*Besides it was almost two days after Ali-ce's memorial service.*"

"*Perhaps. But it took a great amount of courage to offer such a personal sentiment in the company of all those Governors and the Robes too,*" projects Mer-cy while pulling the curtains across the window as well.

"*I do not know any of them except your brother,*" reflects Roy with his jaw relaxing toward his sinking chest. "*I only had to remember our vows. I would have liked to say it in English though. My accent is so bad.*"

"*Perhaps you can after you rest a while,*" projects Mer-cy.

"*Nothing makes me feel closer to her,*" reflects Roy.

"*Go ahead. Nothing would please me more,*" projects Mer-cy.

I began with:

"*Come, my love, in onyx night, for you, my love, will lie upon the ebony night,/*

Brighter than new snow on a raven's coat. And when I shall die, come sable night,/

She joined me with:

Bring me my love, arrange his likeness in a constellation of stars.

And so, dear night, you will make your black face and carbon brow so fine/

That the world will be in love with night and forget its worship of the garish sun."

"*It is just as beautiful in English,*" projects Mer-cy into the dimness of the lightless room.

"*Yes. ALI-CE, ALI-CE, ALI-CE....,*" reflects Roy into a brightness that seems to engulf him. Except for its painless brightness, he welcomes the light and is pleased that the train is finally moving.

"*Roy, Roy....,*" projects Mer-cy as she turns to Roy's bed.

"Mer-cy," responds Sr. Geon while he hurries into Roy's room along with the light through the opening door.

"Is he gone, Sr. Geon?" Mer-cy asks turning to the Succor.

"Yes, I am sorry," deplores Sr. Geon while touching Roy's neck.

"There is no reason to delay the train any longer," judges Mer-cy with her eyes fixed on Roy.

"I will notify the engineer immediately," advises Sr.Geon looking at his timepiece.

"Can Roy be buried next to Ali-ce?" Mer-cy asks without looking up.

"That will not be a problem. Chi-cago is private property. He died in the care of a Succor. I can sign the death certificate. There will be no cause for an autopsy, necropsy or any other delay of his funeral," concludes Sr. Geon ushering Mer-cy out of Roy's Room.

"Thank you for being here," sighs Mer-cy as she sits down.

"The sun is almost gone," observes Sr. Geon reaching for a light switch. "I will turn on some lights."

"Perhaps a little later," whispers Mer-cy looking toward Roy's room.

"*...and he will make the face of heaven so fine that all the world will be in love with night and forget its worship of the garish sun,*" reflects Sr. Geon before pressing the intercom button for the engineer.

The End

Made in the USA
Columbia, SC
07 September 2019